Library of
Davidson College

Conceptions of ether
Studies in the history of ether theories
1740–1900

Conceptions of ether

Studies in the history of ether theories
1740–1900

Edited by
G. N. Cantor and M. J. S. Hodge
University of Leeds

Cambridge University Press
Cambridge
London   New York   New Rochelle
Melbourne   Sydney

Published by the Press Syndicate of the University of Cambridge
The Pitt Building, Trumpington Street, Cambridge CB2 1RP
32 East 57th Street, New York, NY 10022, USA
296 Beaconsfield Parade, Middle Park, Melbourne 3206, Australia

© Cambridge University Press 1981

First published 1981

Printed in the United States of America
Typeset, printed, and bound by Vail-Ballou Press, Inc., Binghamton, NY

Library of Congress Cataloging in Publication Data

Main entry under title:

Conceptions of ether.

Includes index.

1. Ether (of space) 2. Physics – History.
I. Cantor, G. N., 1943–   II. Hodge, Michael
Jonathan Sessions, 1940–
QC177.C67   530.1   80-21174
ISBN 0 521 22430 6

CONTENTS

		page	
	List of contributors		vii
	Preface		ix
	Introduction: major themes in the development of ether theories from the ancients to 1900		1
	G. N. Cantor and M. J. S. Hodge		
1	Ether and imponderables		61
	P. M. Heimann		
2	Ether and the science of chemistry: 1740–1790		85
	J. R. R. Christie		
3	Ether and physiology		111
	Roger K. French		
4	The theological significance of ethers		135
	G. N. Cantor		
5	The medium and its message: a study of some philosophical controversies about ether		157
	Larry Laudan		
6	The electrical field before Faraday		187
	J. L. Heilbron		
7	The quantitative ether in the first half of the nineteenth century		215
	Jed Z. Buchwald		
8	Thomson, Maxwell, and the universal ether in Victorian physics		239
	Daniel M. Siegel		
9	German concepts of force, energy, and the electromagnetic ether: 1845–1880		269
	M. Norton Wise		

10	'Subtler forms of matter' in the period following Maxwell *Howard Stein*	309
	Select bibliography of secondary sources	341
	Index	347

CONTRIBUTORS

Jed Z. Buchwald, Institute for the History and Philosophy of Science and Technology, University of Toronto, Canada M5S 1A1

G. N. Cantor, Division of History and Philosophy of Science, Department of Philosophy, University of Leeds, Leeds LS2 9JT, England

J. R. R. Christie, Division of History and Philosophy of Science, Department of Philosophy, University of Leeds, Leeds LS2 9JT, England

Roger K. French, Wellcome Unit for the History of Medicine, University of Cambridge, Cambridge CB2 3RH, England

J. L. Heilbron, Office for History of Science and Technology, University of California at Berkeley, Berkeley, California 94720 USA

P. M. Heimann, Department of History, Furness College, University of Lancaster, Bailrigg, Lancaster LA1 4YG, England

M. J. S. Hodge, Division of History and Philosophy of Science, Department of Philosophy, University of Leeds, Leeds LS2 9JT, England

Larry Laudan, Center for Philosophy of Science, University of Pittsburgh, Pittsburgh, Pennsylvania 15260 USA

Daniel M. Siegel, Department of the History of Science, University of Wisconsin, Madison, Wisconsin 53706 USA

Howard Stein, Department of Philosophy, University of Chicago, Chicago, Illinois 60637 USA

M. Norton Wise, Department of History, University of California at Los Angeles, Los Angeles, California 90024 USA

PREFACE

This volume collects ten new studies of a topic central to the history of scientific thought in the last three centuries: theories of ether.

Typically conceived in the time of Newton (1642–1727) as subtle media, like air but much rarer and more penetrating, ethers were conjectured to be the mediating agents for such physical effects as light, magnetism, electricity, and nervous impulses. We might say, then, that this volume follows the fate of such conceptions of ether in the period since. For there is a striking contrast between 1751 and 1951. In 1751, Newton's and other ether concepts were dominating almost every branch of scientific theorising from electricity and chemistry to physiology and psychology. By 1951, however, we find an eminent physicist, P. A. M. Dirac, having to argue in the journal *Nature* (*168:*906–7) that although Einstein's 1905 principle of relativity led, reasonably enough, to ether's generally being abandoned, with the new quantum electrodynamics we may be, after all, 'rather forced to have an aether'.

With their many explanatory roles and metaphysical presuppositions, ether concepts introduce us to the broadest themes in the scientific thought of the eighteenth and nineteenth centuries. Ether theorising may even be seen as a unifying theme itself. We have not assumed, however, that it is such a theme. Our concern has been to use the history of this unruly family of concepts to raise and clarify general issues that any comprehensive analysis of those two centuries would have to recognise.

Most obviously, ether theories relate to controversies over the origins of classical field theory, and over the interpretation of Faraday's natural philosophy in particular. Again, ether theories often developed differently within distinctive national schools and styles of scientific work; so they raise the challenge of explaining the cultural differences, most notably in French, German, English, and Scottish science. Furthermore, conceptions of ether under-

Preface

went especially striking transformations in the period from about 1800 to 1840; these transformations may then support the view, favoured by some historians but not all, that a second revolution in science, hardly less fundamental than the famous one of the seventeenth century, separates all the scientific ideas and institutions of 1840 from those of 1800. To take another theme, if we look to the theological uses made of ether concepts, we find plenty of evidence that science and religion have not been – as the older rationalist histories depicted them – always in conflict since the Enlightenment. Finally, we can see from the reflections made on the credentials of ether theories how wrong is the easy, cynical assumption that methodological theory and scientific practise have never learned from one another.

These, then, are some of the issues that surround the history of ether concepts and are sampled in the chapters prepared for this volume. No previous work has approached the subject quite as this one does. We would emphasise, however, that the editors and contributors are often deeply indebted to earlier studies. E. T. Whittaker's classic treatise still deserves mention before all others, even though its approach must often now be questioned. Of the more recent literature, the writings of M. B. Hesse, J. E. McGuire, Ernan McMullin, Kenneth Schaffner, and R. E. Schofield have naturally been invaluable.

The select bibliography at the end of this volume is limited to the more significant secondary literature on ether theories in the period 1740–1900. Here the reader will find all those works referred to throughout this volume by the name–date format (e.g., Whittaker, 1910). Included also are a number of other works not listed elsewhere. References to primary sources, and to those secondary sources that do not provide extensive discussion of ether theories, are contained in the endnotes to the individual chapters. It has at times been difficult to decide which secondary works should be included in the select bibliography, but in general only those have been included that contain comprehensive, detailed, or otherwise noteworthy discussions of ether theories.

In planning this volume we have benefited from advice kindly given by several colleagues, particularly John Brooke, Stephen Brush, Robert Fox, and Mary Hesse. To Richard Ziemacki, our editor at the Press, we are grateful for judicious counsel and encouragement. In writing the introduction and compiling the bibliography we have had very generous help from our contributors and from Michael Duffy. It gives us great pleasure to have this chance to thank them all. Our thanks go, too, to Andrea Charters and Gay Lowe, who prepared the typescripts of successive drafts.

Introduction: Major themes in the development of ether theories from the ancients to 1900

G. N. CANTOR
M. J. S. HODGE

Division of History and Philosophy of Science, Department of Philosophy, University of Leeds, Leeds LS2 9JT, England

We might introduce the historical study of ether theories by stating simply what ether is. But two reflections should make us pause before attempting this. First, although essentialism is now favoured again in reputable quarters, the lessons of Wittgenstein and others still stand. For many kinds of things it is futile to seek a common and distinctive essence. For many terms it is misleading to demand a definition specifying the conditions necessary and sufficient for their application. So it is with *ether*. We cannot usefully indicate properties that all ethers must and only ethers can have. Second, even if we could, we should not wish to make the attempt independently of historical inquiry. For any definitional demarcation of one kind of theory from others may serve merely to separate lines of theorising that in fact developed together, or to conflate those that developed apart.

However, although the complexities of history may make the abstract definitional problem difficult, they do allow solutions for particular purposes. For the purposes of a study of the period from 1740, we can find in the ether theories of Isaac Newton (1642–1727) cases exemplary enough to suggest an introductory characterisation of ether theories in general.

As every schoolboy is traditionally supposed to know, Newton's law of gravitation has bodies such as the moon and the earth attracting one another with a force proportional to the product of their masses and inversely proportional to the square of the distance between their centres. His ether theory of gravitation would explain this tendency as the effect of forces, not of attraction but of repulsion, exerted by the particles of a rarified medium dispersed unevenly throughout the vacuities in the gravitating bodies themselves and throughout the space that separates them. It is the medium itself that Newton calls an 'aether' Again, Newton traced the deflection of light rays passing through a prism to the action of a special medium whose particles exert distinctive forces of their own, and this medium he also calls an 'aether'.

Because Newton's gravitational and optical ethers were regarded at the time as typical of the genre, we may say that an ether was a spatially and temporally extended entity exerting but not merely identifiable with certain forces and supposed to fit most of the following descriptions: It may be present in spaces empty of ordinary solids, fluids, and gases; it is not perceivable as such ordinary materials are; it transmits actions or effects including or like those of magnetism, electricity, heat, and nervous impulses; it can penetrate and pass through ordinary solid, fluid, and gaseous materials; changes in its distribution or its state can cause observable changes in ordinary bodies.

Such a characterisation can and should allow for fundamental variations. For, as emerges from any survey of science from Newton's time on, some ethers approximating to this characterisation have been supposed material, others immaterial; some fluid, others solid; some continuous, others particulate; some conforming to various laws of mechanics, others not. Some, moreover, have been interpreted literally, as truly existing *in rerum natura;* others agnostically, as possible representations of real physical processes; yet others strictly as fictions useful in the correlating of sensible phenomena. However, no matter how radically later ether theorists departed from Newton's ideas, they were always prepared to acknowledge the precedents he had set.

Early ether theories
Aither in the classical traditions

In developing their ether theories, scientists of the eighteenth century – men such as Boerhaave, Hartley, and Newton himself – often saw classical, medieval, and Renaissance authors as setting important precedents for their own conjectures. We should, then, begin by looking to those traditions that they were most concerned to reject or correct, to vitiate or vindicate.

A tidy taxonomy of classical thinking cannot be given, of course, without misrepresenting the character of much of that thinking itself. For many writers, especially in Hellenistic times, were deliberately intent on mixing notions from different schools. To mention one obvious example, the *aither* of Aristotle and the *pneuma* of the Stoics were originally conceived as quite different in their place and action in the universe. Many authors, however, were willing to conflate any *aither* with any *pneuma*. Among the myriad conceptual permutations perpetrated in antiquity, a really diligent student of the ancients, such as Newton was, could eventually find a plausible precedent for almost any innovation he was about to make.

Moreover, to take the uses made of the word $\alpha i\theta\acute{\eta}\rho$ (and its cognate *aether*

Introduction

in Latin) as an exclusive guide would be to risk missing much that is important in understanding the later influence of these traditions. *Aither*, as a word in Plato's dialogues, does not introduce us to his fundamental teachings about knowledge, reality, space, and matter. Nor does *aether* take us to the heart of Lucretius's science and metaphysics. Yet Plato's Platonism, especially through Henry More (1614–87), and Lucretius's atomism, especially through Pierre Gassendi (1592–1655) and Walter Charleton (1620–1707), were to be among the most decisive, indirect influences upon Newton's ether theorising. So, even in starting with the word *aither*, one should recognise that philological must be integrated with philosophical history.

There seems little doubt that the accounts of *aer* and of *aither* found in the Ionian philosophers of the sixth century B.C. derived partly from a commonplace picture of the world already reflected in the poetry of Homer. That picture was primarily a religious one; it dramatised the destiny of the soul. In it the sky is a solid hemisphere, a bowl, a canopy covering a round flat earth. Between the earth and the sky is *aer*, misty air. Beyond this *aer* is an upper part, a higher air, the *aither*. This higher air is shining, blazing, even fiery. Brilliant and pure, it is akin to men's souls. Indeed, it may be a kind of soul, a form of life itself. A human soul could achieve immortality by joining this everlasting heaven after release from the body at death. Appropriately, Socrates' contemporary Aristophanes has Euripides, in the *Frogs*, pray to *Aither* as to a god.[1]

One may see two Ionians, Anaximenes and Heraclitus, finding in this *aer* and in this *aither* the foundations for their cosmologies. Anaximenes indeed considered *aer* the original source from which all else arose: He evidently taught that water arises by the condensation of this air, fire from its rarefaction. He probably also made air the controlling agent in the resultant cosmos. Identifying the soul of an animal with its breadth, or *pneuma*, he may have made air the breath or *pneuma* and so the source of life and regularity for the world at large.[2]

Heraclitus by contrast seems to have associated soul not with air but with fire. Perhaps through this association he was led to make a special, pure form of fire the ordering and animating agent for the whole cosmos. It is probable, too, that he equated this pure, divine cosmic fire with *aither*, the blazing heavenly haven for the soul.

Heraclitus's theory concerns the control, not the generation, of the cosmos. But a conjunction of Anaximenes and Heraclitus could clearly yield an ethereal cosmogony. Equate fire with *aither* and *aither* with *pneuma*, and one has the world arising from the condensation and rarefaction of an original ethereal source. Although such equations were common in late antiquity, there is no such synthesis of *aither* and *pneuma* known among the Ionian authors themselves. But equally there was nothing in the early conceptions of *aither* to preclude it. For the status of *aither* remained very much unsettled in pre-Socratic science. It was still, like the *aer* it was so often compared and contrasted with, sometimes a region, sometimes an agency, and sometimes an ingredient, all according to context. The speculations of Anaximenes or Heraclitus were certainly rich in hints for later ether theories. But they themselves can hardly be said to have had theories of *aither*.

The first theory of *aither* as such is probably not to be found until Aristotle (384–322 B.C.), writing nearly two centuries later. For Aristotle had a general theory of elements and a specific theory of *aither* as an element.

Theories of elements in the two generations before Aristotle had been developed mainly as a response to the radical critique of cosmological theory explicit in the writing of Parmenides of Elea in Italy. Writing after Anaximenes and almost certainly after Heraclitus too, Parmenides had argued that the ultimate object of thought and discourse must be a single, undifferentiated, unchanging being behind all appearances. If accepted, such an argument obviously constituted a crisis for the contemporary science. For its various explanatory entities were all supposed to account for the diversity of the cosmos by underlying and giving rise to it in precisely the ways Parmenides had concluded were inconsistent with genuine reality.[3]

By Plato and his pupil Aristotle's time two very different responses had been made to the Parmenidean impasse. On the one hand, Anaxagoras and Empedocles had proposed theories of a mixture and separation of everlasting elements. According to Anaxagoras, infinitely divisible parts of matter, each containing the 'seeds' of all things, are moved by the universal agency of Mind; whereas according to Empedocles, everything traces to four 'roots', earth, air, fire, and water, worked upon by Love and Strife. On the other hand, Leucippus and Democritus had explained the world as a congregation of imperishable atoms moving in void space; the atoms themselves they supposed differentiated by the geometrical properties of size and shape; they may also have had them differ in weight.[4]

Plato's cosmogony, as found in the *Timaeus*, departed deliberately from both these lines of thought. He followed Empedocles only in making earth, air, fire, and water the basic kinds of matter. He followed Democritus only in seeking a geometrical ground for their qualitative differences. The entirely novel framework for the cosmogony of the *Timaeus* was, naturally, Plato's own solution to the Parmenidean problems: his theory of eternal, transcendent Forms serving as models for the work of Reason. For, in his cosmogonical narrative, the four elemental bodies as we know them arise when the Craftsman takes these Forms as recipes and standards for his work in ordering motion and differentiating space.[5]

Aristotle, of course, had his own forms but rejected Plato's separation of forms from things formed. With a sempiternal cosmos he also had no need for an original informing of the universe as a whole. His five elements – earth, air, fire, water, and *aither* – are thus, as kinds, sempiternal too. They are all, as essences or natures, forms embodied forever.

Although he himself usually called it 'the first body', Aristotle's new element was to be identified by later writers as *quinta essentia,* the fifth essence. In calling it *aither* or *aether* they followed a hint of Aristotle's own.[6] Although but one among five elements, it is distinctive, not to say anomalous, when compared with the other four. For individual bodies compounded of the others can not only change in place; they can change in quality and quantity and can even come into or go out of being altogether. Moreover, samples of these four elements can themselves undergo transmutation. For the nature of each is constituted by a pair of qualities – for example, dry and cold for earth – so that for there to be stuff of that nature in a particular place is merely for the matter there to be so qualified. Now, the prime matter that these qualities ultimately qualify has no nature, no qualification in and of itself. If the qualifications are exchanged, then, in, say, a switch of wet for dry or hot for cold, earth and air may arise from fire and water.[7]

This theory of qualification and transmutation has no application to *aither*. That element is capable only of local motion, change of place, not any other. Indeed, for Aristotle, the circular motion so distinctive of the heavens is above all what calls for the addition of *aither* as an element beyond the traditional four. For those four are distinguished by natural motions up and down, motion in straight lines towards the centre or towards the periphery of the spherical cosmos.[8]

How then can an endless circling of *aither* in the heavens yield motion in the terrestrial realm? For, after all, Aristotle's cosmos is moved, it is driven and is led, from the outside in. In two ways, at least, *aither* can move the rest. First, in moving against fire it ignites fire to give heat and light. From the daily and seasonal variations in the sun's heat and light arise many changes among animals, plants, and minerals. Second, *aither* has a representative, an analogue here on earth: *pneuma*. Not soul as such, but the instrument of animation in plants, animals, and men, *pneuma* is breath and is spirit; it is what brings life to the body at conception and leaves it cold and still on departing at death. The hot air of this *pneuma* is analogous to the *aither* of the heavens. For like the special heat of the sun that fertilises the earth in spontaneous generations, the *pneuma* in semen can initiate the movements of a new life at conception. Moreover, in this action it shows that it has circular motion natural to it: the circling of a life cycle. In the conclusion enthusiastically reiterated by all Aristotelians from the master himself to William Harvey (1578–1657) two millenia later, just as the *aither* has the heavenly bodies endlessly returning as individuals, the *pneuma* ensures that earthly life does so as species. The celestial moves the terrestrial as a permanent object of emulation; thus do both emulate the perfect immutable being of the ultimate, the unmoved, mover.[9]

The Stoics integrated *aither* and *pneuma* even more fully, eventually identifying them explicitly with one another, although the precise details of this and other developments of Stoic thought are beyond reconstruction on the evidence surviving. Zeno of Citium founded the tradition, in the generation after Aristotle, in lecturing (*c.* 300 B.C.) on the Painted Porch (*Stoa Poikile*) in Athens, but the physics of the Stoics must be gathered mostly from untrustworthy biographies and scraps of commentary written much later.[10]

The Stoics made the ability to act and be acted upon the criterion for real existence, and they argued that only corporeal entities meet this criterion. Corresponding to these two abilities are the two ultimate principles of all bodies, the active and the passive principles. These two are thus inseparably present throughout the continuum, the plenum of the universe. The passive principle is matter itself, in so far as it is an inherently inert stuff that Nature is forever at work upon. The active principle, Nature or God, embodied as *pneuma,* a blend of fire and air, sometimes equated with *aither,* is everywhere actively mixing with matter, penetrating and shaping it so as to constitute bodies that can themselves act and be acted upon. This active principle, as God or Nature, is an immanent *logos,* a rational, providential intelligence, the very life of the universe itself from which all plant, animal, and human life draws its diverse animation and faculties. However, its action, through its embodiment as the *pneuma,* the *aither,* is left very unclear. This is mainly because, being continuous, not particulate, and being inseparable from passive matter, it cannot be conceived as one ingredient juxtaposed with others in the composition of a solid or fluid body.[11]

Its primary physical function is to give coherence or cohesion, *hexis,* to bodies. This *hexis,* no mere structural unity, is indeed the source of all the qualities in a body. Its own ground lies not in a static constraint but in a continual motion. With this motion comes a tension, a *tonos.* On one ancient account, 'there is a tensional motion in bodies which moves simultaneously inwards and outwards', the outward motion giving rise to 'quantities and qualities', the inward to 'unity and essence'.[12] The commentators help us little, though, in understanding these simultaneously contrary movements.

The inward one was sometimes associated with the contraction owing to the cold in the air of the *pneuma*, the outward with the expansion owing to the heat in its fire.[13] In our own time such Stoic notions are often glossed with talk of dynamic continua and equilibria, of forces, of energy, and even of fields of force. But, in themselves, these are largely empty, not to say misleading, moves, adding little to our understanding of the Stoic *pneuma* as a physical theory.

That theory had its rationale in the principal problems it was originally meant to solve. In an ultimately continuous corporeal cosmos there are different degrees of activity, coherence, unity, and integration in bodies as one passes from minerals through plants and animals up to man. The Stoics would explain this diversity, and this hierarchy, by referring it to active and passive principles and so to the universal ability to act and be acted upon.[14]

A similar preoccupation with a cosmic hierarchy from merest matter up to God is of course dominant in the neo-Platonic authors, most notably Plotinus (*c*. 205–70 A.D.) and Proclus (410–85). However, although Stoic notions, especially of *spermatikoi logoi* (seminal reasons), were drawn on by these men, there is nothing in their theories truly equivalent to the Stoic *pneuma*, nor indeed to the Aristotelian *aither*. For the neo-Platonists have their hierarchical cosmos arising eternally as a graduated series of emanations ultimately deriving from the One that is above even Mind and the World Soul. All bodies are activated by the World Soul, whose immaterial animation is directly contrasted with the inertness of matter itself. The emanational dependence of bodies on the One, who is also Being and the Good, is understood by analogy with light. For, like a radiant source, this God is in no way diminished even as He brings forth the visible from the darkness that is matter. The neo-Platonists derived their emanationism largely from what they took to be the central themes in Plato's writings. Consistently enough, when elaborating accounts of the elements, they kept closest to Plato's account in the *Timaeus* of earth, air, fire, and water and their movement by the World Soul; and they avoided any special *aither* as a fifth element.[15]

The Stoic and neo-Platonic unifications of heaven and earth contrast radically with one another. Contrasting no less radically with both was the unification presented by the Roman poet Lucretius (*c*. 99–55 B.C.) in his development of the atomism of Leucippus, Democritus, and their follower Epicurus. Lucretius's cosmogony starts with the formation of the earth from a mindless rushing together of atoms driven only by their weight and their collisions with others. As the heaviest atoms making up the earth become packed and linked together, they squeeze out those going to make the heavens: These are smoother, rounder, smaller, and lighter; ejected, then, they rise up as a fiery *aether,* before congealing to form the moon, sun, stars, and planets. Once formed, these heavenly bodies are swept round in their orbits by further streams of air and of *aether* that are themselves deflected into circular paths by surrounding material even further out in space. The *aether* has a special place in Lucretius's cosmos that is determined by the distinctive geometry and gravity of its atoms, but in the matter, and so in the natural motions of those atoms themselves, it is not, ultimately, special at all.[16]

The contrasts among even the few classical cosmologies mentioned here are, needless to insist, striking and real. They may seem to preclude any significant generalisations about the resources classical texts could later provide for early modern ether theorists. There are, however, four that may be worth suggesting, naturally with a warning as to the exceptions one could easily bring against them. First, the conceptions of *aither* often depend directly on ontologies, on theories of being and of substance. Second, it is in integrating accounts of the heavens and of spirits that the classical writers most often faced the problems their theories of *aither* and *pneuma* were to solve. Third, as solutions to those problems, these theories often raised in turn further problems of how these entities were related to the one matter or many matters making up the bodies we see and touch. Fourth, these ideas were given by many writers vague and analogical characterisations that made them plausibly invoked in contexts and in combinations never contemplated by their original authors. For this last reason, if for no others, the classical inheritance has proved as difficult for the modern historian to trace in well-defined lines as it has proved fruitful for medieval, Renaissance, and Enlightenment scientists to develop.

The classical traditions in the Middle Ages and the Renaissance

The medieval centuries saw deployment of the Greek ideas of *aither* and *pneuma* whenever – and that was often – doctrines of spirit or of heaven were being given a physical interpretation.

Two examples must serve to stand for many. First, Christian philosophers elaborated accounts of a 'spiritual body' to go with St. Paul's teaching (1 Corinthians 15:44) that, upon resurrection to everlasting life, a person would be given a new body distinguished from his earthly one by its spiritual nature.[17] Again, problems of conception and incipient animation, in natural, not to mention virgin, births, prompted theories integrating the immortality of the soul, its bodily instruments, and its potential for joining God and the angels in heaven.[18]

Whether among Islamic, Jewish, or Christian writers, however, the Middle Ages saw no radical innovations setting major direct precedents for later ether theorists. One reason for this is familiar enough. With few exceptions, most writers on natural philosophy were consciously continuing traditions tracing ultimately to Plato and to Aristotle rather than to the Stoics or the Atomists. So either their cosmologies followed neo-Platonic modes of unifying the heavens and the earth through an emanationist metaphysics of light rather than *aither* or *pneuma,* or they perpetuated the Aristotelian division of the heavenly and the terrestrial, with its confinement of *aither* to the one and *pneuma* to the other.

The neo-Platonic tradition of light metaphysics found notable expression in the writings of Robert Grosseteste early in the thirteenth century. In Grosseteste's case one may discern at least four appeals to optical metaphors and analogies.[19] First, the intellect's gazing upon the Platonic forms parallels the seeing we do with our eyes; second, God as Christ is viewed as the Light irradiating the soul; third, light is identified as the ultimate form of corporeal substance as such, so that the material creation can be interpreted as if produced by an original point of light propagating itself in three dimensions; fourth, all subsequent action in the physical world may be understood as like the radiation of light. From these last two themes, Grosseteste derives explicitly the requirement that all explanatory causes of natural effects be specified geometrically, in lines, angles, and figures. Descartes was to have his own reasons for reaching similar conclusions four hundred years later. But there are perhaps traces of light metaphysics in his cosmology and in his influential ethereal theory of light itself.

Naturally, Renaissance scholars ensured that all the classical traditions we have mentioned were actively discussed in the generations before Descartes' own. They often studied, also, the Hermetic writings, supposedly deriving from that venerable Egyptian source of priestly wisdom Hermes Trismegistus, but probably in fact composed mostly in Alexandria in the second and third centuries A.D. These writings present a rich blend of ideas, mainly of Stoic, Platonic, Judaic, and Christian origins, as a cosmological foundation for a gnostical faith and magical practice.[20] Two treatises particularly are relevant here. The Latin *Asclepius* matches the grades of soul in plants, animals, and men with the lower and higher elements, so that the soul of animals is sustained by the ceaseless motions of fire and air, whereas the intellect unique to man comes from *aether*. The Greek *Poimandres,* or in Latin, *Pimander,* has the more sluggish elements, earth and water, moved by a spiritual word, *pneumatikos logos,* that provides an obvious link between Stoic physics and Hebrew cosmogony. Many unresolved issues still surround the vexed question of the influence of the Hermetic teachings on sixteenth- and seventeenth-century science. It is often hard to be sure that a concept traces to Hermetic teachings, if only because for most concepts there are several other classical sources.[21]

The scientific tradition most thoroughly permeated with Hermetic views was the alchemical one, notably represented in the Renaissance by Paracelsus (1493–1541) and by later chemists like Van Helmont (1577–1644), who shared his Hermetic interests.

On Paracelsus's view, spirit of various kinds is the true object of the chemical and medical arts. For all minerals, plants, and animals embody and so imprison spirits ultimately of divine origin and celestial nature. From these hidden, invisible spirits they derive the secret powers elicited, concentrated, and coordinated by the scientific adept whose skill and wisdom arise likewise from heavenly powers implanted with the soul in his own body.[22]

The various metaphors and ancient teachings invoked by Paracelsans hardly cohere well enough to make a single, definite theory of *spiritus;* its constitution and relation to bodies, especially, remain obscure, not to say mysterious. For, it seems, no one spirit is common to all bodies, there being, rather, as many as distinct kinds of bodies. In any case the full powers of a body's spirit lie dormant until roused by the right agent. In the alchemists' work, fire is the great awakener; it can purify bodies by liberating their essences, their spirits, because it is itself a ubiquitous embodiment of spirit in nature.[23]

There are strong echoes here of the material Stoic *pneuma* and of the immaterial neo-Platonic *anima mundi*. The interpretation of the corporeal world is hierarchical and dualistic: hierarchical in that bodies are assigned to grades of perfection from minerals to man; dualistic in that these grades are thought to be diverse degrees in the informing and enlivening by spirit or soul of the inchoate and inactive.

Many difficulties in understanding eighteenth-century ether theories arise because those theories drew sometimes on such hierarchical and dualistic philosophies of nature, and sometimes on the most radical alternatives to them. Few texts are more important for those eighteenth-century theories than the long chapters on the element fire in the *Elementa chemiae* (1732) of Hermann Boerhaave (1668–1738). Although Boerhaave's teaching on fire is plainly indebted to chemists in the Paracelsan tradition, it is no less obviously derived from the natural philosophy deliberately developed by René Descartes (1596–1650) as a comprehensive replacement for any hierarchical and dualistic interpretations of the corporeal world.

Such complexities in the origins of the eighteenth-century ether theories necessitate discussion of several broad and deep problems in the historiography of early modern science. Is one to follow Duhem in seeing important continuities between fourteenth-century and seventeenth-century natural philosophy? Or was modern science, as Koyré came close to saying, born in a revolutionary new victory for Plato over Aristotle, won by Kepler, Descartes, Galileo, and Newton? Or did it arise in a gradual mechanisation of the *Weltbild* consummated by Huygens, as Dijksterhuis would suggest? Or was the Aristotelian cosmos destroyed by an incongruous, even unwitting, collaboration between the magical and the mechanical philosophers; or should one say, rather, by the combined forces of the Hermetic, the neo-Platonic, and the Epicurean traditions? Of late the possibilities proffered have proliferated even as persuasive probabilities have remained elusive.[24]

Our aim, of course, can only be to raise such questions here in suggesting their relevance to eighteenth-century ether theories considered in their native contexts. This we can do by concentrating on the extent to which two crucial authors, Newton and Boerhaave, were both, in their very different ways, following and yet departing from, even radically, Descartes and other mechanical philosophers of the previous century.

Descartes, subtle matter, and the mechanical philosophy

Descartes' conjectures for any specific explanatory purposes in physics always invoke his most general conjecture of all: the hypothesis that matter has been differentiated from the beginning of creation into three elements. He usually calls these three simply the first, second, and third elements. However, he also identifies them as 'fire', 'air', and 'earth' respectively, although insisting that they are not to be equated with what have customarily gone under those names.[25]

His avowedly mythic sketch of this cosmogonical differentiation, in his *Le monde* (published posthumously in 1664), begins with the formation of the second element. This element arose everywhere as minute spheres formed like grains of sand rubbed round as they roll about in running water. The other two elements arose as products and residues from this process, for the first element comprises simply the portions rubbed off, and the third comprises any particles left unrounded because too large or packed too firmly in irregular lumps and those formed by conglomeration.[26]

The portions of this first matter were, in their very formation, filling the changing and indefinitely small interstitial spaces. So they have no fixed shape; and by having to move round the spheres of the second element, they have acquired very quick motions. In any large, swirling vortex, the excess first element matter, not needed to fill the interstices between the second element particles, will have moved towards the centre to form there a liquid of perfect fluidity. The sun formed as a vast turning globe of this first element.

Now, as it turns it presses on and moves the surrounding layers of minuscule second element spheres. Light as a physical action is the transmission of this pressure through the second element, whereas movements of the first element constitute heat, fire that is without light.[27]

All Descartes' most influential physical hypotheses deploy second element and hence, too, first element material. In the 1637 treatises, *La dioptrique* and *Les météores,* the two were not distinguished. There, 'matière subtile' includes whatever particles transmit light. However, in *Le monde* and *Principia philosophiae* (1644) – the works later generations took as definitive – Descartes' subtle matter, his celestial matter, what his contemporaries called 'the Cartesian aether', comprises the second element permeated, as always, by the first.[28] His science was thus ethereal throughout, in that all observable changes in the gross, sensible bodies made of the third element are traced to actions upon them of the second and first.

His science offered moreover a comprehensive alternative to all hierarchical and dualistic cosmologies. For its foundations, as Newton and others insisted – as did Descartes himself – lay in Descartes' famous division of all created substances into minds and bodies, together with his conclusion that the essence of bodily substance is simply extension in the geometer's three dimensions of height, breadth, and depth. By confining the mental to the human, the angelic, and the demonic, Descartes made all other creatures purely corporeal and so of a common essence – extension, the one essential property of matter itself and hence of body in general.

Principia philosophiae drew the corollaries decisive for physics. Extension (and so matter) is divisible without limit; there are therefore strictly no atoms, no absolutely indivisible bodies; even corpuscular components of the three elements are ultimately interconvertible if subject to suitable motions. Since there is matter wherever there is extension, the material world is, like space, unlimited in its extent; and within it there is no empty space, no vacuum void of bodily substance. And since this substance is everywhere, the heavens and the earth are all made of the same continuous material. A body is rarefied not by expanding its own matter, but by distension with invading matter, as cotton wool is swollen on absorbing water. This invasion, like all motions, is possible only if there is a circulation wherein every material portion moves instantly into the place left by that next to it, the last filling the place left by the first.[29]

As Descartes emphasised, his metaphysical theory of material substance as essentially extended led very directly to a world of bodies moved in swirling vortices of continuous material.

For Descartes, as for many earlier philosophers, a substance is whatever needs nothing else in order to exist. God is, then, the only substance, strictly speaking. But among creatures, whatever needs God alone in order to exist may be called substance. We know any substance only as having attributes. And its essential nature is constituted by the attributes it must retain to remain in existence. Any portion of matter may lose its colour, its smell, or its solidity. But if it loses all its extension it ceases to exist altogether. The extension of a body and of the space it fills are the same extension. So, if the extension of the body is the extension of a substance, then presumably the extension of the space is, too, even though there is no ground for deciding whether it is the extension of the same substance. We know matter and we know space as possessing the same attribute. So, in thinking of either, we are clearly and distinctly perceiving the same essence.[30]

This conclusion depends on Descartes' application of the category of substance to space as well as to God and to matter. Newton, disagreeing, will argue that spatial extension, as a condition of all substantial being, even God's, is no substance itself; whereas Leibniz will propose that, as a relational order among contemporaneous bodies, space depends on unextended monadic substances being apprehended by human minds through those extended appearances that constitute bodies. Where Descartes reckons space a substance of the same essence as matter, Newton judges it more, and Leibniz less, basic than the substance of any created entity.

Descartes claimed that with extension and motion he would make the world. These two are what God created in the beginning, and all natural phenomena have arisen from God's subsequent conservation of them. Motion is introduced into matter by God, but not as another substance with another essence. For motions are, as shapes are, merely modes of extension, ways for extended substance to exist, requiring it only to be extended. A body of a particular shape is present wherever all the matter within a volume of that shape is at rest relative to itself and in motion relative to adjacent matter. In Descartes' natural philosophy nothing active, nothing substantial – nothing like the spirits of the Paracelsans, much less the *anima mundi* of the neo-Platonists – comes between the intrinsically active substance that is God and the intrinsically inactive substance that is matter. God moves this matter directly; no hierarchy of intervening created agents stands below God but above the common stuff of creation.

The ultimate and universal laws of nature are, then, the laws that God has prescribed for himself as an immutable supplier of motion to matter and so conserver of motion in matter, and hence of shape and motion in bodies. Left alone by other bodies, alone with God, a body would keep its shape and its state of motion or of rest; and, if moving, would continue with constant speed in a straight line. In the collisions between bodies, likewise, God conserves the total quantity of motion measured as the product of the speed and the volume of the matter moving.[31]

In the world of Descartes' plenum, of course, no body is ever free from the action of others. Every portion of matter is continually agent and patient. The necessary and sufficient condition for all such interactions is motion and contact. To affect the shape or motion of another now at a distance, a body must move so as to collide with the other or move some intermediate that will do that; to affect one already adjacent it must move so as to press upon it. These effects can be wrought and suffered by bodies because they are only changes in the modes of their extension. Their extension itself cannot be changed; none of it can be lost or exchanged for another attribute. The power of a moving body as a mover, upon contact, of others is a force; but this force is strictly the effect of the mover's motion. In Descartes' physics, the forces possessed by a moving body are the consequences, not the causes, of its motion.[32]

That bodies owe all their differentiations and motions to God and his conserving laws for their actions as movers Descartes claims to know with certitude from the very natures of God, matter, and motion themselves. But what of the general hypothesis of the three elemental forms of matter? And what of specific hypotheses about the shapes of bodies and figures of motions that may be responsible for the particular phenomena of heat, light, gravity, magnetism, muscular contraction, and so on? Here, by contrast, Descartes insists that we must settle for plausible conjecture constrained only by agreement with the fundamental laws of conservation and with the phenomena themselves.[33]

It is customary to stress how highly speculative all Descartes' ethereal hypotheses were, and to suggest that his main influence was in convincing people of the coherence of mechanical explanation in general, not of the probability of any conjectured mechanism in particular. But Descartes was influential at both levels. His restriction of physical explanation to homogeneous matter, local motion, and contact action was, of course, very influential as a general ideal. But so too were his particular hypotheses. One has only to look at Huygens on gravity, Newton on colours, Leibniz on planetary motion, or Boerhaave on animal heat to see that whether later physicists, chemists, and physiologists accepted or rejected, modified or replaced, Descartes' proposals, they often took them very seriously in their own efforts to solve those problems.

The planets, mostly composed as our own earth is of the third element, have stable orbits, Descartes suggests, because the second element is more concentrated away from the centre of any vortex; the density of each planet therefore determines a distance from the centre where it has no tendency to spiral inward or outward. The proportion of terrestrial (third element) to celestial (second and first element) matter explains also the weight and falling of heavy bodies near the earth. With its greater speed, the ambient celestial matter has greater centrifugal force. So, rising from below, it displaces a terrestrial body downwards with a force dependent on the proportion of solid and celestial matter in it.[34]

A prism can produce the spectrum of different colours only if its surface is struck obliquely by the incident ray of white light. To explain this, Descartes conjectured that the light particles are thereby acquiring different rotational tendencies and so the power to initiate in our eyes and nerves the motions causing different sensations of colour.[35]

The motions causing such sensations are those in the 'spirits' – matter predominantly subtle – in the nerves and in the brain itself. Likewise, in the muscular responses to such sensations, the motions are transmitted by surges of spirits in nerves and other tissue.[36]

Fire – the motion in first element matter constantly agitating second element spheres – can initiate burning by detaching and expelling as vapour the solid parts of a combustible body of wood, say, or paper. Air is needed for combustion to allow the expelled vapour to escape, just as wine cannot leave a tapped barrel unless air is free to enter it.[37]

Our contributors trace some of the influence such Cartesian conjectures had in later generations. What is needed here is a clarification of the sense in which Descartes' ethereal hypotheses were mechanical explanations. The word *mechanical* was so much used then, and has been so much used by historians in our own time, that without some settled sense for it we will be liable to misunderstand the whole development of ether theorising ever since.

The clarification must naturally be historical. The best hope lies, then, in seeing why Robert Boyle (1627–91) explicitly included 'the Cartesians' among the 'mechanical philosophers' in his instructive essay *Excellence and grounds of the corpuscular or mechanical philosophy* (1674).[38] Boyle's principal disagreements with Descartes are important. After all, he denies that extension is the one property essential to matter, distinguishes solid matter from empty space, and admits that much of space is void. In the essay, however, Boyle is careful not to rest his characterisation of a 'mechanical account' of phenomena or of the 'mechanical philosophy' on any theses separating him and Descartes. That characterisation is deliberately rested on the common ground.

The common ground is that a mechanical philosophy admits but two 'grand' principles: matter and motion; and it restricts the 'affections' of material particles to the two 'mechanical' ones of figure or shape and motion. Moreover, it restricts the powers of bodies to act on others – whether large bodies or minute 'corpuscles' – to those they acquire from being in local motion, and it restricts the actions they can exert to changing the local motion of others upon contact – in impact or pressure – with them.[39] A mechanical philosophy, Boyle insists, is corpuscular, if it has the same laws of motion holding for the indefinitely small parts of bodies as for those bodies themselves. A corpuscular philosophy is mechanical, if those laws are restricted to the conservation of local motion and to the powers and actions that arise in particles from their local motions.

Boyle can, then, find the Cartesian physics, ether and all, within these restrictions. Indeed, the Cartesian ether is, he argues consistently enough, in principle detectable through its mechanical effects; although his 'attempt to examine the motions and sensibility of the Cartesian Materia Subtilis, or the Aether' – with a pair of bellows in a receiver exhausted of air – gave only negative results.[40]

More generally, Boyle takes the Cartesian ether to show that any apparently unmechanical agent may well be given an underlying mechanical interpretation. In its constitution and operation the Cartesian subtle matter is explicitly corporeal and mechanical, of course. But it is hardly, Boyle urges, any less ubiquitous and active than the 'universal spirit of some spagyrists [i.e., Paracelsans], not to say, the *Anima Mundi* of the Platonists'. Like these agents, it is, in Boyle's words, an 'active principle'.[41]

Boyle's point, though obvious, is indispensable for the historian. Far too often even scholarly specialists in this period have tried to establish that some entity, one of Newton's ethers, for example, derived historically from some other entity, the Paracelsan *spiritus* perhaps, on the grounds that the general effects and so explanatory roles of the two were similar, as widespread, hidden, penetrating, and activating agents.[42] Boyle reminds us, however, that almost all the traditions of natural philosophy had some such entity filling some such role. What distinguished these traditions was not the cosmic work these various entities did, but how they were essentially constituted and how they ultimately operated. Boyle is right: There are manifest overall functional analogies between Descartes' ether and the Platonists' *anima mundi* – and all the other entities in that explanatory family, too, from the Stoic *pneuma* and the Epicurean *aether* to the Paracelsan *spiritus*. Equally, though, Boyle shows us that what the historian of scientific traditions in this period needs to concentrate on is not those surface resemblances, but the deep differences in the theories of matter, soul, substance, essence, causation, providence, and law

Introduction

that divide those traditions from one another in their fundamental conceptions of these and all other entities.

Boyle saw himself and Descartes as two mechanical philosophers standing close together and separated by vast metaphysical gulfs from any Platonists or, equally, any Paracelsans. Newton, on the other hand, was, from early on, deliberately to distance himself from the metaphysical foundations of Descartes' science and so from any form of the mechanical philosophy. There is a striking contrast – between the treatment of Cartesian metaphysics, largely but not always implicit, in Boyle's essay on the mechanical philosophy and that, totally explicit, in Newton's extremely revealing manuscript 'De gravitatione', written about 1670.[43] To understand this contrast we must consider briefly the criticisms being made by Platonists, at the time, of the Cartesian mechanical philosophy.

A new Platonism versus the new mechanical philosophy

The mechanical philosophy was opposed, often vehemently, by many who drew their inspiration either from ancient neo-Platonists like Plotinus and Proclus or from Plato himself. Nor was the opposition confined to metaphysical abstractions. It touched directly on the explanation of observable phenomena. The action of magnets provides an especially good illustration.

At the opening of the century, in his pioneering study of magnetism, *De magnete* (1600), William Gilbert went beyond the phenomena to articulate a conception of nature drawing heavily on neo-Platonic ideas. He considered that magnetism was an incorporeal power or agent, indeed, the very soul of the earth. Without such a soul there would be, he insisted, 'neither life, nor primary activity, nor motion, nor coalition, nor controlling power, nor harmony, nor endeavour, nor sympathy'; instead, 'the whole universe would fall into wretchedest Chaos, the earth in short would be vacant, dead, and useless'.[44]

Gilbert argued that the action of a magnet on a lodestone could not be explained by material effluvia but was instead owing to the soul associated with each and with the earth whose nature they share; for to this soul and common nature they owe their tendency to come together like one body in a single action.

Following Gilbert himself, Kepler – a devotee of Proclus – concluded that magnetism was active as such a spiritual power throughout the world. In explaining the elliptical orbits of the planets, he traced their motions to the magnetic action of the rotating sun.[45]

Descartes, of course, offered a quite different explanation of magnetic action in his *Principia philosophiae*. He conjectured that the magnet caused an effluvium of subtle matter to circulate through the body of the magnet and the surrounding space in a closed loop. This flux of subtle matter rarefied the air between the magnet and a block of iron so that they were forced together by the pressure of the external air. In order to account for the specific interactions between two magnets – the positions of their poles determining whether attraction or repulsion occurred – Descartes postulated that the particle streams were aligned, that the particles had screw threads, and that channels in the magnets were similarly threaded.[46] Whatever the difficulties with this conjecture, Descartes' intention was clear; he would eliminate from natural philosophy the type of soul envisaged by Gilbert and Kepler and instead reduce magnetic action to the motion of particles of matter.

In the latter half of the century the neo-Platonic position was most extensively, if not coherently, elaborated by Henry More, Ralph Cudworth, and their Cambridge associates. Initially an admiring correspondent of Descartes, More subsequently, in his *Divine dialogues* (1668) and *Enchiridion metaphysicum* (1671), made a direct attack on Cartesian physics and metaphysics as inconsistent with true theology.

Central to More's critique was the extension, and action in the world, of spirits.[47] For Descartes, of course, all spirits, and even animal and plant souls, though not the rational human soul, were material and therefore extended only as any other matter and as space itself is. In reply, More argued that the natures and extensions of space, of matter, and of spirit are entirely unalike. The extension of a body is physically divisible. By contrast, the extension of a spirit, like the extension of space itself, is not divisible into physically separable parts. Matter, moreover, is in itself inert, whereas spirit is naturally active and initiative of motion. To move our material bodies, our souls must be extended so as to be indivisibly present and active throughout our bodies. The body does not, therefore, fill its place in space to the exclusion of extended spirits. These can penetrate, expand, and contract within bodies and the places they fill. Body and spirit, as substances of entirely different natures and different extensions, can be present together without thereby ceasing to be distinct. Although one piece of matter excludes another from its place in space, many spirits can be present together there. The density of spiritual substance can vary with the number and concentration of the spirits present. The one spirit present everywhere is God; indeed space, in Cudworth's words, is 'the infinite extension of an incorporeal Deity'.[48]

So, in More's and Cudworth's philosophies of nature, any region of space, empty and void of material particles, as much of space is, may be permeated not only by God himself but by many created, active, spiritual substances.

The Cartesian position was ostensibly upheld by Henry Power – a student at Cambridge originally – in his *Experimental philosophy* (1664). He there argued strongly against magnetism's being an immaterial principle and, drawing explicitly on Descartes' own conjecture, referred it to the motion of particles of subtle matter. Elsewhere, however, he identified animal spirits as 'the very top and perfection of all Nature's operations, the purest and most aetherial particles of all Bodies in the World whatsoever (and so consequently of nearest alliance to the Spiritualities) and the sole and immediate instrument of the Soul's operations here'. He speculated, moveover, that this pure ether was the Pauline 'spiritual body'. From these passages doubt may remain whether Power was thoroughly Cartesian or whether, as Rattansi has claimed, he was partially influenced by the Cambridge Platonists. The decisive question, difficult to answer with confidence, is whether his perfect and pure ether particles are supposed to possess any activity except that deriving from the motion given them by God or by other creatures.[49]

The extended, intrinsically active, immaterial spirits of More and Cudworth certainly could move themselves and bodies in precisely the ways denied any mechanical ether like Descartes'. Accordingly, these Cambridge Platonists attributed the action at a distance of magnets, for example, to a spiritual substance that acts 'fatally, magically and sympathetically' (meaning, presumably, without volition, that is, fatalistically; secretly and powerfully; and in accord with the inclinations of like natures to act and suffer similarly).[50] In such an action, needless to say, there is no mechanical impact or pressure, those being actions possible only between bodies, and so not between bodies and spirits nor between one spirit and another.

This clash – between the mechanical philosophies of men like Descartes and Boyle and the neo-Platonic alternatives deliberately opposed to them, especially by More and Cudworth during Newton's early years at Cambridge – provides the indispensable background for Newton's physics and metaphysics, and so for his ether theorising. We cannot expect here to establish once and for all precisely what he did or did not owe to these two traditions. We can only show why no fuller analysis of Newton's philosophy of nature could avoid that question.

Newton's ethers

Exegesis of Newton's ethers is fraught with difficulties. He constructs several different, even incompatible theories, but works none out in detail. The many, sketchy conjectures, scattered through his published and unpublished writings alike, do show clearly, however, that at two periods ethers were among his central concerns. The first period was the 1670s. For it was then that he composed two documents destined to have notable influence when eventually printed: (1) his second paper on light and colours, read at the Royal Society in 1675/6 but not published until 1757, when Thomas Birch's *History of the Royal Society* appeared, and (2) his ether letter to Boyle written in 1678/9 and printed by Birch in his edition of Boyle's *Works* in 1744.[51] The second period begins about 1710 and is signalled by new texts included in the second edition of the *Principia* (1713) and, even more so, by those in the second English edition of the *Opticks* (1717).

Recent historical research has made possible a coherent account of Newton's involvement with ethereal media at these two periods, especially by clarifying the problems those ethers were to solve and, equally, those they were to raise.[52]

Newton followed the mechanical philosophers, at least in supposing ordinary gross bodies to be composed of hard, impenetrable particles. In conformity with the 'analogy of nature' he conceived that aeriform fluids, light rays, and ethereal fluids were likewise formed of such particles, the ether particles being the smallest. Discussing the constitution of ether in the manuscript 'De aere et aethere' (written in the 1670s) he suggested that 'just as bodies of this Earth by breaking into small particles are converted into air, so these particles can be broken into lesser particles [comprising the ether] by some violent action'.[53] Newton could not, however, easily reconcile this conception of the microscopic realm with his general philosophy of nature. The central problem was that this conception invoked inert matter, whereas nature was indubitably active. His solution, as Heimann (this volume) discusses, was to insist that as well as inert matter there must be 'certain active Principles, such as that of Gravity, and that which causes Fermentation, and the Cohesion of Bodies'.[54] He was thus led to reject Descartes' identification of extension as the essence of matter and space alike, and hence to reject also the Cartesian material plenum and restriction of action to contact action. For these 'active Principles' involved forces that were causes and not mere consequences of motion: They were principles of active causation among bodies. It is in describing such interactions that Newton sometimes invokes, for example, the metaphor of 'sociability' used by alchemists and also by the neo-Platonists.[55]

As to what constituted these 'active Principles', Newton gave several different accounts. At times he considered them the immediate exertions of God's will, as if He were directly supplying all activity in the universe. At other times, particularly in writings during the quarter century beginning early in the 1680s, it seems, rather, that the various forces between particles were themselves the active principles. However, throughout much of the remainder of his working life, Newton attributed this activity to one or more ethers (McMullin, 1978:75–110).

Although Newton's presentations differ, on his main account ether consisted of very minute particles that (1) repelled one another, and (2) repelled and were repelled by particles of gross matter. The first of these forces accounted for ether's great elasticity; thus, Newton often claimed that ether was 'much of the same constitution as air, but far rarer, subtler, and more strongly elastic'.[56] In framing this theory of the elastic ether predicated on the notion of interparticulate forces, Newton had, therefore, to reject Descartes' contact-action plenal ether. In support of this rejection he appealed to theology, to methodological arguments about the role of hypotheses, and to the physical inadequacy of the Cartesian theories of planetary motion and of light; he even cited, too, the 'oldest and most celebrated Philosophers of *Greece* and *Phoenicia*'.[57]

The question arises where the activity of this ether resided, since at first sight it appears to be merely a subtle form of aeriform fluid that would therefore have no intrinsic activity. Newton provides two solutions, both of which are problematic. He appears to have thought the activity of the medium related to its elasticity; ether being highly elastic, it was therefore exceedingly active. Moreover, since the elasticity itself depended on the short-range repulsive forces between particles, the activity of ether was ultimately owing to the great strength of the repulsive forces. However, in some writings he suggested that the minuteness of the particles themselves provided a further explanation of activity. For example, in query 21 of the 1717 *Opticks* he argued that since 'attraction is stronger in small Magnets than in great ones in proportion to their Bulk . . . [the] exceeding smallness of its [ether's] Particles may contribute to the greatness of the force by which these Particles may recede from one another, and thereby make that Medium exceedingly more rare and elastick than Air'.[58] Yet with either source of activity, particularly the second, Newton seemed to be attributing to subtle matter that very property he wanted to deny to *all* matter, namely, activity. This was Newton's dilemma, and one he failed to resolve (McMullin, 1978).

As for the repulsion between ether and gross matter particles, Newton's model implied that ether is less dense inside bodies than in the ambient space. Moreover, he assumed that at any interface ether density does not alter discontinuously but that there is a density gradient extending a small distance either side of the interface. This hypothesis he used to explain several sorts of phenomena, including the refraction of light.

A ray of light passing, say, from air into glass and so from a region of denser to one of rarer ether would, Newton reasoned, be 'most pressed, urged, or acted upon by the medium on that side towards the denser aether, and receive a continual impulse or ply from that side to recede towards the rarer [i.e., it is deflected towards the normal], and so is accelerated'.[59] Similarly, in passing from glass into air, the ray is both deflected away from the normal and – in conformity with the dynamic account of refraction – retarded. These hypotheses could also explain the deviation (inflection) of light towards any body close to where it is passing. However, although Newton employed this explanation in the 1670s, by the time he wrote the *Opticks* experimentation had convinced him that light rays were, rather, deflected away from such bodies.[60]

Newton also attempted to explain gravitational attraction by this kind of ether theory (although in the 1670s he had employed a very different model that utilised different sizes of ether particle). The postulated repulsion between ether and gross matter suggested, if it did not directly imply, that the ether density increased progressively with distance from the sun. So a body such as a planet would be, as it were, squeezed from the denser ether towards the rarer ether closer to the sun. At the more fundamental microscopic level this analogy needs to be replaced by an analysis of the forces acting on the spherical body of the planet. Ether being more dense on the side of the body away from the sun, the repulsive forces acting on that side would exceed those acting on the other hemisphere. Hence the body would experience a net force towards the sun and move in that direction (McGuire, 1977:110–11).[61]

These examples illustrate two of several classes of phenomena that Newton considered an ether theory could explain. There was, however, one obvious physical problem that the existence of a ubiquitous ether raised. Since Newton had shown in the *Principia* that a resisting medium would affect the motion of bodies, ether should produce secular accelerations in the planets and so spiral trajectories towards the sun. As these expectations did not agree with astronomical observations, Newton resorted to arguing that ether should produce only unobservable effects on the planets. In query 22 of the *Opticks* he assumed that both the elasticity and the rarity of ether were 700,000 times greater than those of air. The resulting resistance, he claimed, would be so small as 'scarce[ly to] make any sensible alteration in the Motions of the Planets in ten thousand Years'.[62]

The exact nature of Newton's ether and of its relation to specific subtle fluids is obscure. In his second optical paper, he suggested that ether was not composed of 'one uniform matter' but was a mixture of 'the main phlegmatic body of aether', which was inactive, together with active and more subtle 'aethereal spirits'.[63] Moreover, he considered that this mixture of ethers could, as many chemists believed, be condensed to produce diverse forms of matter. In the same paper, Newton cited the ability of electrical action to be propagated through glass as evidence for very subtle effluvia capable of penetrating the pores of solid bodies. It was by equating these effluvia with the 'aethereal wind' that Newton reached an account of electrostatic phenomena fairly close to many seventeenth-century electrical theories.

Guerlac (1963, 1967) has suggested that Newton's interest in electrical phenomena helps explain his renewed interest in ether theory during the early 1700s. At that time, Francis Hauksbee, the curator of experiments at the Royal Society, was, with Newton's encouragement, carrying out attrition experiments apparently indicating that subtle streams of matter were emitted from a rotating glass globe. With this further strong evidence for ethereal matter that was highly active, Newton began to prepare a second part of book 3 of the *Opticks* devoted to the role of ether, particularly in electrical effects; indeed, there are manuscripts showing that he considered electricity to be the subtle spirit that produced all natural phenomena, including animal motion. However, in the event, he suppressed publication of this addition to the *Opticks* and instead cautiously tailored some of his material into the seven ether queries added to the 1717 edition.

In that edition he published further experimental evidence favouring ether's existence. This is the famous two-thermometer experiment showing that a vacuum posed no resistance to the propagation of heat. Heat thus appeared to be 'convey'd through the *Vacuum* by the vibrations of a much subtiler Medium than Air, which after the Air was drawn out remained in the *Vacuum*'.[64] This experiment was performed at the Royal Society under Newton's direction but this time by J. T. Desaguliers in the autumn of 1717 (Guerlac, 1967).

Like other ether theorists, such as Descartes and Power, Newton also employed ethers to account for both animal motion and perception. In his second optical paper he gave a detailed explanation of muscular action in which he differentiated the 'aethereal animal spirit' from the 'common aether'. He saw the first of these acting as an intermediary capable of binding the 'muscular juices' and the more subtle common ether. Thus, when the soul injected a small quantity of the 'aethereal animal spirit' into the muscles, this radically altered the degree of combination of the other two fluids and so changed the muscular tension. Likewise, in the *Opticks* of 1717, Newton's account of vision has rays (i.e., streams of corpuscles) of light striking the retina and there setting the medullary ether into vibration, the ether vibrations then travelling along the optic nerve to the 'place of Sensation'.[65]

For Newton, then, ethers were at least putatively the cause of a wide variety of phenomena; and in almost all cases ether's role was principally as an active agent initiating in bodies new motions that they would not otherwise have acquired. Obviously unlike Descartes' subtle matter in not being mechanical in Boyle's sense, Newton's ethers are not quite like More's spirits either. For his ethers are often, if not always, explicitly material, their constituent particles arising indeed in the division and dispersion of standard nonethereal material. It was not their matter as such, then, that distinguished them, but rather the short-range forces those particles possessed. Indeed, the closest analogues to More's spirits in Newton are not ethers themselves, but these active forces, forces, that is, not reducible to the force of inertia. It is these forces, these penetrating, immaterial sources of motion, after all, that enable Newton to count ethers as active principles of motion. Thanks to the repulsive forces between its particles, the gravitational ether, for example, can serve as the source of the power in heavy bodies to acquire new motions and not merely to retain old ones.

Even if some such interpretation of them is correct, however, much remains unclear about Newton's ethers, his ethereal spirits. In resolving this unclarity, it is tempting to interpret them as akin to the spirits of various earlier writers. But any arguments for their conceptual affinities and historical connections with those other spirits must respect their most distinctive features: They are active and material – but not mechanical in the sense of the mechanical philosophy.[66]

Boerhaave and fire as a subtle imponderable fluid

As Heimann (this volume) explains, the lectures on fire and heat of the Dutch physician and chemist Hermann Boerhaave (1668–1738) had a remarkably pervasive influence upon the ether theorising of the eighteenth century.[67]

The lectures had a curious career as public documents. In 1724 a publisher printed, without permission, as Boerhaave's *Institutiones et experimenta chemiae,* texts of the lectures given annually since 1718. Peter Shaw and Ephraim Chambers soon put this work into English as *A new method of chemistry* (1727), credited to Boerhaave and with extensive notes added, especially from Boyle's and Newton's works. By 1732, Boerhaave had himself arranged for an authentic version of the lectures to appear as *Elementa chemiae,* and Timothy Dallowe published an English version in 1735. Not deterred, however, Shaw produced in 1741 his own English version of the *Elementa,* but once again under the old title, *A new method of chemistry,* and with almost all the same notes as before.[68]

As Love has emphasised, Boerhaave's account of the material constitution of fire as an element is ultimately closer to Descartes' and Boyle's than to any other previous ones.[69] Like Descartes' first element, fire, it is a ubiquitous, imponderable, penetrating, and active material. Consistently, too, with the general constraints of Boyle's corpuscular philosophy, the particles, naturally indivisible and not transmutable into particles of other elements, are solid, hard, smooth, and rounded. However, Boerhaave was familiar with the attractive and repulsive forces judged by Newton as important for chemistry as for optics. Recognising that such powers were not 'mechanical', Boerhaave explicitly denied trying to explain all chemical operations by mechanical principles; and insisted that powers of attraction and repulsion between complex, compound particles were needed in accounting, for example, for the action of solvents.[70]

Placing Boerhaave's fire in relation to the natural philosophies of Descartes, of Boyle, and of Newton is therefore no straightforward task, as Shaw recognised in his notes. On one principal point, Shaw rightly found Boerhaave at odds with Boyle and Newton and in agreement with several Continental chemists and with Newton's Dutch follower s'Gravesande: Fire was formed as such at the beginning of things and has not been 'mechanically producible' since from other kinds of bodies. The physical cause of sensible heat is therefore not conceived, à la Boyle, as consisting merely of the movements of the parts of hot bodies; for then, of course, it would be producible mechanically.[71]

But does this mean that fire as a material substance incorruptible and inaugmentable since creation is itself not a mechanical agent in Boyle's sense? Boerhaave appears never to face this issue directly. What he says about fire would seem to leave it perhaps deliberately unsettled. The particles of fire have no one direction of motion natural to them. Fire is thus without gravity or levity. Left to itself, a pure sample of elementary fire expands and disperses itself in all directions. However, whether this dispersal arises from forces of repulsion acting between the particles or from mere collisions in their jostling motions is unclear. The rarefaction of the gross bodies it acts upon is a universal and reliable sign of fire as an agent. And this rarefaction is possible because fire is repelled by the corpuscles of the invaded, rarefied body. In being reflected as light is from solid surfaces, streams of fire particles are not merely bouncing back upon contact, but are, Boerhaave seems to imply, actively repelled by the particles of the surface, as Newton's light particles are. Unlike any other material, fire is equably distributed throughout the world except where concentrated or dispersed. And the convergent motions of ordinary bodies that are attracting one another are the main cause of its concentration, but again for reasons left obscure. Apart from these repulsions and expansions, the motions of the fire particles and their actions on each other and on other particles remain, then, also obscure.

What is explicit and emphatic in Boerhaave is that without the continual motions of fire particles all nature would fall into a cold, stiff, dead stillness. So once again we see that this fire plays the role played in different ways by the neo-Platonic *anima mundi*, Stoic *pneuma,* Cartesian subtle matter, and Newtonian active principles. It can play this role on earth, tirelessly, because it is continually circulating through the sun; there the parallel streaming of its particles is restored and so, too, is its power as an active mover of other bodies.

As Shaw's notes stress, Boerhaave's fire is a subtle, active, ubiquitous, conserved, spontaneously dispersive, expansive, imponderable fluid. But it is not presented by Boerhaave within a general theory of fluid bodies as compared with aeriform and contrasted with solid bodies. It is an ether, that much is clear. What is not clear, as Shaw urged, is how Boerhaave or Newton himself would have related it to Newton's ether. The challenge of making such connections was one to be often and fruitfully taken up later in the century.

Newton and Boerhaave were, needless to say, not the only sources for ether theories at that time. Our contributors have introduced prominently many others. But we may allow Newton and Boerhaave to raise the major issues that any comprehensive history would have to encompass. For these issues epitomise, in turn, many challenging complexities in the relationships between the scientific developments of the eighteenth and what it is now customary to call the scientific revolution of the seventeenth century.

Major issues *c.* 1740–*c.* 1905
The diversity of ether concepts

If we ask what is new and what is old, in the ether theorising of the 1740s, the answer is hard to give with precision. On the one hand, we have several publications that greatly increased knowledge of Newton's ether theories and interest in them. Bryan Robinson, in 1743, published a *Dissertation on the aether of Sir Isaac Newton,* drawing mainly on the *Principia* and *Opticks*. Then in 1744 the letter, already mentioned, of Newton to Boyle was published for the first time in Birch's *History of the Royal Society;* it was quickly included by Robinson in 1745 in a new book entitled *Sir Isaac Newton's account of the aether.*[72] As Heimann (this volume) has explained, there were several reasons for these publications' being of special interest in the 1740s. He stresses three: the growing emphasis on the role of repulsive forces and on a balance in nature between repulsive and attractive actions; the spreading influence of Boerhaave's conception of fire; and the increasing preoccupation with electrical experiments.

On the other hand, however, we must always recognise that several older lines of thought continued to sponsor ether theories. Most obviously, the Cartesian tradition in natural philosophy was represented by numerous effluvial explanations especially for optical, electrical, and magnetic phenomena. These explanations kept, more or less strictly, within the constraints originally placed by the mechanical philosophy on the understanding of subtle matter. Again, among physiologists the Galenists' spirits were very much alive and so, too, the Stoic conceptions of *pneuma* and *aither* that Galen himself had drawn upon.

Clearly, then, the ethers of mid-eighteenth-century science were highly diverse in their constitutions and operations. A brief glance at three active sites of ether theorising in this period can provide a general indication of the diversity.

1. By mid-century it was agreed that electrified bodies can, like magnets, sometimes attract and sometimes repel other bodies. Theories to explain these attractions and repulsions abounded. Some theories attempted to follow the precedent set by older Cartesian explanations of magnetic action. Streaming effluvia were posited whose parts acted on one another and on particles of gross bodies solely by contact.

On these theories the action of the electrified body is propagated by an intermediary, the effluvium, that is itself in motion as a whole. As Heilbron (this volume) notes, the contrast is thus direct and fundamental with a theory such as Benjamin Franklin's highly influential one. On Franklin's theory the repulsive or attractive action is propagated by a stationary medium surrounding the electrified body. The medium, following the precedent of Newton's ethers, consists of particles repelling each other with forces acting across the short distances between them. Although these particles may be subject to various motions, the medium propagates the electrical action without moving as a whole.

2. For a second example of ether theorising, we may turn to that crucial area of overlap between the physics and the chemistry of the late eighteenth century: the caloric theory of the three states of matter. Caloric itself, the matter of heat, was supposed to consist of small particles that, once again, repel one another across the distances between them. The caloric introduced into a body therefore distributes itself so as to surround each of the particles of that body. Consider now the net force exerted upon any two particles each coated with particles of caloric. If rather few caloric particles are present, the original mutual attraction between the ordinary matter particles will predominate and the body will remain solid. Add more caloric, more heat, however, and the body will become fluid when there is a net repulsive tendency but one

still small enough to be counteracted by the pressure of the atmosphere. Add still more and there will eventually be a net repulsion great enough to overcome that pressure; the fluid now will become aeriform or, as one said by the end of the century, gaseous.

Recent scholarship has increasingly emphasised how central this conception of bodily states was to the revolution in chemistry associated with Antoine Lavoisier and epitomised in his *Traité élémentaire de chimie* of 1789.[73] Perhaps, however, as Christie (this volume) suggests, its full significance for Lavoisier's whole career as a chemical theorist has yet to be appreciated. One issue may indicate how fundamental it was. For nothing is more important to Lavoisier's chemistry than a distinction that has no equivalent in earlier chemists: his distinction between two ways in which two elements may leave a solid or fluid and enter the surrounding atmosphere. They may leave separately: That is, all the particles may have each their own coating of caloric particles, in which case the gas produced is a mixture of the two elements. Or they may leave together: That is, particles of the elements may be associated in clusters with each such cluster coated with caloric particles, in which case the gas formed is a compound of the two elements. We see, then, that Lavoisier's very conceptions of principles, elements, mixtures, and compounds as applied to gases in general, and to oxygen, hydrogen, and water in particular, depend directly on this caloric theory of the states of matter. So, too, therefore, did his innovations in the theory of acidity and combustion, directly dependent as they were on his conceptions of principles, elements, mixtures, and compounds in the gaseous state. The evidence that Lavoisier's chemical researches were knowingly guided by this ethereal theory must take on major significance once one appreciates how it eventually provided the main unifying foundation for the reorganisation of chemical science presented in the *Traité*.

3. In physiology and in theology, as French and Cantor (both in this volume) bring out, the main problems often involved the transmission of action not between bodies but between body and soul or mind. Early in the eighteenth century, one such issue was raised directly by Georg Stahl and Friedrich Hoffmann. Stahl insisted that spirits were either material or immaterial; so they must fall on one side or other of the gulf between soul and body and cannot therefore bridge it. As he concluded, subtle spirits could not be used to eliminate souls from physiology by taking over their traditional role as active sources of bodily motions. To this conclusion Hoffmann responded, as French explains, by supposing that whereas most matter was essentially inert, the animal spirits were by contrast active sources of motions, although quite distinct from the mind itself. Such controversies highlighted a problem met

Introduction

with in many forms in the eighteenth century. If one admitted anything like the Cartesian dualistic distinction between mind and body, then the question naturally arose how any of the various ethereal entities under discussion would fit in with that distinction, or whether they should rather be thought to refute it.

We cannot understand the ether theorising of the eighteenth century unless we appreciate its diversity. Equally, we cannot understand the science of that century unless we recognise the central place in it of such diverse ether theorising.

A tentative taxonomy of ethers

One possible way of distinguishing between ether models is to examine how they solve some particular problem, such as electrical attraction. However, ether theories tend to be problem relative and therefore the choice of any one problem is likely to exclude many popular ether theories. Yet all the ether theories discussed in this volume do in some way or other claim to account for the interaction between two entities, either the action of body A on body B, or the influence of mind (M) on body B. In either case we are concerned with some 'change' in body B mediated by the ether that in some way or other has been 'disturbed' by body A or mind M. The kinds of 'disturbance' are what we intend to classify, but there is a difficulty in deciding when an account of 'change' is required, since this decision is theory dependent. Newton, for example, unlike Kepler, considered that no account was required to explain a body's constant rectilinear motion. It should also be noted that if we were to confine our discussion to the accounts of observable motion we would be omitting a number of subject areas in which no manifest motion occurred. For example, the 'change' in body B could be an increase in its temperature or illumination.

Five types of ether model are here described and illustrated by appropriate examples. Before discussing these models it will be helpful to remind ourselves of two ways in which changes in B might be discussed without appealing to an ether. First, an extreme kind of 'positivist' might account for the change in B by appealing to the regular correlation between the parameters of A and B.[74] Thus, for example, he would claim that the acceleration of B towards A could be described only in terms of the 'masses' of A and B and their distance apart. Second, for others, including less extreme positivists and several of the writers discussed in the next section, the motion of B might, instead, be attributed to some 'power' or 'force' associated with A.

1. According to the first ether model body A (or mind M) must determine the motion of a stream of ether particles that carry B along with it. If space is

considered a plenum this flux of ether would form part of a larger circulatory current, but otherwise the ether particles might merely be projected like a volley of arrows. In the first case ether particles would be in contact with B, but in the second interaction might involve short-range forces. Examples of the first include the Cartesian explanation of magnetic action and planetary motion; LeSage's explanation of gravitation by the flux of ultramundane particles provides an example of the second (Aronson, 1964; Laudan, this volume). These illustrate what might be called projectile models of ether.

2. If body A (or mind M) were responsible for altering the density of ether so that, for example, its density increased with distance from A, then the density gradient in the neighbourhood of B would produce its motion. Body B would, as it were, be squeezed from the denser ether to the rarer closer to A. In his queries to the *Opticks*, Newton utilised this kind of model, and it may be termed a density gradient model of ether action.[75]

3. A paradigm for the third model is Franklin's theory of electrostatic action in which the force between A and B is principally due to the forces associated with the ethereal atmospheres surrounding the two bodies. Thus, among other forces, the ether particles around A repel the ether particles surrounding B.[76] According to this model – the interactive atmospheric model – the attractive and repulsive powers of the ether particles are of central importance.

4. Franklin's discussion of electricity also involved a very different model to account for electrodynamic phenomena; when two differently charged bodies were brought close together, ethereal matter flowed from the more to the less highly charged one. The short-range repulsive forces acting between the ether particles accounted for the fluidity of ether and its tendency to flow from a more densely packed region to one with a lower surface density of ether particles. This kind of model was also used extensively in heat theory to explain heat 'flow', but in this case the fluid of heat, caloric, permeated the whole of a body's volume and did not principally accumulate at its surface. Implicit in this model are the assumptions that the quantity of the fluid is conserved, and this in turn had strong empirical implications.[77] Partly for this reason this ether model and to a lesser extent the previous one were fairly successful during the later eighteenth century. Owing to the close analogy between this ether and well-known hydrodynamical principles, it should be called the hydrodynamical ether model.

5. Ethereal media might also transmit action from A to B but without translational motion; instead, the ether elements or particles themselves merely undergo a minute vibratory motion, a rotation or deformation or, perhaps, no motion at all. Moreover, the elements or particles are able to affect only their

neighbours and do not therefore act over measurable distances. A paradigmatic example of this kind of ether model is Descartes' theory of light, which attributed light to a pressure across the plenum. Another is Fresnel's wave theory incorporating an elastic solid ether consisting of vibrating particles constrained by short-range forces.[78] Paradoxically then, according to this classification, Descartes and Fresnel employed similar ether models; however, on the grounds of this single similarity we should not associate these two writers who faced very different problems and proposed radically distinct ether theories. A variant kinematic ether would be Faraday's early theory of electrostatic action according to which particles of the medium closest to a charged body become polarised and thus in turn polarise the 'contiguous', but spatially separated, particles. Action is thus propagated along a line of particles, just as magnetic action can be propagated along a line of freely suspended magnets. This theory, as Heilbron (this volume) shows, had many earlier precedents. It was, however, criticised by Robert Hare on the ground that it still involved action at a distance. Initially at least, Faraday did not fully appreciate Hare's criticism, and as Gooding argues, he saw no difficulty in the interaction between neighbouring, though separated, particles.[79]

Opposition to ethers

Although the chapters in this volume concentrate on the uses to which ethers were put, we should not neglect the body of opinion that rejected ethers from science. French (this volume) shows, for example, how physiological animists attributed activity to living matter and thus rejected the option of employing subtle fluids as the source of activity. Likewise, Laudan (this volume) discusses Mill's reaction to Whewell, which exemplifies a form of inductivism incompatible with ether theories. There are many other contexts in which the legitimacy of ethers formed a major explicit issue in scientific controversy; indeed, the question of ethers has repeatedly been a subject of contention dividing different schools of philosophers and scientists. Just as there were certain philosophies of science that either permitted or necessitated ether theories, there were others that precluded them on ontological or epistemological grounds.

A number of opponents of ethers have held reductionist philosophies of nature founded on the notion of force, or later energy. In the eighteenth century many attempts were made to explain all the activity of matter, and often mind also, in terms of forces, such theories frequently drawing on query 31 of the *Opticks* rather than the preceding ether queries. A good example of this position is provided by Joseph Priestley (1733–1804), who employed the notion of force to explain all physical activity. Moreover, this move allowed him to overcome, or rather circumvent, the traditional mind–body problem by making both mind and body aspects of force and therefore not distinct and antithetic entities. Paradoxically, one of Priestley's major sources for his account of mind was David Hartley, who had proposed a physiological account of perception and mental action employing the vibrations of a medullary ether. Priestley's debt to Hartley was, however, partial, for although he adopted much of Hartley's theory of mind, and particularly his associationist theory, he neglected, or rather rejected, his ethereal physiology.[80]

Priestley was not, of course, unique in adopting forces without employing ethers. Roger Boscovich, John Michell, John Robison, Henry Cavendish, and William Herschel provide further well-known examples of late eighteenth-century force theorists who accounted for activity in the physical world by forces but without evoking ethereal fluids.[81] In adopting this position they argued that perfectly adequate and simple explanations were obtained by referring observable motions to the forces that produced them. Moreover, they claimed, there was positively no need to evoke any ad hoc explanatory entities like ethers. These writers also articulated theories of matter employing centrally directed forces. It should be noted, however, that they shared no consensus over whether those forces were centred on points – as in Boscovich's point atomism – or on hard material particles of finite volume or whether, as in the case of Priestley, discussed subsequently, the forces alone were sufficient. They also differed over the general form of the force 'curve'. This diversity, together with some indications of the different routes taken in formulating their force theories, suggests that these writers cannot simply be labelled 'Boscovichian atomists'.

A further set of arguments against ether theories came from those who adhered to a particular interpretation of ontological dualism. Thomas Reid, for example, held the traditional distinction between matter, which was inert, and the active incorporeal mind, yet his account of this dualism was related to his interpretation of Newton's first rule of philosophy. Reid considered that any cause evoked by the scientist had to be true in the sense that its existence was observable. Thus the ether, not being open to direct sensory examination, could not count as a legitimate cause, and he dismissed it as a mere phantom of the imagination. Reid was specifically concerned with Hartley's ether, which was supposed to fill the nerves and thus act as the immediate cause of animal motion. However, Reid complained, such a hypothetical entity did not explain how the mind and body interacted, and instead he attributed this interaction to some 'power' beyond our comprehension. Likewise, the mutual attraction between two masses was due to the 'power' of gravitation, which could be expressed as a law but whose cause was unknown. Reid rejected both the kind of materialism that he attributed to Hartley and also the immaterialist ontology associated with Berkeley's writings. However, in discussing the nature of matter he drew heavily on the sensationalist epistemology set out in Berkeley's early writings and employed this approach to reject ethers.[82]

In the nineteenth century sensationalism often became incorporated into the broader current of positivism, which likewise was antipathetic to ethers. Auguste Comte, for example, was cited by those who rejected ethers, since like Reid he considered that science should not be concerned with uncovering hidden causes, but should instead aim at discovering the exact invariable laws that governed phenomena.[83] At one end of the spectrum are writers like Friedrich Engels, for whom gross matter alone was acceptable; at the other are instrumentalists who had no use for ethers.[84]

Symbolic of the attitudes of many towards ethers was Lord Salisbury's somewhat uninformed comment in his presidential address before the British Association in 1894; he claimed that ethers occupy 'a highly anomalous position in the world of science'. He proceeded to describe ether as 'a half-discovered entity' that had been hypothesised merely to provide a medium for the transmission of light. However, his address is probably best remembered for his jibe that 'the main, if not the only function of the word ether has been to furnish a nominative case of the verb "to undulate" '.[85]

Just as certain late eighteenth-century writers attempted to explain all physical, and often also mental, activity in terms of forces, so a century later energy came to fulfil a similar role. By that time the principle of the conservation of energy had become a pillar of physics and was employed, as discussed by Siegel and Wise (both in this volume), by ether theorists as a necessary postulate governing the propagation of light or electrical disturbances in ether. Though it was used mathematically in these physical theories, there were a number of writers, particularly in Germany, who sought in 'energetics' a general metaphysical principle. Thus, Wilhelm Ostwald, for example, considered that science should be remodelled so as to avoid the untenable hypothesis of material atoms; instead, the mathematical formalism of thermodynamics should be given a physical interpretation solely in terms of the ontology of energy.[86]

The opponents of ether theories have, at least until this century, been confronted by a battery of arguments why ethers are acceptable, even necessary, in science. Although the chapters in this volume discuss these pro-ether arguments in some detail, it may be helpful to list them, or at least the more common ones, briefly:

1. One argument is that science should be concerned primarily with causes, as well as laws. Ethers are therefore required, since they provide causal accounts of phenomena; see the chapter by Christie.
2. A variation on (1) was propounded by those writers who attempted to explain all physical activity in terms of mechanics and who therefore insisted that the propagation of light, for example, be explained by ethers acting according to the principles of mechanics; see 'The rise of "mechanical" ethers' in this Introduction and the chapters by Laudan and Buchwald.
3. Many scientists have also claimed that a well-formulated ether theory will have great explanatory and predictive value and may also aid the conceptualisation of complex phenomena (Siegel).
4. Potentially, at least, ether theories offer the possibility of unifying disparate phenomena and diverse branches of science (Heimann; Wise; Siegel).
5. Ethers can be employed in extra-empirical contexts, for example, to solve theological problems (Cantor) or to account for physiological activity (French).
6. Ethers have been demanded by certain philosophies of nature. For example, whereas many field theorists filled space with ether, some German Kantians even required space to be in some sense active (Wise).

So far in this section the division has been drawn rather too sharply between the supporters and opponents of ether theories. The position, however, is far more complex: There can be many shades of opinion among both ether theorists and their opponents, and some scientists have had no strong commitment either for or against ethers. Again, ether theorists probably form a continuum between those who believe ether really to exist and instrumentalists who consider ethers merely useful but fictitious hypotheses by which phenomena may be explained and predicted. Henri Poincaré provides us with an example of the latter position, since, although he bequeathed to the metaphysicians the problem of whether ether existed, he asserted that 'what is essential . . . is, that everything happens as if it existed'.[87]

The range of attitudes is illustrated by the British and Irish responses to Fresnel's wave theory of light. George Airy asserted that the wave theory was true, but he conceived it solely as a mathematical theory of wave propagation and did not commit himself on either the existence of the ether or the value of ether models. By contrast, David Brewster strenuously denied the existence of ether and rejected the wave theory as the true physical theory; nevertheless, he acknowledged the theory as a useful mathematical hypothesis that, owing to its high degree of corroboration, must include some true law of nature. At the other end of the spectrum, William Whewell adopted the realist position, equated the wave theory with ether, and asserted that ether was a necessary constituent of the physical universe. Less extreme positions were adopted by John Herschel, Baden Powell, and Humphrey Lloyd, all of whom held that the wave theory, although not perfect, was certainly superior to any rival. Moreover, they accepted the ether hypothesis as a useful one that was in some limited sense about a real entity, and they encouraged research into its precise constitution (Cantor, 1975).

Forces, ethers, and fields

With these developments, especially in the work of Fresnel, Cauchy, and, later, Maxwell, we are brought to face some very broad issues concerning the relations between the science of the nineteenth and that of the preceding century. In taking up these issues systematically we would, naturally, have to go even further back to consider the place of eighteenth-century natural philosophy in the whole history of modern science.

The ether theories of that century can help here, provided, of course, that a familiar truism is kept in mind: namely, that innovations in those theories could often be made as conscious departures from seventeenth-century positions but never as deliberate approaches to nineteenth-century ones.

A sustained attempt to interpret the eighteenth, quite properly, as the sequel to the seventeenth century is made in Robert Schofield's monograph *Mechanism and materialism: British natural philosophy in an age of reason*. Of its three parts, the second, 'Aether and materialism, 1740–1789', offers as careful and detailed an examination of that century's ether theorising in physics, chemistry, and physiology as is available anywhere.

However, Schofield's highly informative book does not succeed in providing an adequate, general framework of interpretation for its subject. To simplify severely, there are three major difficulties.

1. His fundamental distinction contrasts (*a*) explanations that refer phenomena to particles of homogeneous matter acting upon one another with attractive and repulsive forces with (*b*) those referring them to various material substances – often ethers – characterised by special qualities. Thus, according to the first – 'mechanism' – in heating a body one merely gives its parts increased motions; according to the second – 'materialism' – one is introducing a special matter of heat. Now this contrast may seem to work well enough for many cases; although, as Schofield acknowledges, it would appear defeated by any theories tracing phenomena to motions, but motions whose production and propagation require the action of a special material

agent, as in Rumford's theory of heat, for example, where heat arises as a vibratory motion, caused only in and by a special ether.[88] However, in reality, the contrast is fundamentally unsound. For, as the very case of Newton himself shows clearly, there was no sense in which ethereal explanations were alternatives to explanations by forces; the forces possessed by the ether particles were essential to ethereal explanations, as the decisive sources of the motions putatively responsible for the phenomena.

Certainly, Schofield is right to emphasise that the new ethereal theories of heat as a special material substance implied a fruitful conception of a conserved quantity of heat quite unlike any quantitative conservationary conception implicit in older theories equating heat with motion. However, ethereal matters were taken to stand in many very different relations to the observable qualities and measurable quantities they were to explain. Granted, caloric really was supposed to be heat even if it was not thought to be sensibly hot; but the gravitational ether was not weight itself nor even heavy, any more than optical ether was light or illuminating. Certainly, Schofield is right to stress that an electrical ether and an optical ether, say, were often conceived as being, in an important sense, different kinds of matter. But he is mistaken in his insistence that the differences were like the differences distinguishing Aristotelian elements. For the distinctness of one ether from another depended, in many explanations, on the assumption that the particles of each exerted no forces on the particles of the other. Such a distinctness of ethereal matters presupposed, then, a distinctness among ethereal forces that had no equivalent in any Aristotelian theory of elements. Tracing the phenomena to forces as causes of motion in corpuscular particles and tracing them to the distinctive properties of ethereal matters cannot, then, be seen as contrasting explanatory strategies in Enlightenment science.

2. Schofield's developmental application of his distinction is open to many objections. He finds two legacies from Newton: 'mechanism' from the *Principia;* 'materialism' from the *Opticks* and other ether writings. Then he has post-Newtonian natural philosophy going through three stages: 'mechanism' up to about 1740, 'materialism' for the next generation, and, finally, a return to 'mechanism'. However, this proposal involves three unacceptable corollaries. First, there is no evidence that Newton or his followers saw his legacy as divided in the way Schofield suggests; second, the natural philosophy of the 1740s (as exemplified by Franklin, say) saw no revival, such as Schofield proposes, of 'qualitative', even 'Aristotelian', modes of explanation dominated by taxonomic aims; third, the natural philosophy of the late eighteenth century in no sense marked a return to that of its opening years. In particular, the suggestion that with Joseph Priestley, a key figure for Schofield, we are

going back pretty much to Boyle represents a radical misunderstanding of Priestley, as McEvoy for one has emphasised.[89]

3. Schofield's scheme misconstrues the relationship of eighteenth-century natural philosophy with both earlier and later thinking. On his account, the differences between Newton (in the *Principia*) and the mechanical philosophers, like Boyle and Huygens, are ultimately not very important (they are all 'mechanists'); and, on the other hand, the gap between his 'neo-mechanists', such as Priestley, and nineteenth-century 'field theorists', such as Michael Faraday (1791–1867), is also implied not to be very wide, being bridged by such natural philosophers as Humphry Davy (1778–1829).

Overall, then, too many of Schofield's leading assumptions and conclusions are problematic for his interpretative scheme to be acceptable as a whole.

Another, very different proposal may hold more promise for any attempt to map the place of prominent eighteenth-century natural philosophies between those of the preceding and succeeding centuries. It has been developed by Heimann and McGuire in several papers and is perhaps most crisply expounded in McGuire's 'Forces, powers, aethers and fields'.[90] Their analysis, concerning metaphysics as much as physics, is abstract and complex, not to say convoluted in its arguments; but we do not need here to decide for or against the analysis as a whole. It is mainly their specific suggestions about Priestley and Faraday that are of direct concern.

Consider, once again, Boyle, Newton, Franklin, Priestley, and Faraday as representative figures, writers any broad generalisations must fit. On the Heimann–McGuire analysis, Newton departs radically from Boyle; Franklin departs rather little from Newton; Priestley, however, departs radically from Newton; and Faraday develops but hardly rejects the position held by Priestley. So, as they see it, with the two big moves away from Boyle's mechanical philosophy, the first by Newton and the second by Priestley, one is getting close to Faraday's 'field theory of matter'.

Now, this use of the phrase *field theory* requires us to make a brief excursus on its possible ambiguities. For, too often, scholarly specialists have risked being at cross purposes with one another through failure to disentangle these ambiguities.

It may help to distinguish three very different reasons a historian of natural philosophy might have for using the word *field*.[91]

1. He might be identifying a set of theories by indicating the *explicanda* they were about. So – granted quite unhelpfully – he might simply refer to Aristotle's, Descartes', Newton's, and Einstein's fundamentally different explanations for the downward falling of unhindered heavy bodies near the

earth's surface as their respective theories of the earth's gravitational field. To refer to them thus would presumably imply nothing definite about what sorts of explanations are given by these theories, but would merely designate the specific kind of effect they were to explain and note that it is an effect explained today by a field theory. Whether such a usage ever serves a useful historical purpose is dubitable; but in any case we can and will do without it here.

2. He might be discussing various late eighteenth-century theories explaining, for example, the attraction of suitably electrified bodies for bits of paper and chaff at a distance from them. Among these explanations, he might distinguish some as field theories of electrostatic action. And how would he decide when a theory of electrostatic action is and is not a field theory? This question, it turns out, is not a straightforward one at all, for neither the physicists themselves nor the philosophers of physics have ever settled on a clear and coherent explication for the notion of field theory.

We would not presume here to offer such an explication. However, we would like to provoke further discussion, and to do this by pointing to some ambiguities in the notion of field theory that are especially troublesome for historians of physics.

To look ahead briefly, there would seem to be a tension between two historical conclusions: First, there is no disputing that the English word *field* was first made part of the vocabulary of physics in the 1850s and 1860s by William Thomson (Lord Kelvin) and Maxwell. Nor is there much difficulty in discerning why they wanted to use the word; as we shall see, they wanted to draw attention to some particular features of certain electromagnetic theories, and they made it pretty clear which features these were. The trouble is that these features do not serve by themselves to demarcate those theories as a family from theories that admit action at a distance: Hence, then, the seeming conflict with another historical conclusion. For, second, it does seem (as Wise, this volume, emphasises) that these theories were deliberately constructed in order to avoid action at a distance in physical theory. So the question arises: Were the features of these theories that made it appropriate for Thomson and Maxwell to call them field theories not, in fact, the same features that made it reasonable for these theories to be taken as alternatives to action-at-a-distance theories?

We would suggest that this may well be so. We would suggest, too, that if it is so, then one may have a resolution of a further historical difficulty. For there does seem a tension between two other historical conclusions: The first theories to have the word *field* used of them by Thomson and Maxwell were novel and unprecedented in many respects; however, as Heilbron (this volume) brings out, in the features that made it appropriate to use that word of them, they hardly seem novel at all, having, rather, plenty of precedents in the eighteenth century, if not earlier.

Perhaps, then, we may do well, on combining all these historical points, to ask whether the first theories to be called field theories were indeed novel and were indeed deliberate alternatives to action-at-a-distance theories, but, equally, were neither novel nor demarcated from action-at-a-distance theories in those features that made it appropriate to call them field theories.

Consider next how Thomson and Maxwell introduced the word *field* to refer to the region of space beyond the boundaries but in the 'neighbourhood' (Maxwell's word) of an isolated magnet or electrified body. Consider, more particularly, the space in the neighbourhood of an isolated, spherical electrified body. Any point in this space has, of course, a property that it would not have were the electrified body removed to an infinite distance, namely, that a suitable test body placed there moves and moves, let us suppose, towards the electrified body. The region is then the field of the body's electrical action, the field of the force the body exerts – just as (although no Victorian physicists seem to have invoked this lexical precedent explicitly) the area of ground that a regiment would move to defend, if entered by the enemy, was traditionally identified as its field of action.

Now, to ascribe such properties to these points in space would not commit us to any explanation, any account of the causes, for these motions in the test body. Nor would one make any such commitment in going on to map the paths, the directions and accelerations, of these motions – for all the points on a plane in this space intersecting the centre of the body – with two sets of lines: (*a*) straight lines, converging on the centre of the electrified body, with arrows on them to represent the directions of motion, and (*b*) circular lines, drawn like contours about the electrified body as a centre, to mark points of equal acceleration in a test body. Moreover, even to call this mapping a representation of the field of force around the electrified body would hardly be to venture an explanatory, even less a causal, theory for the apparent actions at a distance. For this would be little more than another way of referring to what is still only a mathematical representation of the effects, the accelerated motions, themselves. Suppose, further, that we introduced lines marking points not of equal acceleration but of equal potential energy, the potential energy at a point being defined as the work done in bringing the test body there starting from an infinite distance away. Then, of course, in some cases involving more than a single electrified sphere, we might have points where the potential energy was not zero, but where a test body was subject to no net force and so no acceleration. However, even with this mapping of equipotentials, we would still be describing the form of the apparent attractions at a distance, rather than considering what their underlying causes may be. Nor would we have a causal theory of the field if we added the assumption that the actions of the underlying causes are propagated at finite velocities, so that an instantaneous annihilation of the electrified body would be followed only after an interval of time by a change in the potential energy at a point beyond its former boundaries.

Suppose now, though, that a theory is proposed concerning the constitution and operation of a physical agency whose noninstantaneous causal activity is presumed to be distributed as this mapping of its effects implies. Then surely we would have passed to a physical theory of the field, the spatial region, of that action. Notice, however, that we would be calling it a field theory not because it invokes one kind of physical agency rather than another, but because it is a theory constructed as a causal explanation of the field of action as understood to be mappable and noninstantaneously changeable in these ways. It counts as a field theory not because its *explanans* is of one sort rather than another, but because the *explicandum* is conceived as coming under a distinctive sort of description.

Such a use of the word *field* not only respects Maxwell's and Thomson's original intentions; it also requires the historian to face a number of decisions. For various *apparent* actions at a distance have been recognised, of course – and physical explanations for them offered – from time immemorial. So the origin of field theories of these actions must have awaited not the recognition of the effects, but the conclusion – and the historian has to decide when that may have been reached, implicitly or explicitly – that they can be mapped in some special way: either as lines of variably accelerated motion in a test body or as lines of continuously variable force, or rather, perhaps, as lines of equal potential energy. And as for potential energy as a concept: Again, the historian would have to decide whether it presupposed the general concept of energy that emerged with the earliest specifications, in the 1850s, of the first and second laws of thermodynamics.

There are thus two historiographical implications in this sense of field theory. First, as Hesse (1961, 1967a) has stressed, in considering the relevant eighteenth-century developments in natural philosophy and in mathematics, it may be wrong to assume that one must have been the cart and another the horse. Innovations in the conceptual analysis and experimental measurement of forces and in the algebra of potentials all helped to promote the quest for field theories of specific physical actions. Their precise, respective roles need still to be traced in detail, however. Second, in this sense of field theory we *cannot* contrast field theories as a family with action-at-a-distance theories.

Introduction

The reason is already plain. Any such theory provides an explanation for *apparent* action at a distance. Indeed, being primarily explanations specifically of electrostatic action, or of magnetic action or of gravitational action and so on, we may call them field theories of specific bodily actions or forces. They do not, as such, presuppose a denial of the everyday distinction between the region of space beyond the active body and the space that this body occupies. For although the field is not taken to have an edge, a boundary, the body may well be so taken. There is, moreover, no presupposition made about whether the *apparent* action at a distance is mediated with or without *underlying* action at a distance – whether, for example, by an ether comprising particles acting upon one another over short or even long distances, or by one with interactions only between contiguous particles, or, again, by a continuous medium.

Now, this last point does not resolve but may clarify the historical difficulties we noted earlier. For we must ask, surely, how there arose the traditional association between field theories and opposition to all actions at a distance. We have seen that the way the word *field* came into physics does not ensure the association as a matter of definition. Why then was the association there from the start as a matter of fact? And why was it so close an association as to have become, since, almost a matter of definition?

This question is again far from a straightforward one, as Siegel, Heilbron, and Wise (all in this volume) show (and as Wise has made even clearer in a personal communication). There is one consideration, however, that makes an obvious point of departure. Physical theories for the field of a specific bodily action were to explain a transaction between two bodies that was presumed to be mediated and so to be propagated with finite velocity. So, naturally, as Faraday and Maxwell both reflected, one did not want to duplicate in the medium exactly the sort of action at a distance apparently exerted by the two bodies, since then the theory would hardly offer an explanation of that apparent effect: Hence, therefore, the historical importance of ethers that admitted action at a distance only between neighbouring or contiguous particles. Whatever other difficulties there might be with such mediation of macroscopic bodily action, at least it would not merely and precisely duplicate these actions at a smaller scale. More generally we might say that since a main rationale for mediation theories was to avoid such duplication, then, as *physical* theories, theories of fields naturally tended to reduce and limit actions at a distance even if they did not necessarily aim to eliminate them. This tendency would, moreover, be reinforced in *mathematical* analyses of the field considered not simply as the field of the body's action but as constituted by the actions of the medium itself. For, with the whole action transmitted with finite velocity, the component actions falling within any region of the field can be analysed mathematically as actions exerted by one part on another and so without separate consideration of the action of the body itself upon the field. In this way the field's actions and energy can be treated as having an independent existence. Further, in considering the transmission of activity and changes in energy distribution over distances much greater than any separating the physical parts of the medium (if it is supposed to have such separate parts), any discontinuities may be, as Wise (personal communication) puts it, ignored or smoothed over. Such an ether as a physical entity may still be assumed to involve actions at a distance; but once the object of mathematical analysis is taken to be the field itself considered as a set of spatial elements, characterisable in terms of the energy distribution over them, then actions at a distance are dispensed with if not rejected.

The extent to which a historian of physics understands field theories of specific bodily actions by contrast with action-at-a-distance theories may well depend, then, on whether he is concentrating on the sorts of physical entities introduced by the theories or whether he is, rather, concentrating on the sorts of mathematical analyses for which the theories were providing physical representations. If his emphasis is on the first he will also incline (as Heilbron, this volume, does) to stress the eighteenth-century precedents for nineteenth-century field theories like Faraday's. If his emphasis is on the second he will be likely to stress (as Wise does) their novelty.

3. A third use for the word *field* would imply a direct contrast with actions at a distance. For the historian might wish to characterise a theory of matter – as both McGuire (1974) and Hesse (1967a) have characterised the one Faraday developed in his later years – as a field theory of matter. Now, on a field theory of matter one cannot distinguish where a body is from where it acts, for it is wherever it acts. So there is denial of bodily action at a distance. For, the body being denied a boundary, there is no spatial neighbourhood beyond that boundary and so in that sense no field of force beyond the body.

For on a field theory of matter, such as Faraday seems to have favoured from the late 1840s on, the forces constituting bodies are not conceived as contributing to the making of a body by being added to solid extended particles. Nor are they conceived as constituting a body by acting with intensities that diminish with increasing distances from points where their intensity is infinite, because forces of continuously varying finite intensity over space are all that constitute bodies. So with matter present wherever any such forces are present and with such forces varying continuously, there are no points in space where no matter is present. On those theories that we might call field theories of material substance, space is, then, full of matter, of continuously variable force fields.

Returning now to the historical problem of the developments separating Boyle and Faraday, we can see why complexities naturally arise in any quest for the 'origins of field theory'. Even insofar as the quest is a search for the eighteenth-century roots of Faraday's thinking, we would have to distinguish between sources for his field theories of electrical or magnetic actions and sources for his field theory of matter. And we would have to distinguish, accordingly, among the very different relations each might bear to earlier conceptions of action at a distance, both apparent and underlying action at a distance. We would have to distinguish also, therefore, among the very different relations each might have to earlier ether theories.

Heilbron (this volume) explores precedents for Faraday's field theory of electrical action. We may usefully here consider one of the Heimann–McGuire suggestions about his field theory of matter: They suggest that successive movements in thought, from Descartes through Boyle, Newton, Priestley, and Faraday, be seen as a decline for inert extension and a rise of active forces in the ontology of material substance. We have already seen something of the early steps in this progression. Boyle added solidity to Descartes' extension in marking material corpuscles off from space, whereas Newton had active forces of attraction and repulsion added to solid, atomic corpuscles that still retained their shapes and so volumes.

Newton's step was of course of great significance. In introducing these forces, he knowingly repudiated the metaphysical foundations of the mechanical philosophy. What is more, by bringing into his science of mechanics entities, attractive and repulsive forces, that fell outside the ontology of the mechanical philosophy, Newton invited a change in the very understanding of the words *mechanical* and *mechanism*. Isaac Watts acknowledged as much in asking in the next generation whether plants and animals can be formed 'by the *Mechanical Motions and Powers of Matter'*, and answering:

> If by the word mechanical, we mean nothing else but those motions and powers, which proceed from the essential properties of matter considered as mere solid extended substance; then I cannot allow the proposition to be true: But if we conclude in the word mechanism, all those additional powers and motions also, which arise from the original laws of motion, which God imposed upon matter at first, such as gravitation or mutual attraction, and others of the same kind, then I allow that all things in the successive ages of the world are formed mechanically; always supposing the divine agency preserving all the atoms of matter and their motions accord-

ing to these Laws. And it is my opinion, that all beyond this is miracle.[92]

To understand the next step, as Heimann and McGuire reconstruct it, we must appreciate why many eighteenth-century writers argued that, with these active Newtonian forces admitted to the ontology of science, the shapes of the corpuscular particles had become explanatorily otiose and their edges metaphysically embarrassing. Boscovich, most notably, urged that nothing explanatory would be lost and metaphysical coherence would be gained if the various forces were conceived as added to otherwise empty spaces – by being made to act, according to various force laws, around point-centres in those spaces – and not to tiny volumes of solid matter. Going on beyond Boscovich, Priestley and some others late in the century argued that such punctiform centres of force are themselves explanatorily redundant; they are moreover epistemologically unwelcome to an empiricist, for a point where a force acts with infinite intensity is no possible object of experience. An empiricist conception of material substance as constituted by its sensible active powers may and need admit only, they concluded, forces whose intensity varies continuously over regions of space.

Many authors, British and Continental, contributed to the emergence of such conceptions of matter, and their contributions involved arguments at many levels: theological as much as methodological, transcendental as much as empirical. Hume's scepticism about direct sensory experience of active powers had to be refuted, for example, as did charges of materialism and pantheism. The very presence of these arguments should convince one that whether or not the Heimann–McGuire analysis is correct as a whole, it is surely sound in its insistence that an impressive conceptual distance separates many late eighteenth-century natural philosophies from Newton's atoms and superadded forces.

How far those natural philosophies take us toward Faraday and so Thomson and Maxwell is, of course, another issue. That issue is one we may clarify usefully by considering the general fate of ether theories in the century separating Franklin and Maxwell.

The rise of 'mechanical' ethers

Eighteenth-century ethers. Even a cursory glance at the scientific literature indicates how radically physics changed in the period 1800 to 1840. Yet it is easier to identify this change than to characterise it accurately, and most simplistic general comparisons fall short of the mark. This situation is complicated by the differences in the historical development of the various branches of physical science. In this connection, Kuhn, among others, has

Introduction

suggested that the physical sciences divide into two classes that follow different developmental patterns.[93] The 'classical' sciences – astronomy, mechanics, and geometrical optics – reached a high degree of theoretical and mathematical sophistication well before the beginning of the eighteenth century. By contrast, at that time the second group, the 'Baconian' sciences, were still not governed by any strong theory, although a great deal of diverse empirical data had been collected. Among these areas were chemistry, heat theory, and electricity; physical optics does not fit very neatly into either of Kuhn's categories. During the latter half of the eighteenth century the data base of these sciences was extended considerably, and they were given explicit theoretical foundations that not only helped direct experimentation but also brought them somewhat closer to, although by no means into line with, the 'classical' sciences. It is significant that, in the period beginning around 1740, it was in these 'Baconian' sciences, rather than in the 'classical' ones, that ether theories were primarily employed. The precise relation between ethers and developments in these branches of science is, however, a question requiring further research.

Ether theories as applied to both electricity and heat were particularly fruitful, principally in respect to models (4) and (3) in the taxonomy of ethers given in a preceding section. Single and two-fluid theories of electricity were used extensively with a fair degree of success in explaining electrostatic action quantitatively, whereas the hydrodynamic model proved adequate to explaining the 'flow' of both heat and electricity. A rather late but particularly impressive example of this type of theory was Sadi Carnot's use in 1824 of the waterfall analogy in developing the theory of heat engines.[94]

In these cases ether theories not only aided conceptualisation but also provided explicit theories that made predictions and directed experimentation and the design and use of apparatus. Yet these examples should not blind us to the extensive use of ethers in nonquantitative contexts during the eighteenth century. Often ethers were proposed to explain, in a rather loose sense, some general class of phenomenon – such as cohesion or electrical attractions and repulsions – with little concern for the detailed experimental results. Moreover, a number of these less concise ether theorists were concerned primarily with biblical exegesis and the role of subtle fluids in the divine economy of nature.[95] Perhaps for these reasons ether theorising became somewhat disreputable in the eyes of some experimentalists.

Very often eighteenth-century ether theories were founded on direct analogy with the phenomena exhibited by known fluids such as water or air. Thus early vibration theories of light sought analogies between the phenomena of light and sound: Just as sound was propagated by vibrations of air particles, so light was a vibration transmitted by the ether particles. Another example is the analogy between heat and the behaviour of macroscopic fluids, which allowed heat flow to be visualised in terms of conserved particles of matter and, in this instance, enabled a small range of phenomena to be discussed quantitatively. However, it was argued that the matter of heat was invisible and yet seemed able to penetrate ordinary matter, and thus it was considered to be subtle. It also had to be weightless and able to spread from a source, and thus it was necessarily an elastic fluid. Moreover, although these largely hydrodynamic models of ethers were able to illustrate phenomena in an easily conceptualised manner, they were inadequate at explaining any but a limited range of phenomena in detail.

To take the important example of chemistry, ether theory offered an explanation of how particles might combine, but it was inadequate to explain, let alone predict, the elective combination of specific substances. Again, to cite the examples of refraction and inflection discussed by Newton, a general account could be given of how differential ether density might explain the deviation of a ray; however, this model was totally unable to deal with dispersion or diffraction patterns. No contemporary theory, it should be noted, was manifestly successful in these areas. One of the advantages of ether theory was that it offered, potentially at least, a programme by which recondite phenomena could be explained. Moreover, it seemed to provide a unified causal account of diverse types of phenomena. The writings of many early theorists make it clear that they considered speculative guesses in order, as if the object were to make somewhat plausible guesses about the underlying structure of nature.

Ether theory c. 1840. If we now examine theories of light, heat, and electricity in 1840 we find a major change. The earlier theories had by then been almost entirely replaced by high-level theories involving not direct analogies but either purely mathematical models or 'mechanical' models expressed in mathematical terms. Moreover, these deductive models provided precise predictions that were tested against experiment and that proved, at least in this respect, greatly superior to their eighteenth-century predecessors. These fields, unlike perhaps chemistry, had evolved into 'classical' physical sciences, in Kuhn's terms.

The example of physical optics will be used to illustrate this change, since it was of primary historical importance in the development of ether theories. At the turn of the century, Thomas Young advanced a vibration theory of light explicitly analogous to his theory of sound. He conceived ether as similar in structure to air but much rarer and more elastic. Just as vibrations in the air produced sound, so vibrations in this ether produced light. Young employed almost no mathematics, nor did he have a worked-out theory of wave propagation. Even his famous principle of interference was discussed simply in terms of the path difference between two rays, and he considered only the two simplest cases: when the phase difference was either 0° or 180°.[96]

If we now examine the state of the wave theory of light in the 1830s we find that men like Cauchy, Challis, Airy, and Powell were articulating the theory through mathematical models. As Buchwald (this volume) discusses in detail, many of these writers were concerned with devising 'mechanical' models of ether that could account for such complex phenomena as double refraction and dispersion.

These changes affected other branches of physics, albeit in somewhat different ways, during the period 1800–40. Largely responsible for these changes was a rather diverse group of scientists, principally French and often associated with Pierre Laplace, who greatly extended, rather than originated, a programme in scientific problem solving.[97] They conceived mechanics as the basic science to which all physical problems were to be reduced. Any problem had thus to be analysed into its simplest elements in terms of the particles and forces involved. A model would be formulated in mathematical terms, in agreement with the laws of mechanics, and from this model testable predictions could be made and compared with experiments. This method of analysis and synthesis, often involving calculus and sophisticated mathematical techniques, was used extensively and successfully in most branches of physics during the opening decades of the century, whereas previously its main area of application had been celestial mechanics. The programme as discussed so far leaves open to question the precise models employed and the further conditions imposed on them. At this level there was, of course, much disagreement. For example, while Laplace, Biot, and Malus adopted a particulate theory of light based on dynamical principles, Fresnel formulated his wave theory of light in terms of a mechanical model of ether.[98] It was indeed in the context of Fresnel's wave theory that new and exacting conceptions of ethers emerged.

'Mechanical explanation'. Before discussing further this change in ether theorising, it is necessary to clarify the concept of *mechanical explanation,* which has been the source of much confusion. In contrast to the mechanical philosophies of the seventeenth century, Newton introduced *forces;* and, in his three laws of motion, he postulated a very specific relation between the behaviour of a body – in terms of its 'quantity of matter' and its motion – and the forces acting on it. The first law states that when there is no external force acting,

the body's momentum (*mv*) will be unaffected. In the second law, a proportional relationship is specified between the impressed force and the change produced in the body's momentum, the direction of the change being in the direction of the force. Finally, when two bodies act on one another, the forces of interaction will be equal in magnitude but opposite in direction.

Of direct concern here is the insistence by nineteenth-century physicists (together with some earlier writers) that an explanatory model had to employ, or at least be consistent with, Newton's laws of motion in order for it to be mechanical. The paradigmatic example of a mechanical theory was Newton's own theory of gravitation (as distinct from his ethereal hypotheses concerning gravitational attraction). It did not employ contact action and therefore would not have counted as a mechanical explanation in Descartes' sense. Instead it involved bodies acted on by forces and behaving in accordance with the laws of motion.

This example does, moreover, illustrate further characteristics of mechanical explanation as understood in the nineteenth century.[99] Gravitational theory is very specific about the magnitude and direction of the forces acting on material bodies. It states not only that the force is attractive, acting along the line joining the bodies, but that its magnitude is given by the equation

$$F = G \frac{m_1 m_2}{d^2}$$

Mechanical explanations in general include among their premises mathematical expressions for the force vector. The example of gravitation allows us to identify other typical features of such force laws. First, they specify some property of the bodies involved – in this case their 'quantity of matter' (or masses), m_1 and m_2. (It could, of course, be their electric charges or viscosities). Second, the formula just given includes a spatio-temporal parameter – the distance (d) between the idealised points. Third, there is a constant – the gravitational constant (G). In other instances the constant(s) and variable parameters might, of course, be very different. Likewise, although some writers have suggested that only a single form of force function is permissible, there seems no reason in the present context to restrict the form of the equation. Finally, although usually implicitly, these nineteenth-century mechanical models assumed that both matter and energy were neither created nor destroyed.

The significance of mechanical ethers. The considerations just outlined specify the minimum necessary conditions of a mechanical explanation as employed in nineteenth-century physics. Since these conditions were of general applicability, they were, of course, applied to ether models. This conception of a mechanically adequate ether theory, though not unknown in the eighteenth century, allows us to draw some fairly sharp contrasts between the speculative ethers of the eighteenth century and the mechanical ethers central to certain branches of physics, particularly optics, in the nineteenth century.

In physical optics, the turning point came in about 1820 with Fresnel's version of the wave theory. In this field, as discussed in a preceding section, physicists developed *mechanical* and *mathematical* ether theories that not only explained in exact terms a number of phenomena but also made quantitative predictions that were impressively confirmed. During the next few decades a number of different ether models were proposed – one writer in 1849 counted fourteen different ether theories[100] – often with the specific intent of accounting for some of the more intractable phenomena, such as dispersion and absorption, both of which involved postulating an interaction between ether and gross matter. Yet despite these differences and the remaining problems in some parts of the theory, ether theories had to be mechanical, in the sense just discussed. Moreover, for these ether models to be of any use in the context of contemporary optical theory, they had, through their mathematical formalism, to account for at least a moderate range of optical phenomena in quantitative terms.

These requirements severely limited the range of ether theories that were scientifically worthy of consideration after about the first quarter of the century. William Rankine, for example, in 1855 distinguished the luminiferous ether from caloric and the electrical and magnetic fluids on the ground that, because the former was founded on the laws of mechanics, it was 'a probable representation of a state of things which may really exist'. By contrast, he considered that the latter fluids functioned 'merely as a convenient means of expressing the laws of phenomena', and, moreover, they often retarded the progress of science, since the unwary adopted them as true causes.[101]

One implication of the rise of mechanical ethers was that the diversity of subtle fluids disappeared from the literature, albeit more slowly from the popular press. In the eighteenth century it was frequently supposed that space was permeated by a number of noninteracting ethers, each performing a distinct function; one might account for electrical phenomena, another for heat, and a third, say, for magnetism. Although there was much conjecture that these fluids might somehow be related to one another, this situation needs to be contrasted with that of the nineteenth century, when ether theorists tended to concentrate on a single ether, which might perform a number of different functions and which was conceived in mechanical terms. The new physicists' conception of the ether lost the pictorial appeal of the early ether theories founded on commonplace analogies such as the flow of water or water waves. Instead, in order to account for polarisation, theorists suggested that ether was like some strange elastic solid, which could thus transmit transverse undulations. Furthermore, ether theory was comprehensible only to those who understood mechanics and analytical mathematics: It was no longer susceptible to direct analogical arguments but was dependent on the more abstract dynamical and mathematical theory.

A further implication of this change in ether theory concerned its domain, for instead of using it to unify diverse branches of natural philosophy, theorists initially applied the new luminiferous ether solely to optics. Even questions of photochemistry and the role of light in plant growth usually were considered outside the domain of the wave theory of light. Yet the dream of a universal ether never faded completely, and in a few areas connected with optics, and thus the luminiferous ether, some progress was made. The first of these was heat theory. Following the successes of the wave theory of light a number of writers, beginning around 1830, suggested that since radiant heat appeared to be reflected, refracted, polarised, and so on, it too was a vibration in ether (Brush, 1970). This theory, which was extended by Ampère to include heat conduction, was adopted by many scientists in preference to caloric theory.

Maxwell and post-Maxwellian ethers. Maxwell's theory of electromagnetism provided the focal point for ether theory in the latter half of the nineteenth century. The relation between electromagnetism and ether theory was, as Siegel (this volume) argues, important for Maxwell himself, since he acknowledged the need for adequate mechanical models. He came closest to achieving this aim in the early 1860s when, in his paper 'On physical lines of force', he developed a molecular vortex model. Here he considered the ether as forming closed vortex cells which are in rotation and between which are small particles functioning as idle wheels. He was able to show that the equations governing the behaviour of this mechanical model corresponded to those of electromagnetism, the rotation of the cells corresponding to the magnetic component and their translation to the electric current. Moreover, Maxwell suggested that vibrations of this ether constituted light, and he subsequently extended his theory to account for dispersion.

Since Maxwell's theory applied to both electromagnetism and optics, ether theorists were faced with the problem of constructing a model that not only would be internally consistent with mechanical principles, but also would be able to explain a wide range of electromagnetic and optical phenomena.

By 1864, Maxwell had abandoned the specific model just outlined, principally because, as Siegel (this volume) suggests, the notion of idle wheels appeared somewhat contrived and could not be translated into a specific physical particle. As Whittaker (1910) discusses in detail, Maxwell's theory led to a plethora of ether models combining rotational and translational motions of ether – one motion corresponding to the electric component, the other to the magnetic. Yet although such models did achieve considerable popularity, German physicists in particular, drawing on neo-Kantian philosophy, often propounded a very different type of ether theory. One example, which is discussed by Wise (this volume), is Helmholtz's theory of 1870 that ether is not composed of particles in motion but may be characterised by its state of polarisation, which itself is propagated through the medium.

One of the most important extensions of Maxwell's theory was Fitzgerald's 1878 account of reflection and refraction.[102] In this he drew on the ether model developed some forty years earlier by James MacCullagh, who in turn was responding to a model of Fresnel's.[103] In order to account for the polarisation of light the ether vibrations had to be transverse, not longitudinal, and therefore ether, Fresnel claimed, must be like an elastic solid. Whereas some writers criticised this conception of ether as incompatible with the free motion of the planets, MacCullagh's main criticism was that the requisite optical phenomena could not be explained by assuming ether to be both incompressible and not susceptible to distortion. What he instead proposed was an elastic solid ether that admitted distortion (Schaffner, 1972:59–68). While MacCullagh utilised this model in optics, Fitzgerald employed it in a wider domain by showing that MacCullagh's equations corresponded to those for Maxwell's theory of electromagnetism.

Stein (this volume) discusses Fitzgerald's model of ether and the responses to it from physicists, particularly Thomson, Larmor, and Lorentz, who developed further ether models. Of the complex developments of the last three decades of the century, which have been discussed by Stein and others (e.g., Whittaker, 1910; Doran, 1975), three in particular deserve attention. One of these was the renewed interest in the theory that the atoms composing material bodies were local disturbances – vortex rings – in an incompressible ether (Silliman, 1963). Helmholtz had shown that vortex rings were stable and indivisible and rebounded after collision, just as atoms were conceived to behave. Moreover, these properties of vortex rings had been demonstrated experimentally using smoke rings. This theory also had the advantage of removing one of the difficulties raised by electromagnetic field theory in that it eliminated the distinction between matter and ether by positing that matter was composed of ether.

A second development involved what some historians have considered a major departure both for ether theory and indeed for the foundations of physics (McCormmach, 1970). In place of mechanical theories of ether, Lorentz and others at about the turn of the century suggested that the ultimate reality was constituted by the electromagnetic ether and electric particles (e.g., electrons). Indeed, they contended that all the laws of nature were reducible to ether, which obeyed the electromagnetic field equations. Doran (1975) argues that this conception was by no means novel but involved an essentially Leibnizian theory of matter held by many British physicists, who conceived ether as a continuous medium to which electromagnetic properties were ascribed. As Stein (this volume) points out, however, there are some difficulties with this historical thesis.

Third, aberration experiments have been a recurrent source of empirical problems for ether theories. To take a couple of early examples, attempts were made to reconcile the observations of James Bradley, published in 1728, and those of later writers by proposing specific relationships between the earth and the ether in its neighbourhood. Young, for example, considered that ether was stationary in respect to the sun and that the earth passed through the ether without disturbing it; on the other hand, Fresnel, in responding to an experiment reported by Arago, suggested that material bodies carry with them excess ether in proportion to the square of their refractive indices (Schaffner, 1972:20–39). However, the most famous aberration experiment was that of Michelson and Morley in 1887, in which they tried to detect the expected second-order difference between the velocity of light along the line of the earth's motion and the velocity perpendicular to it. According to Fresnel's theory of partial drag, which had been revived by Lorentz in conjunction with the electromagnetic theory of light, a detectable difference was expected. However, the Michelson–Morley experiment indicated that the 'relative velocity of the earth and the ether is probably less than one sixth of the earth's orbital velocity, and certainly less than one fourth'; in other words, considerably less than predicted.[104]

One response to the Michelson–Morley experiment was to propose a new version of ether theory that would 'save the phenomena', or more exactly, save the theory. In the 1890s, Fitzgerald and Lorentz suggested the initially ad hoc hypothesis that moving bodies contract in the direction of their motion through the ether by a factor of $\sqrt{1-(v^2/c^2)}$.[105] This formula became central to Einstein's special theory of relativity and Einstein readily acknowledged his debt to Lorentz. Though Holton (1969) has argued that the Michelson–Morley experiment was not of crucial importance to the development of Einstein's theory, as has sometimes been claimed, Einstein was directly concerned with the problem of matter–ether interaction (Hirosige,

Introduction 53

1976). His reading of Lorentz, and thus his familiarity with earlier ether theories, provides a route by which ether theory may be connected with the theoretical core of twentieth-century physics.

Epilogue: the twentieth century

There is a widely held view that ether theories suffered a dramatic and sudden demise with the rise of the special theory of relativity, since a stationary ether permeating space seems incompatible with Einstein's two postulates. In other words, after 1905 ethers became the concern of the historian rather than the physicist. To endorse this view would allow the present volume to end on an incisive note. However, conceptually rich theories, unlike the individuals who frame them, do not suddenly disappear from science but instead often survive in some modified form. This is certainly the case with ethers. Concern with ether continued after 1905, and it has been suggested by Goldberg (1970) that, particularly among British physicists, ether theorising contributed to the initial lack of interest in Einstein's theory. Most important, however, the view that ether theories did not survive 1905 is incorrect because there have been, and still are, many ether theories that, in principle, are perfectly compatible with special relativity and even general relativity. Moreover, quantum theory has led to new conceptions of ether, and not a few physicists have urged the *necessity* of some form of ether theory.

Like E. T. Whittaker, we have devoted a single volume to a survey of ether theories in the eighteenth and nineteenth centuries. As he required a second volume for the early twentieth century – and even wrote a third that has never been published – so we must suggest that to do justice to the recent and highly technical history of the topic at least one further volume would be required. Although ether theorising has continued during the present century, the rise of Einstein's theories of relativity and other developments have raised a number of new and difficult problems that, in turn, have placed further constraints on ether models. Of the many recent physicists who have been concerned with ethers we can select only two, Einstein and Dirac, to illustrate the continuing history of the subject beyond the period covered in the present volume.

Einstein, in an address delivered at the University of Leyden in 1920, stated:

> More careful reflection teaches us, however, that the special theory of relativity does not compel us to deny ether. We may assume the existence of an ether; only we must give up ascribing a definite state of motion to it, i.e. we must by abstraction take from it the

last mechanical characteristic which Lorentz had still left it. . . . [There] is a weighty argument to be adduced in favour of the ether hypothesis. To deny ether is ultimately to assume that empty space has no physical qualities whatever. The fundamental facts of mechanics do not harmonize with this view. . . . According to the general theory of relativity space without ether is unthinkable; for in such space there would not only be no propagation of light, but also no possibility of existence for standards of space and time (measuring-rods and -clocks), nor therefore any space–time intervals in the physical sense.[106]

Our second example, Dirac, submitted a short but famous letter to *Nature* in 1951, which included the following:

Physical knowledge has advanced very much since 1905, notably by the arrival of quantum mechanics, and the situation has again changed. If one examines the question in the light of present-day knowledge, one finds that the aether is no longer ruled out by relativity, and good reasons can now be advanced for postulating an aether.[107]

Notes

1 G. S. Kirk and J. E. Raven, *The Presocratic philosophers* (Cambridge, 1964), 10–14. See also W. K. C. Guthrie, *The Greeks and their gods* (London, 1950), 207–8, 263, 324; W. K. C. Guthrie, *A history of Greek philosophy* (Cambridge, 1962), 1:270–3; O. Gilbert, *Die meteorologischen Theorien des griechischen Altertums* (Leipzig, 1907), 17–65, 662–701.

2 For the texts of Anaximenes and Heraclitus and commentary, see Kirk and Raven, *The Presocratic philosophers*, 143–62, 182–215. A recent, incisive survey of pre-Socratic philosophy is given in J. Barnes, *The Presocratic philosophers*, 2 vols. (London, 1979). Valuable articles are collected in D. J. Furley and R. E. Allen (eds.), *Studies in Presocratic philosophy*, 2 vols. (London and New York, 1970, 1975); and in A. P. D. Mourelatos (ed.), *The Presocratics* (New York, 1964).

3 The historical significance of the Parmenidean impasse, clearly appreciated by Aristotle, is made a central theme in Barnes, *The Presocratic philosophers*.

4 Texts and commentary for these authors are given in Kirk and Raven, *The Presocratic philosophers*, 263–85, 320–95, 400–26. Our summary draws on Hesse (1961), 35–45.

5 For Plato on matter and the elements, see G. Vlastos, *Plato's universe* (Oxford, 1975). Important collections of articles on Plato include G. Vlastos, *Platonic studies* (Princeton, N.J., 1973); R. E. Allen (ed.), *Studies in Plato's metaphysics* (London, 1965); and G. Vlastos (ed.), *Plato: a collection of critical essays*, 2 vols. (New York, 1971). See also the article 'Plato' in *Dictionary of scientific biography*, ed. C. C. Gillespie, 15 vols. (New York, 1970–8).

6 See the comprehensive and detailed article 'Quinta essentia', by P. Moraux, in A. F. von Pauly, *Realencylopädie der classischen altertumswissenschaft*, ed. G. Wissowa,

Introduction 55

(1963), 47:1181–1263. We have been reminded (by an anonymous referee) that a clear statement of the theory of five elemental bodies, including *aither*, is to be found at 981c of the *Epinomis*, a work plausibly ascribed to Plato or a close associate. Moraux discusses whether this statement anticipates Aristotle's teaching on the five elements.

7. See F. Solmsen, *Aristotle's system of the physical world* (Ithaca, N.Y., 1960). A valuable collection of articles is J. M. E. Moravcsik (ed.), *Aristotle* (New York, 1967). See also the article, by G. E. L. Owen, D. M. Balme, and L. G. Wilson, on Aristotle in Gillespie, *Dictionary of scientific biography*.

8. Solmsen, *Aristotle's system*, 287–301; F. Solmsen, 'The vital heat, the inborn pneuma and the aether', *Journal of Hellenic Studies 77* (1957), 119–23.

9. Solmsen, 'The vital heat, the inborn pneuma and the aether'; A. L. Peck, *Aristotle: generation of animals*, rev. ed. (Cambridge, Mass., and London, 1953), app. B.

10. On Stoic physics, generally, see S. Sambursky, *Physics of the Stoics* (London, 1959); D. Hahm, *The origins of Stoic cosmology* (Columbus, Ohio, 1977); H. A. K. Hunt, *A physical interpretation of the universe: the doctrines of Zeno the Stoic* (Melbourne, 1976). This last translates a number of passages into English for the first time. On *pneuma*, see G. Verbeke, *L'Evolution de la doctrine du pneuma* (Paris, 1945).

11. Sambursky, *Physics of the Stoics*, 21–48; Hahm, *Stoic cosmology*, 157–74; Hunt, *A physical interpretation of the universe*, 17–59.

12. Nemesios, *De natura hominis*, quoted, in translation, in Sambursky, *Physics of the Stoics*, 128.

13. A. A. Long, *Hellenistic philosophy* (London, 1974), 157.

14. M. Lapidge, 'Stoic cosmology', in *The Stoics*, ed. J. M. Rist (Berkeley, Calif., 1978), 161–86.

15. See especially Plotinus, *Enneads*, I, iv, and V, i.

16. Lucretius, *De rerum natura*, V, lines 416–649.

17. Here Origen's *De principiis* (written in the third century) was an important source for later writers. For other such early sources, see Moraux's article cited in n. 6.

18. M. A. Hewson, *Giles of Rome and the medieval theory of conception* (London, 1975).

19. Our summary here follows D. Lindberg, *Theories of vision from Al-kindi to Kepler* (Chicago, 1976), 94–103.

20. F. A. Yates, *Giordano Bruno and the Hermetic tradition* (London, 1964), 1–20.

21. R. S. Westman and J. E. McGuire, *Hermeticism and the scientific revolution* (Los Angeles, 1977).

22. W. Pagel, 'Paracelsus', in Gillespie, *Dictionary of scientific biography*.

23. W. Pagel, 'Paracelsus and the neo-Platonic and gnostic tradition', *Ambix 8* (1960), 125–66; J. R. Partington, *A history of chemistry*, 4 vols. (London, 1959–70), 2:chap. 3; A. G. Debus, *The chemical philosophy: Paracelsian science and medicine in the sixteenth and seventeenth centuries*, 2 vols. (New York, 1977).

24. See, for example, M. Bonelli Righini and W. Shea (eds.), *Reason, experiment and mysticism in the scientific revolution* (New York, 1975). V. L. Bullough (ed.), *The scientific revolution* (New York, 1970), collects many such interpretations.

25. N. Kemp Smith, *New studies in the philosophy of Descartes* (London, 1952), 103–14.

26. The text of *Le monde* is in vol. 11 of R. Descartes, *Oeuvres de Descartes*, eds. C. Adam and P. Tannery, 12 vols. (Paris, 1897–1910). An English translation of some opening sections is in R. Descartes, *Descartes: selections*, ed. and trans. R. M. Eaton (New York, 1927).

27. Kemp Smith, *New studies*, 118–22.

28 *Le monde* is usually thought to have been completed by 1633. Love shows that in its treatment of subtle matter, at least, it is closer to *Principia* than to earlier writings: R. Love, 'Revisions of Descartes's matter theory in *Le monde*', *British Journal for the History of Science 8* (1975), 127–37.
29 Our summary partly follows A. Kenny, *Descartes: a study of his philosophy* (New York, 1968), 204–5.
30 *Principia philosophiae*, bk. 1, §§ 51–64. The Latin text (1644) and French translation (1647) are in vols. 8 and 9 of the Adam and Tannery edition of Descartes' works cited in n. 26. For a partial translation into English, see R. Descartes, *The philosophical works of Descartes*, trans. E. S. Haldane and G. R. T. Ross, 2 vols. (Cambridge, 1968) 1:201–303. For a recent analysis of Descartes' arguments for extension as the essence of matter see R. J. Blackwell, 'Descartes' concept of matter', in *The concept of matter in modern philosophy*, ed. E. McMullin, rev. ed. (Notre Dame, Ind., 1978), 759–75.
31 *Principia philosophiae*, bk. 2, §§ 36–45.
32 For the complexities in Descartes' conceptions of motion and force, see Westfall (1971), 56–98; A. Gabbey, 'Force and inertia in seventeenth-century dynamics', *Studies in History and Philosophy of Science 2* (1971), 1–68;
P. H. J. Hoenen, 'Descartes's mechanicism', in *Descartes*, ed. W. Doney (New York, 1967), 353–68; P. Machamer, 'Causality and explanation in Descartes' natural philosophy', in *Motion and time, space and matter*, eds. P. Machamer and R. Turnbull (Columbus, Ohio, 1976), 168–99; W. E. Anderson, 'Cartesian motion', in ibid., 200–23; and even more recently, D. Clarke, 'Physics and metaphysics in Descartes' *Principles*', *Studies in History and Philosophy of Science 10* (1979), 89–112; and G. C. Hatfield, 'Force (God) in Descartes' physics', ibid., 113–40.
33 See P. J. Olscamp's introduction to R. Descartes, *Discourse on method, optics, geometry and meteorology* (Indianapolis, 1965); G. Buchdahl, *Metaphysics and philosophy of science* (London, 1969).
34 Aiton (1972), 49–57.
35 A. Sabra, *Theories of light from Descartes to Newton* (London, 1967), 46–69.
36 Kemp Smith, *New Studies*, 130–7.
37 Ibid., 105–11.
38 The complete text of this essay is reprinted in M. B. Hall, *Robert Boyle on natural philosophy: an essay with selections from his writings* (Bloomington, Ind., 1965), from R. Boyle, *The works of the Honourable Robert Boyle*, ed. T. Birch, 2nd ed., 6 vols. (London, 1772), 4:68–78.
39 See §§ 3–5 of Boyle's essay.
40 R. Boyle, 'About an attempt to examine the motions and sensibility of the Cartesian Materia Subtilis, or the Aether, with a pair of bellows, made of a bladder, in the exhausted receiver', reprinted in Hall, *Robert Boyle on natural philosophy*, 358–63, from Boyle, *Works*, 3:250–2.
41 Boyle, in Hall, *Robert Boyle on natural philosophy*, 198, 209. It is thus an error to think that mechanical philosophers were distinctive in not admitting active principles.
42 For such an attempt, see P. Rattansi, 'Newton's alchemical studies', in *Science, medicine and society in the Renaissance: essays to honor Walter Pagel*, ed. A. G. Debus, 2 vols. (New York, 1972), 1:167–82; P. Rattansi, 'Some evaluations of reason in sixteenth and seventeenth-century natural philosophy', in *Changing perspectives in the history of science: essays in honour of Joseph Needham*, eds. M. Teich and R. Young (London, 1972), 148–66.
43 For the text (with translation) of 'De gravitatione', see A. R. Hall and M. B. Hall

(eds. and trans.) *Unpublished scientific papers of Isaac Newton* (Cambridge, 1962), 89–156.
44 W. Gilbert, *On the magnet, magnetick bodies also, and on the great magnet the earth: a new physiology, demonstrated by many arguments & experiments* (London, 1900), 209–10.
45 E. J. Dijksterhuis, *The mechanization of the world picture*, trans. C. Dikshoorn (Oxford, 1961), 313.
46 Descartes, *Principia philosophiae*, bk. 4, §§ 133–180. Cf. J. F. Scott, *The scientific work of René Descartes* (London, 1952), 188–92.
47 Our summary follows that in J. Passmore's article on More in *Encyclopedia of philosophy*, ed. P. Edwards, 8 vols. (New York and London, 1967), 5:387–9.
48 R. Cudworth, *The true intellectual system of the universe: wherein all the reason and philosophy of altruism is confuted, and its impossibility demonstrated*, 3 vols. (London, 1845), 3:231–2.
49 H. Power, *Experimental philosophy in three books: containing new experiments microscopical, mercurial, magnetical* (London, 1664), 71–2, 152–61. Cf. Rattansi, 'Newton's alchemical studies'.
50 Cudworth, *True intellectual system*, 1:249.
51 I. Newton, 'An hypothesis explaining the properties of light', in T. Birch, *The history of the Royal Society of London* 4 vols. (London, 1756–7), 3:247–305; I. Newton to R. Boyle, in R. Boyle, *The works of the Honourable Robert Boyle*, ed. T. Birch, 5 vols. (London, 1744), 1:70–3. P. 74 of this work contains a letter from Newton to Oldenburg concerning ether. These texts are reprinted in facsimile in *Isaac Newton's papers and letters on natural philosophy*, ed. I. B. Cohen (Cambridge, Mass., 1958).
52 See, in particular, Aiton (1969); Guerlac (1967); Hall and Hall (1967); Hawes (1968*b*); McGuire (1968); McMullin (1978); Rosenfeld (1969). Only after completion of this introduction did the editors become aware of Corson (1974), which contains a detailed discussion of the chronological development of Newton's ideas about ethers drawing on both published and manuscript documents. Corson's interpretations are generally though not invariably in accord with those suggested here.
53 I. Newton, 'De aere et aethere', in Hall and Hall, *Unpublished scientific papers of Isaac Newton*, 227.
54 I. Newton, *Optice* (London, 1706), 344. The translation is that of the *Opticks*, 4th ed. (London, 1730; reprinted, New York, 1952), 401.
55 For example, Newton 'An hypothesis', 253; Newton to Boyle, 71.
56 'An hypothesis', 249.
57 Cotes's preface and the 'General scholium' added to the 2nd (1713) ed. of the *Principia*. See also query 31 of the *Opticks*.
58 Newton, *Opticks*, 2nd ed. (London, 1717), 326. See also McGuire (1970).
59 'An hypothesis', 255. See also Newton to Boyle, 70.
60 *Opticks* (1717), 292–313. See R. H. Stuewer, 'A critical analysis of Newton's work on diffraction', *Isis 61* (1970), 188–205.
61 *Opticks* (1717), 324–6. However, Howard Stein has drawn to our attention a passage in the *Principia* (bk. 3, prop. 6, cor. 2) where Newton states that ether itself is subject to gravitational attraction. There remains the problem of how these passages can be reconciled.
62 *Opticks* (1717), 327.
63 'An hypothesis', 250–1.
64 *Opticks* (1717), 323–4.
65 'An hypothesis', 252–5; *Opticks* (1717), 328.
66 Some of Newton's accounts of ether, especially in manuscripts he did not publish,

even raise the question whether, after all, ether is mechanical in the sense of conforming to his own laws of motion, his own science of mechanics. At least once – in a passage quoted by Heimann (this volume) in his n. 17 – Newton seems to describe ether as lacking resistance and so without inertia, and as not acting, therefore, in accord with his laws of mechanics. But how is inertia to be avoided in an ether presumably formed from particles of the same matter as air? If the particles had negligible volume, then, perhaps, the medium would have no measurable inertia. But could God still be conceived as endowing such particles with attractive or repulsive forces and, if so, with what consequences for the medium as a whole and for the bodies floating in it? As McMullin (1978) implies, in discussing such texts, a unified and coherent interpretation of all Newton's ether theorising, published and unpublished, still eludes even recent scholarship.

67 On Boerhaave's career see G. A. Lindeboom, *Hermann Boerhaave: the man and his work* (London, 1968). For the chemistry, H. Metzger, *Newton, Stahl, Boerhaave et le doctrine chimique* (Paris, 1930), remains indispensable.
68 Lindeboom, *Hermann Boerhaave,* 179 ff.
69 Love (1974); H. Boerhaave, *A new method of chemistry,* trans. P. Shaw and E. Chambers (London, 1727), 220–38.
70 Love (1974); Boerhaave, *A new method of chemistry,* trans. P. Shaw, 2 vols. (London, 1741), 1:386–7, 397.
71 Boerhaave, *A new method* (1727), 222, 233.
72 On Robinson and his career, see Schofield (1970).
73 Fox (1971). H. Guerlac gives a bibliography in his article on Lavoisier in Gillespie, *Dictionary of scientific bibliography.* This article is reprinted, with minor changes, as H. Guerlac, *Antoine-Laurent Lavoisier, chemist and revolutionary* (New York, 1975).
74 For example, E. Mach, *The principles of physical optics,* trans. J. S. Anderson and A. F. A. Young (London, 1926).
75 *Opticks* (1717), 324–5.
76 For example, I. B. Cohen (ed. and intro.), *Benjamin Franklin's experiments: a new edition of Franklin's experiments and observations on electricity* (Cambridge, Mass., 1941), 302; Cohen (1956); D. H. D. Roller, 'The development of the concept of electric charge: electricity from the Greeks to Coulomb', in *Harvard case histories in experimental science,* eds. J. B. Conant and L. K. Nash, 2 vols. (Cambridge, Mass., 1957), 2:541–639.
77 Cohen, *Benjamin Franklin's experiments;* Cohen (1956); D. Roller, 'The early development of concepts of temperature and heat: the rise and decline of the caloric theory', in Conant and Nash, *Harvard case histories,* 1:117–214.
78 Sabra, *Theories of light from Descartes to Newton;* A. Fresnel, *Oeuvres complètes d'Augustin Fresnel, publiées par Henri de Sénarmont, Emile Verdet et Léonor Fresnel,* 3 vols. (Paris, 1866–70).
79 D. Gooding, 'Conceptual and experimental bases of Faraday's denial of electrostatic action at a distance', *Studies in History and Philosophy of Science 9* (1978), 117–49; R. Hare, 'A letter to Professor Faraday, on certain theoretical opinions', *American Journal of Science and Arts 38* (1840), 1–11, and *41* (1841), 1–14. Faraday replied in ibid. *39* (1840), 108–20.
80 D. Hartley, *Observations on man, his frame, his duty, and his expectations* (London, 1749); J. Priestley, *Hartley's theory of the human mind, on the principle of association of ideas: with essays relating to the subject of it* (London, 1775).
81 R. Boscovich, *Theoria philosophiae naturalis,* 2nd ed. (Venice, 1763); W. Herschel, *The scientific papers of Sir William Herschel,* ed. J. L. E. Dreyer, 2 vols. (London, 1912); J. Priestley, *The history and present state of discoveries relating to*

vision, light and colours (London, 1772); Schofield (1970), 235-76; R. McCormmach, 'Henry Cavendish: a study of rational empiricism in eighteenth-century natural philosophy', *Isis 60* (1969), 293-306.
82 T. Reid, *Essays on the intellectual powers of man* (Edinburgh, 1785), esp. essay 1, chap. 3, and essay 2, chap. 3. See also Laudan (1970).
83 H. Martineau (ed.), *The positive philosophy of Auguste Comte*, 2 vols. (London, 1853), 1:225. Cf. W. R. Grove 'On the correlation of physical forces', *Literary Gazette, and Journal of Belles Lettres* (1844), 25.
84 F. Engels, *Herrn Eugen Dührings Umwälzung der Wissenschaft* (Stuttgart, 1921).
85 Lord Salisbury, 'Presidential address', *Report of the sixty-fourth meeting of the British Association for the Advancement of Science* (London, 1894), 8.
86 W. Ostwald, *Vorlesungen über Naturphilosophie* (Leipsig, 1902). See also E. Hiebert, 'The energetics controversy and the new thermodynamics', in *Perspectives in the history of science and technology,* ed. D. H. D. Roller (Norman, Okla., 1971), 67-86.
87 H. Poincaré, *Science and hypothesis* (London, 1905), 211.
88 Goldfarb (1977). Cf. Fox (1979).
89 See R. E. Schofield, 'Joseph Priestley, natural philosopher', *Ambix 14* (1967), 1-15; and the criticism of that article in J. G. McEvoy, 'Joseph Priestley, natural philosopher: some comments on Professor Schofield's views', *Ambix 15* (1968), 115-21. See further J. G. McEvoy and J. E. McGuire, 'God and nature: Priestley's way of rational dissent', *Historical Studies in Physical Sciences 6* (1975), 325-404; J. G. McEvoy, 'Joseph Priestley, "aerial philosopher" ', *Ambix 25* (1978), 1-55, 93-116, 153-75, and *26* (1979), 16-38.
90 McGuire (1974). This article draws in turn on Heimann and McGuire (1971), Heimann (1971), and Heimann (1970). See also J. E. McGuire and P. Heimann, 'The rejection of Newton's concept of matter in the eighteenth century', in McMullin, *The concept of matter in modern philosophy.*
91 On the historical and conceptual complexities surrounding the notion of field, see Stein (1970); Tonnelat (1959).
92 I. Watts, *The works of Isaac Watts,* eds. D. Jennings and P. Doddridge, 6 vols. (London, 1753), V, 594.
93 T. S. Kuhn, 'Mathematical versus empirical traditions in the development of physical science', *Journal of Interdisciplinary History 7* (1976), 1-31; reprinted in T. S. Kuhn, *The essential tension* (Chicago, 1979).
94 S. Carnot, *Reflexions sur la puissance motrice du feu et sur les machines propres a développer cette puissance* (Paris, 1824).
95 For example, John Hutchinson, Samuel Pike, and William Jones of Nayland. See Heimann (1973); Heimann and McGuire (1971); the forthcoming dissertation by Chris Wilde of Cambridge University.
96 See, principally, T. Young, 'On the theory of light and colours', *Philosophical Transactions 92* (1802), 12-48; T. Young, *A course of lectures on natural philosophy and the mechanical arts,* 2 vols. (London, 1802), 1:457-71.
97 R. Fox, 'The rise and fall of Laplacian physics', *Historical Studies in the Physical Sciences 4* (1974), 89-136.
98 E. Frankel, 'The search for a corpuscular theory of double refraction: Malus, La Place and the prize competition of 1808', *Centaurus 18* (1974), 223-45; Buchwald (this volume).
99 E. Nagel, *The structure of science: problems in the logic of scientific explanation* (London, 1961), 153-202.
100 R. Moon, *Fresnel and his followers: a criticism, to which are appended outlines of theories of diffraction and transversal vibration* (Cambridge, 1849), ix.

101 W. Rankine, 'Outlines of the science of energetics', *Proceedings of the Royal Philosophical Society of Glasgow 3* (1855), 381–99.
102 G. F. Fitzgerald, 'On the electromagnetic theory of the reflection and refraction of light', *Philosophical Transactions 170* (1880), 691–711. See Schaffner (1972), 204–7.
103 J. MacCullagh, 'An essay towards a dynamical theory of crystalline reflexion and refraction', and other related papers reprinted in J. MacCullagh, *The collected works of James MacCullagh,* eds. J. H. Jollett and S. Haughton (Dublin, 1880).
104 A. A. Michelson and E. W. Morley, 'On the relative motion of the earth and the luminiferous ether', *Philosophical Magazine 24* (1887), 449–63.
105 Several papers on this subject have been reprinted in vol. 4 of H. A. Lorentz, *Collected papers,* 9 vols. (The Hague, 1935–9).
106 A. Einstein, 'Ether and relativity', in *Sidelights on relativity* trans. G. B. Jeffery and W. Perrett (London, 1922).
107 P. A. M. Dirac, 'Is there an aether?' *Nature 168* (1951), 906–7.

1

Ether and imponderables

P. M. HEIMANN

Department of History, Furness College, University of Lancaster, Bailrigg, Lancaster LA1 4YG, England

The physics of active, ethereal imponderable fluids represented the main current of speculation among British natural philosophers in the second half of the eighteenth century. These 'fluids' were envisaged as being composed of particles that mutually repelled each other, this property being referred to as 'elasticity'. The 'elastic fluids' were conceived as being 'subtle' (being able to penetrate the empty spaces between the particles of ordinary matter in bodies), as being weightless (or at least with no measurable weight), and as being attracted by the particles of ordinary matter. Frequently postulated to explain the phenomena of electricity, magnetism, optics, heat, and chemistry, the operations of the repellent particles of the imponderable fluids were thus traced to interparticulate forces of attraction and repulsion. These characteristic features of the imponderable fluids suggest that their articulation was influenced by Newton's concept of ether, postulated as a 'subtle', 'elastic', particulate substance in the queries appended to the second English edition of the *Opticks* (1717), and by Newton's theory of interparticulate forces of attraction and repulsion, especially as explicated in query 31 of the 1717 *Opticks* (Cohen, 1956; Schofield, 1970; Heimann and McGuire, 1971; Heimann, 1973).

The extent to which the theory of ethereal substances proposed by eighteenth-century British natural philosophers can be analysed in relation to Newtonian categories has, however, been questioned (Home, 1977*a*). The theory of imponderable fluids may seem to echo the Cartesian doctrine of the all-pervading subtle ether that, though in decline in cosmology (Aiton, 1972:244–56), provided the model for the fluid theories of fire and magnetism advanced by Daniel Bernoulli and Euler in the 1740s. Nevertheless, there are important differences between the Cartesian and Newtonian ethers. The Cartesian ether was a plenum; its operations arose from the contact action of its component particles. By contrast, the Newtonian ether was composed of particles that were separated by void space and that acted on gross bodies by means of their repulsive forces. The 'Newtonian' character of the elastic fluids postulated by British natural philosophers was explicit, and cannot be regarded as a Newtonian veneer disguising an essentially Cartesian vortex model. Nevertheless, the development of the imponderable fluid theories cannot be ascribed solely to the influence of the ether concept, though in attempting to clarify the diverse conceptual origins of the imponderable fluid theories this chapter emphasises the primary importance of the Newtonian concept of ether. An analysis of the development of Continental imponderable fluid theories would show different patterns of thought, with 'Newtonian' concepts exercising a less dominant influence; and although understanding imponderable fluid theories in eighteenth-century British natural philosophy in relation to Newton's theory of ether does not completely characterise their intellectual provenance, that theory does provide an illuminating context for historical analysis.

Newton's ether theory aroused little interest until the 1740s, and a central aim of this chapter is to analyse the reasons for its adoption at that time and to clarify the manner in which the concept of ether provided the paradigm for the imponderable fluid theories. Further, the incorporation of Boerhaave's 'fire' in theories of an ethereal 'electrical fire' and of Stahl's 'phlogiston' (the chemical principle of inflammability) as a modification of ether led to a broadening and transformation of the theory of ether. By the end of the century attempts were made to formulate a unified theory of ether, reducing the diversity of phenomena to the modifications of a single ethereal active substance. This chapter attempts, then, to clarify how Newton's theory of ether was transformed into the concept of an inherently active substance, a theory of nature that contradicted Newton's own doctrine of the intrinsic passivity of material entities, all activity being for him ultimately grounded in divine agency. For these later theorists, active powers were held to be intrinsic to some or even all material substances, and ether functioned as the source of the activity of nature. The transformation of Newton's theory of ether was associated with a blurring of the distinction between the categories of activity and passivity in Newton's natural philosophy, and a rejection of its theological implications.[1]

However, despite the dominance of the theory of ethereal fluids in late eighteenth-century British natural philosophy, alternative conceptual schemes continued to be canvassed. The speculative and qualitative cast of the theory of imponderable fluids was unacceptable to those natural philosophers interested in a mathematical theory of nature, to whom Newton's *Principia*

(1687), with its stress on the role of the forces quantitatively determining natural phenomena, appeared a more appropriate paradigm for physical theory. The Scottish natural philosophers Robison and Playfair opposed the supposition of interstitial particles of ether to explain gravity and of imponderable 'ethers' to account for the phenomena of electricity and heat (Cantor, 1971; Olson, 1975:157–224). Cavendish attempted to develop Newton's theory of the unity of matter and the Newtonian programme of the quantification of the interparticulate forces; for Cavendish, the fluids of light, electricity, and phlogiston were ponderable, and heat was regarded as the motion of the particles of ordinary matter. Ether played no role in his theory of nature, and he rejected the supposition of imponderable fluids and anomalous forms of matter.[2]

The disjunction between the theory of imponderable fluids and the quantitative theory of interparticulate forces as alternative programmes for physical explanation does illuminate the aims of some eighteenth-century British natural philosophers. However, it has been argued by Schofield that there was a fundamental opposition in eighteenth-century British natural philosophy between (1) any theory of atoms and interparticulate forces and (2) any theory of an ether or of imponderable fluids. On Schofield's account, although these two lines of theorising both derived from Newton's natural philosophy, they employed quite distinct principles of physical explanation.[3] This interpretation imposes a distorting analytic framework, simplifying Newton's natural philosophy and the complexities of eighteenth-century interpretations of Newton's theory of nature. In Newton's natural philosophy the mode of action of ether was grounded on the agency of repulsive forces; and eighteenth-century ether theorists stressed the role of the repulsive forces associated with the ethereal imponderable fluids, arguing that there was a balance in nature between the attractive force associated with ordinary matter and the repulsive force of the ethereal fluids. To make a contrast between force and ether concepts the analytic framework for the interpretation of eighteenth-century British natural philosophy is to ignore the relation between Newton's ether and his concept of 'active principles', natural agents that he held to be distinct from the 'passive' principles that characterised the properties of matter. In this chapter it is argued that the status of ether as an active principle was fundamental to the development of theories of active, ethereal imponderable fluids. In place of a disjunction between 'force' and 'ether' concepts, it is suggested that a more appropriate analytic framework for characterising the 'Newtonian' origins of British imponderable fluid theories is to be found in the relations among the Newtonian concepts of forces, active principles, and ether. For Newton's speculations were ambiguous, and these categorial complexities were reflected in the conceptual structures of eighteenth-century ether theories, in which 'Newtonian' concepts of force, active principles, and ether were conflated in the theory of inherently active ethereal imponderables endowed with repulsive forces (Heimann and McGuire, 1971).

Newton's ether theory: the published sources

Newton provided several sources from which eighteenth-century natural philosophers could learn of his ether theory. His first published discussion of ether was his introduction of a subtle and elastic ether in the queries to the 1717 *Opticks* to explain optical reflection and refraction and to provide a causal explanation for gravity.[4] But his earliest speculations on ether, introduced in his 'Hypothesis explaining the properties of light' written in 1675 (though not published until 1757)[5] and in a letter to Oldenburg of 1676 correcting this paper (published in 1744),[6] placed these optical speculations in the context of a cosmology based on the circulation and transformation of ethereal spirits. Newton qualified this ethereal cosmology in a letter to Boyle written in 1679 (though not published until 1744),[7] where he expressed the possibility of an explanation of gravity in terms of the differential densities and sizes of ether particles. The problems of interpreting Newton's theory of ether derive both from the different emphases of these discussions and from the ambiguity of Newton's arguments.

Newton's ether was thus most familiar to the eighteenth-century natural philosophers in the account given in the 1717 *Opticks,* which introduced the physical model of ether as composed of mutually repelling particles. The most extended account of the mode of action of ether was given in query 21, where Newton argued that

> the exceeding smallness of its Particles may contribute to the greatness of the force by which those Particles may recede from one another, and thereby make that Medium exceedingly more rare and elastick than Air, and by consequence exceedingly less able to resist the motions of Projectiles, and exceedingly more able to press upon gross Bodies, by endeavouring to expand it self.[8]

In suggesting this model to explain the agency of gravity, Newton was not invoking the Cartesian concept of pressing or impact of contiguous particles constituting a plenum. In the *General scholium* to the second edition of *Principia* (1713) he had argued that gravity 'must proceed from a cause that penetrates to the very centres of the sun and planets' and that 'operates not according to the quantity of the surfaces of the particles upon which it acts (as mechanical causes are accustomed)'.[9] In referring to the 'pressing' of the ether, Newton was not suggesting the Cartesian impact model; nor was he envisaging a fluid medium acting by hydrostatic pressure. As he pointed out in query 28, at any point in a fluid the pressure acts equally in all directions, whereas 'Gravity tends downwards';[10] hence fluid pressure is irreconcilable with the directionality of gravity. Though Newton's ether has sometimes been viewed as providing some form of contact-action explanation of gravity, Newton's own statements emphasised the irreducibility of gravity to 'mechanical' or contact-action theories.[11]

Newton apparently envisaged ether acting by a differential density arising from the repulsive forces exerted by the minute particles of ether, while the great 'elastick force' of the ether, its tendency to 'expand it self', enabled it to 'press upon gross Bodies' and to cause planets to approach or recede. Arguing analogically, in query 21, Newton claimed that just as the small size of light corpuscles implied forces of great intensity in relation to size, the particles of ether, which were 'exceedingly smaller than those of Air, or even than those of Light', were endowed with the strongest forces with respect to their sizes. Ether was composed of 'Particles which endeavour to recede from one another', its agency manifested through the differential density of the ethereal 'medium'.[12] So Newton's gravitational ether did not act by contact action; its particles were separated by void space and acted on one another by their repulsive forces. It has been argued that in attempting to reduce the force of gravity to the repulsive forces of the ether particles, Newton's ether embodied the problem of action at a distance that it purported to explain (McGuire, 1968:187; Westfall, 1971:395). In questioning the intelligibility of ether as an explanation of gravity this interpretation ignores an important characteristic of the ether, its status as a Newtonian active principle.

Newton distinguished between gravity and those physical properties – hardness, extension, inertia – that he held to be intrinsic to the nature of matter. As he emphasised in a famous statement to Bentley (published in 1756), gravity was not to be considered as 'innate, inherent and essential to Matter'.[13] In query 31 of the 1717 *Opticks,* Newton contrasted passive principles such as *vis inertiae* (force of inertia) and the 'passive Laws of Motion as naturally result from that Force' with 'active Principles, such as that of Gravity'.[14] In Newton's natural philosophy active principles were agents that were not reducible to the passive principles of matter. As a putative explanation of gravity the ether of the 1717 *Opticks* was, by implication, an active principle establishing the intelligibility of the phenomenon of gravity (Heimann and McGuire, 1971:240–5). This interpretation is in consonance with the theological function of ether. One likely motive for Newton's inclusion of ether in the 1717 *Opticks* was to refute Leibniz's criticism of Newton's concept of gravity as a 'miracle', an 'occult quality' and a 'fiction'.[15] Arguing for a contact-action or 'mechanical' explanation of gravity, Leibniz expounded his views at length in his correspondence with Clarke in 1715–16. Newton conceived active principles as manifest in certain 'general Laws of Nature'. Regarded as the manifestation of God's lawful, causal agency in nature, they functioned as the cause of motion and gravity.[16] Rejecting the reducibility of gravity to a contact-action model, Newton conceived the ether of the 1717 *Opticks* as an active principle communicating God's causal agency and as a physical model (though not a contact-action model) establishing the intelligibility of the distance force of gravity.

This interpretation of ether as an active principle implies that ether had an ambiguous conceptual status in Newton's natural philosophy. Composed of particles of matter, ether would ostensibly appear to fall under the category of passive principles. However, Newton's manuscript references to ether between 1706 and 1717 show that he sought to distinguish between the passivity and inertia of *ordinary* matter and the active properties of ether.[17] Despite these reflections on the anomalous nature of ether particles, Newton did not characterise their properties in a systematic way, and his only published hints about these issues were in his discussion of the 'greatness' of the 'elastick force' of ether, and its implied status as an active principle. Nevertheless the conceptual status of ether as an active principle was to exercise considerable influence on the development of the imponderable fluid theories. In Newton's natural philosophy active principles were distinct from the passive properties of matter, and their activity was dependent on God's causal agency; but the ambiguous conceptual status of Newton's ether as both an active and a material principle led many eighteenth-century theorists to interpret ether as a substance endowed with inherent activity, conflating the active – passive dualism of Newton's natural philosophy.

In query 31 of the 1717 *Opticks*, Newton linked active principles with the 'great and violent' processes of chemistry, questioning the reducibility of chemistry to the 'passive Laws of Motion'.[18] Taking ether as an active principle, some eighteenth-century chemists assimilated 'fire' and 'phlogiston' to ether and emphasised the irreducibly chemical properties of ether. The association between ether and chemical active principles was heightened by the publication of Newton's 'Hypothesis on light' and his letter to Oldenburg, where the operations of ether were linked to chemical processes, and where ether served the same function as active principles in maintaining the activity of nature. Newton suggested that 'the whole frame of nature may be nothing but aether condensed by a fermental principle'. He supposed that nature

> may be nothing but various Contextures of some certaine aethereall Spirits or vapours condens'd as it were by precipitation, much after

the manner that vapours are condensed into water or exhalations into grosser Substances and after condensation wrought into various formes, at first by the immediate hand of the Creator, and ever since by the power of Nature . . . Thus perhaps may all things be originated from aether.

Newton added that 'nature is a perpetuall circulatory [the word *circulatory* was omitted from the published version] worker'; by the chemical transformation of ethereal spirits and by 'nature making a circulation', the activity of the cosmos was conserved.[19] In stressing that ether was an underlying first principle from which all things originated, in supposing that the activity of the cosmos was conserved by the circulation of ethereal spirits, in arguing for the generation of all things from ether, and in relating the operations of ether to chemical processes, Newton echoed seventeenth-century alchemical and neo-Platonist writers.[20] In a manner analogous to his disjunction between active principles and the passive principles of matter, this chemical, active ethereal cosmology lay outside the framework of the 'passive Laws of Motion'.[21]

The emergence of imponderable fluid theories

Newton's ether theory aroused little interest among natural philosophers until the 1740s. A contemporary assessment of the ether hypothesis of the 1717 *Opticks* as 'something new in the latest edition of his *Opticks* which has surprised his physical and theological disciples'[22] hinted at the reason for this lack of interest. Newton's discussion of the mode of action of ether in query 21 was far from clear, and Newtonian natural philosophers had been schooled to be wary of contact-action or 'mechanical' explanations of gravity. Natural philosophers such as John Keill and theologians like Clarke had taken up the cudgels in public defence of Newton's theory of the attractive force of gravity and his theological argument in explicating gravity as the effect of the divine will. Ether appeared to have a questionable status in this 'Newtonian' world view. Indeed, the review of the Latin edition of the 1717 *Opticks* in the *Acta eruditorum* regarded Newton's introduction of ether as providing a contact-action explanation of gravity, reflecting Leibnizian criticisms of the 'occult' nature of attractive forces.[23] In England, Robert Greene claimed, in his *Philosophy of the expansive and contractive forces* (1727), that Newton had proposed ether in the 1717 *Opticks* in response to the criticisms of Newton's theories of atoms, the void, and the passivity of material entities in Greene's *Principles of natural philosophy* (1712) (Heimann and McGuire, 1971:255–61). Despite these responses, Henry Pemberton, who had edited the third edition of *Principia* (1726) for Newton, mentioned Newton's 'subtle and elastic substance diffused through the universe' in his *View of Sir Isaac Newton's philosophy* (1728).[24] Nevertheless, until the 1740s, Newtonian popularisers avoided ether in favour of Newton's theory of short-range interparticulate forces as developed in the queries to the *Opticks* (Schofield, 1970:19–62; Thackray, 1970:8–82).

The new interest in Newton's theory of ether among British natural philosophers was related to the burgeoning enthusiasm for electrical studies. As Benjamin Martin noted in his *Philosophia Britannica* (1747), though Newton had discussed ether 'he seem'd not at all delighted with the thought, nor ever laid any stress upon it'; by contrast, contemporary theorists 'are arriv'd at great dexterity since Sir Isaac's time . . . [and] can now almost prove the existence of this aether by the phenomena of electricity'.[25] Four important developments in the period 1717–46 fostered this shift in opinion: an increasing stress on the role of repulsive forces and on the balance in nature between attractive and repulsive forces; the impact of Boerhaave's concept of 'fire'; the publication of Newton's early letters to Boyle and Oldenburg on ether, and of Bryan Robinson's *Dissertation on the aether of Sir Isaac Newton* (1743); and the interest in electrical effluvia (Cohen, 1956:205–362; Heimann, 1973:10–17). These four points will be discussed in turn.

The early Newtonians focused on the description of attractive forces, which was consonant with the gravitational paradigm for the theory of interparticulate forces, but in his *Vegetable staticks* (1727), Stephen Hales developed the implications of Newtonian repulsive forces. Hales was concerned to discuss the production of gases in chemical and biological processes and, following Newton, to explicate the properties of gases in terms of repulsive forces. In the *Opticks,* Newton had associated the 'repelling Power' of particles with the gaseous state: A 'true permanent Air' contained 'particles receding from one another with the greatest Force'. Moreover, Newton had argued that '*Aether* (like our Air) may contain Particles which endeavour to recede from one another', the 'elastick force' of the ether being traced to the 'exceeding smallness of its Particles' and hence the 'greatness of the force by which those particles may recede from one another'.[26] Hales considered that chemical processes were maintained by the production and absorption of gases by chemical substances, which he supposed brought about by attractive and repulsive forces. Associating 'air' with the alkali principle, he interpreted the interaction between acidic and alkaline substances in terms of the interaction of opposing forces. In expounding a theory of 'air' based on the Newtonian concept of repulsive forces, Hales developed the implications of Newton's theory of ether, postulating that the order of nature was dependent on a balance between attractive and repulsive forces. Nature would become 'one inactive cohering lump' if matter were 'only endued with a strongly attracting power'; so intermingled with 'attracting matter' there was 'a due proportion of strongly repelling elastic particles [air], which might enliven the whole mass, by the incessant action between them and the attracting particles'.[27]

Hales envisaged a fundamental balance of nature, a balance between attractive and repulsive forces. Associating attractive and repulsive forces with different material entities, an attracting matter and an 'elastic', ethereal repelling matter, he considered nature as inherently active, its activity maintained by the 'incessant action' of attractive and repulsive forces. Hales's two-substance theory of attractive ordinary matter and repulsive 'air' was to exercise considerable influence on the development of Benjamin Franklin's theory of electricity in the 1740s, which posited a dualism of ordinary matter and a repulsive, ethereal electric 'fluid'; the concept of a balance of powers between attractive matter and ethereal repelling 'fluids' was to be a characteristic feature of the theory of unified ethereal substances developed by James Hutton in the 1790s; and Hales's arguments were to have a significant influence on the development of the imponderable fluid theories.

Boerhaave's concept of 'fire', the theoretical kernel of his *Elementa chemiae* (1732; translated into English by Dallowe in 1735 and Peter Shaw in 1741), was a major influence on the development of Franklin's theory of electricity. For Boerhaave, fire was a physical instrument, the cause of chemical change and 'the instrumental cause of all motion', being 'the great changer of all things in the universe, while itself remaining unchanged'.[28] Boerhaave's 'active element' of fire was not an elastic fluid, and there is no evidence that Boerhaave's formulation of the concept owed anything to Newton's ether. Nevertheless, ether and fire did have some properties in common. Boerhaave stressed the 'excessive minuteness' of the particles of fire, and emphasised that fire was a space-pervading substance that was immutable and not subject to the laws of gravity. Fire had the 'property of penetrating all solid and fluid bodies'; it 'exists always and everywhere'. Fire maintained the activity of the universe, for through the agency of the 'active element' of fire the 'whole universe might continue in perpetual motion'.[29] These outward similarities between Boerhaave's space-pervading substance of fire and Newton's ether led natural philosophers to conflate the two concepts.

This affinity to Newton's speculations was emphasised in Shaw's footnote commentary to his translation. Shaw implied a relation between Boerhaave's quasi-material fire and Newton's active principles; as a substance possessing inherent activity, the source of the activity of nature, fire was interpreted as being analogous to Newton's ether, itself an active principle maintaining the activity of nature. The conflation of ether–fire as an inherently active substance was fostered by the ambiguous conceptual status of Newton's ether as both an active principle and a material substance. The dualism between ether–fire and ordinary matter posited by British natural philosophers in the 1740s can be traced to Boerhaave's view that fire 'has a power of expanding everything else', counteracting the contractive power of 'the remaining bodies which have a virtue implanted in them, whereby they constantly resist this separation of their elements'. Thus there were 'two principles' in nature, 'one of expansion, the other of attraction',[30] a dualism analogous to Hales's theory of attractive and repelling substances, which was explicitly linked to the Newtonian dualism between ordinary matter and ether. By the 1740s natural philosophers had developed dualistic theories in which *'elementary fire'* was considered to be an ethereal substance 'endowed with active Powers distinct from those of other Matter'.[31]

The publication of a systematic treatise on Newton's ether, Bryan Robinson's *Dissertation on the aether of Sir Isaac Newton* (1732), and the publication of Newton's letters to Oldenburg and Boyle in 1744 and of Robinson's pamphlet *Sir Isaac Newton's account of the ether* (1745), which included a reprint of the letter to Boyle and of extracts from the ether queries of the 1717 *Opticks,* brought Newton's theory of ether to the attention of natural philosophers. In his *Dissertation on the aether,* Robinson emphasised that Newton had considered ether to be composed of particles acting on each other at a distance by their repulsive forces. Although his quantitative treatment of optics, gravitation, and capillarity in terms of the sizes and forces of ether particles was not echoed in the writings of the imponderable fluid theorists, his espousal of the Newtonian ether had a considerable impact on the work of the electrical theorists of the 1740s. Benjamin Wilson's identification of ether with an electrical substance in his *Essay towards an explication of the phenomena of electricity deduced from the aether of Sir Isaac Newton* (1746) was made with an explicit acknowledgment to Robinson.

The interest in electrical studies in the late 1740s can be linked to the new concern with Newton's ether; it was supposed that investigations in electricity would enable natural philosophers to 'discover the nature of that subtle elastic and etherial medium, which Sir Isaac Newton queries on, at the end of his *Opticks'*.[32] Boerhaave's ethereal fire, Hales's dualism of ordinary matter surrounded by repelling ethereal air, and Newton's theory of the ethereal elastic fluid were to help shape the hypotheses of the electrical theorists, notably the work of Benjamin Franklin. Franklin had read the work of Hales, Boerhaave, and Wilson, as well as Newton's *Opticks,* and his theory of the electrical fluid brought the concept of material electrical 'effluvia', which electrical theorists had employed as an explanation of electrical phenomena, within the framework of the Newtonian concept of ether. The focus of Franklin's theory of electricity was to seek an explanation for the charging and discharging of bodies, especially the Leyden jar, a device capable of delivering electric shocks, which had been discovered shortly before he began his researches in 1747 (Cohen, 1956:285–478; Home, 1972). In Franklin's influential theory electrification was represented by the permeation of the electrical fluid through the interstitial pores of the electrified body. Franklin considered this 'electrical matter' to consist of 'particles [which were] extremely subtile', differing from 'common matter in this, that the parts of the latter mutually attract, those of the former mutually repel each other'. In Franklin's theory, 'though the particles of electrical matter do repel each other, they are strongly attracted by all other matter', by ordinary matter, that is; and he argued that surplus 'electrical matter' was held in place around electrified bodies as an 'electrical atmosphere' by an attraction between the particles of these bodies and those of the electric fluid.[33] In explicating a theory of electrification, Franklin posited a dualism of ordinary matter and the electric fluid, developing Hales's dualistic theory for the explanation of electrical phenomena. Boerhaave's influence can be seen in his reference to the electrical fluid as an 'electrical fire'. The electrical fluid was conceived as analogous to Newton's ether, being composed of mutually repelling particles. The electrical matter was envisaged as an electrical ether, and Franklin suggested that fire, ether, and electricity were 'different modifications of the same thing',[34] though he ultimately distinguished between the properties of fire and those of the electric fluid (Schofield, 1970:172).

Franklin was concerned to explain electrification rather than to formulate a systematic theory of nature, but his ideas illustrate the impact of the Newtonian ether, and the associated theories of Hales and Boerhaave, on the electrical imponderable fluid theories of the 1740s. A systematic treatment of the identification of Newton's ether with a repelling, active, material substance that was applied to explain a diversity of phenomena can be seen in Gowin Knight's *Attempt to demonstrate, that all the phaenomena in nature may be explained by two simple active principles, attraction and repulsion* (1748). Knight claimed that the activity of nature required the supposition of 'some Active Principle or Principles capable of producing and continuing Motion in the Universe', these active principles being the forces of attraction and repulsion.[35] Knight associated the forces or active principles with two different material entities, and he argued that 'attraction and repulsion cannot both, at the same time, belong to the same individual substance, being contraries', concluding that 'there are in Nature two kinds of Matter, one attracting the other repelling'. Echoing Hales's dualistic theory, Knight held that the repellent ethereal matter clustered around the particles of attracting matter, and argued that the phenomena of light and magnetism could be reduced to the motions of the repellent matter. Light was explained as the propagation of a vibrational tremor along a chain of mutually repellent particles, and magnetism (Knight's major interest) was explained as a 'circulation of the repellent fluid' between magnetic poles.[36]

Although Knight's treatise illustrates the transformation of the Newtonian concepts of force, active principles, ether, and matter in the 1740s, he was careful to emphasise that the active principles of attraction and repulsion were not inherent in material substances but were themselves the manifestation of divine activity in nature. However, the further development of the imponderable fluid theories led some natural philosophers to the enunciation of a theory of nature in which matter was endowed with inherent activity. The transformation of the Newtonian dualism between active and passive principles into a dualism of ordinary attracting matter and an active, ethereal repelling substance, and the conflation of ether–fire as an inherently active substance, led to the conflation of the dualism between active and material or passive principles that was a seminal feature of Newton's natural philosophy. Ether functioned as an inherently active substance endowed with repulsive force, and ordinary matter was considered as possessing an inherent attractive force. This theory contradicted Newton's concept of the intrinsic passivity of material entities; activity was subsumed in the inherent powers of material substances. These arguments were fully articulated in James Hutton's theory of nature in the 1790s, but although many natural philosophers did not explicitly formulate the philosophical and theological implications of this position, an early statement of this argument in Cadwallader Colden's *Principles of action in matter* (1751) illustrates the conceptual transformation associated with the interpretation of ether as an inherently active substance. Asserting that it was 'unphilosophical' to ascribe activity to divine agency, Colden claimed that motion was inherent in matter: Motion 'exists in something, which has in itself the power of moving'. All material substances, including ordinary matter, were regarded as 'acting principles', as being endowed with inherent activity. Ether was a 'species of matter', an acting principle maintaining the activity of nature.[37]

The interpretation of ether as an inherently active substance threatened the theological foundations of Newton's philosophy of nature. For Newton ether was an active principle, a manifestation of God's causal agency in nature. Newton's view of ether was maintained by Colin Maclaurin in his *Account of Sir Isaac Newton's philosophical discoveries* (1748), which echoed Newton's voluntarist theology in pointing out that if gravity were 'produced by a rare and elastic *aethereal medium'*, then 'the whole efficacy of this medium must be resolved into his power and will, who is the supreme cause', for God is 'the source of all efficacy'. Maclaurin's interpretation of the theological status of ether was being called into question by the emergent view of ether as an inherently active substance endowed with repulsive force. This theory of nature was sustained by the view that the phenomena of nature could be reduced to the interaction of two material principles, attractive and repulsive, and that the diversity of phenomena could be explained in terms of a single ethereal substance: The 'terms *Fire, Electricity, electrical Aether, aetherial Spirit*' were 'synonymous'.[38] This interpretation of ether was to be fundamental to the development of the unified theory of ether in the second half of the eighteenth century.

Chemistry and the development of unified ether theories

In query 31 of the 1717 *Opticks*, Newton provided hints about his theory of chemistry. Despite his appeal to a programme of chemical explanation in terms of a quantified science of interparticulate forces (Thackray, 1970:18–42), he associated chemical phenomena with the operation of active principles, questioning the reducibility of chemistry to 'passive Laws of Motion'. The chemical connotations of ether and active principles were heightened by the strongly chemical ethereal speculations in Newton's 'Hypothesis explaining the properties of light' and his letter to Oldenburg. The revival of ether in the 1740s and the development of a dualistic natural philosophy of ordinary matter and ethereal fluids led some British chemists to invoke ether for the explanation of chemical processes, the chemical resonances of ether being strengthened by the assimilation of Boerhaave's concept of fire to the ether concept. The chemical theories that employed ether, and that were enunciated from the late 1740s on, proposed a dualism between ordinary matter and the irreducibly chemical properties of ether, with fire and phlogiston being assimilated to ether.[39] This expansion of the ether concept led to the emergence of unified ether theories in which the diversity of phenomena were reduced to the operations of an ethereal substance.

Peter Shaw's translations of Stahl's *Philosophical principles of universal chemistry* (1730) and of Boerhaave's *New method of chemistry* (1741) made an important contribution to these developments. Shaw followed Stahl in questioning the reducibility of chemistry to the mechanical philosophy, claiming, in his *Chemical lectures* (1734), that 'genuine chemistry' preferred to leave to 'other philosophers the sublimer disquisitions of primary corpuscles or atoms' and urging that chemistry should avoid such 'metaphysical speculations'.⁴⁰ In commenting in his edition of Boerhaave's treatise on Boerhaave's assertion that chemical change 'is effected by means of motion alone', Shaw questioned the reducibility of chemical phenomena to the motion of corpuscles, pointing out that the 'common laws of sensible masses . . . will not reach to those more remote, intestine motions of the component particles'. Shaw claimed that 'besides the common laws of sensible masses, the minute parts they are composed of seem subject to some others, which have as yet been but lately taken notice of, and are yet more than guessed at'. He was here referring to the 'new mechanics' that Newton had based on the power of 'attraction', a quantified science of interparticulate forces 'not reducible to any of those in the great world'. Nevertheless, Shaw questioned the applicability even of this 'sublimer mechanics' to chemistry, going on to quote almost verbatim from Newton's remarks on active principles in query 31, and arguing that active principles were the fundamental agents in nature: 'The author of nature has added to bodies certain active principles to be the sources of motion'. For Shaw the ultimate principles in nature were irreducibly chemical, to be found in the study of the chemical role of active principles rather than in the motion of corpuscles:

> [It is] by means of chemistry that Sir *Isaac Newton* has made a
> great part of his surprising discoveries in natural philosophy; and
> that curious set of queries, which we find at the end of his optics,
> are almost wholly chemical. Indeed chemistry, in its extent, is
> scarce less than the whole of natural philosophy.⁴¹

These arguments helped to shape the chemical doctrines advanced by William Cullen in his chemical lectures at Glasgow in the late 1740s. Cullen was concerned to pinpoint the sources of chemical change, and he followed Newton and Hales in arguing that there were 'two great principles, the attractive & repulsive,' which were 'the source of motion and change'. In emphasising the importance of the theory of elective attractions, Cullen followed Newton's own stress on the quantification of interparticulate forces, but the influence of Hales, Boerhaave, and Shaw is apparent in Cullen's stress on repulsion, on the expansive force of fire as the source of repulsion, and hence on the irreducibility of chemistry to universal forces of attraction and repulsion. In Cullen's view 'fire pervades all bodies and keeps their parts asunder', and hence 'attraction & fire were to be considered as the primary causes of motion'. He associated fire with electricity and ether, reflecting contemporary work in natural philosophy: 'Fire [is] an elastic fluid . . . [we know there is] an aether in bodies from the reflexion &c of light, from electricity. The same with fire & present everywhere'. Cullen developed the dualistic theory of ordinary matter and ether to incorporate the chemical role of ether. Lecturing in 1757–8, he argued that 'there are only two elements, one of them gravitating matter, the other a subtile aether', going on to note that according to this hypothesis, 'its [*sic*] probable that light and all other phaenomena of Fire depend upon this Subtile Aether in all Bodies'. Cullen explained the crucial chemical phenomenon of combustion by supposing 'that inflammable bodies are of such a particular texture as to recover these particles of Aether' as 'phlogiston'.[42] Cullen's introduction of Stahl's principle of inflammability, phlogiston, to denote the chemical manifestation of ether broadened the conceptual framework of the theory of ethereal substances.

These ideas were further developed in the Edinburgh lectures of Cullen's student Joseph Black. In appealing to the theory of ether, Black was concerned to explain the combustion of chemical substances, ether serving to justify the intelligibility of the phlogiston concept. In a lecture in 1768, Black declared:

> Sir Isaac Newton is of the opinion that there is in nature a certain fluid of exceeding elasticity subtility and density, that pervades all nature, the different modifications of which produce the phenomena of Electricity, magnetism and Gravity and the cohesion of the smaller parts of bodies to each other . . . I am therefore of the opinion that this is the inflammable principle. [Talbot, 1967: ch. 13, 42]

Phlogiston was thus a 'modification' of ether, its supposition justified by appeal to the Newtonian pedigree. In a lecture delivered in the 1780s, Black argued that the principle of inflammability, 'phlogiston' or 'subtle aether', was a 'subtile & active fluid', and he affirmed the dualistic theory of nature in denying that this ether or phlogiston was a 'gravitating substance'; rather, 'this matter is exempted from the laws of gravitation'.[43] Questioning the unity of matter, Black stressed the chemical connotations of ether and the disjunction between chemical principles and the laws of ordinary matter. In Black's view, 'heat may be considered in nature as the great principle of chemical movement and life', and he supposed that phlogiston was a manifestation of heat: 'Heat or light are the principles of inflammability'.[44] In Black's theory of nature chemical and thermal phenomena were closely associated, and heat, light, and phlogiston were imponderable substances, modifications of ether and distinct from ordinary matter; chemical and ethereal principles were not reducible to the passive laws of ordinary matter.

Black's lectures expanded the theory of ether to include chemical, optical, and thermal phenomena, providing hints towards the formulation of a unified ether theory. The chemical role of ether, in its 'modifications' as heat, light, and phlogiston, was emphasised in several works published in the 1770s, often reflecting Black's influence.[45] In his *Philosophical inquiry into the causes of animal heat* (1778) the Edinburgh-educated physician P. D. Leslie followed Black in arguing that 'phlogiston is fire and light, or a certain subtle elastic fluid, upon the modifications of which the phenomena of heat and light depend'. This ethereal substance 'is the chief cause and principle of activity in the universe' and is 'exempted from the common laws of gravitation'. Leslie explicitly identified phlogiston with the *'Newtonian* ether, the electrical *aura, materia subtilis,* fire and light'.[46] Similar ideas were proposed by Bryan Higgins in his *Philosophical essay concerning light* (1776). Higgins supposed that light, phlogiston, fire, and electricity were different modifications of ether: 'Fire is not considered as a homogeneal body different from light and phlogiston; and I am unwilling to admit the Electric Fluid as an element different from these'.[47]

The phlogiston concept thus played a central role in eighteenth-century British natural philosophy. However, many chemists considered phlogiston an ordinary chemical substance rather than an ethereal imponderable. British chemists who adopted this interpretation included Kirwan, Cavendish, and – for a time – Priestley, all of whom equated phlogiston with some form of 'inflammable air [hydrogen]'. Nevertheless the view of phlogiston as an ethereal active substance represents an important British chemical tradition: Phlogiston was identified with or viewed as a 'modification' of fire, light, electricity, and ether.[48] This unified ether theory stressed the unity and activity of nature; avoiding a superfluity of diverse imponderable fluids, these theorists reduced the operations of nature to a dualism of gravitative matter and ether.

The systematic *Dissertations on different subjects in natural philosophy* (1792) by Black's associate James Hutton provided a full statement of this unified ether theory, with an emphasis on its philosophical and theological implications. Hutton supposed a dualism of gravitational (attractive) matter and the 'emanation of [repelling] matter from the sun';[49] these two kinds of matter maintained the operations of nature, 'the opposite powers . . . continually balancing one another, or alternately prevailing'. If the gravitational matter were to prevail then 'gravitation would soon bring all the matter of this machine [the universe] to rest, and would lock up every body in a state of the most absolute inactivity'. The emanation of matter from the sun was thus a 'necessary cause of vital motion', for 'without the influence of the sun this world would remain an useless mass of inert matter'. This solar substance was envisaged as an active principle embracing all the imponderable fluids: 'Light, heat and electricity appear to be three different modifications of the same matter'.[50] In explicating this theory of the unified ether or solar substance, Hutton appealed to a wide variety of phenomena. He was especially concerned to explain the interrelationships among heat, light, electricity, and chemistry. The relationship between light and heat, the association of a loss of heat with the emission of light, was explained by arguing that light and heat were different 'modifications' of the repulsive solar substance. Phosphorescence was explained by the claim that light is 'arrested and detained in a certain modification' within bodies forming a 'phlogistic substance'; this matter may be emitted from its connection with a body and 'resume its former character of light'. The phlogiston theory thus implied 'the union of the matter of light and heat with some of the chemical substances of bodies' to form a 'phlogistic substance'. Combustion involved the loss of light and heat and the decomposition of phlogistic substances.[51] Hutton regarded electricity as related to light and thought heat a modification of the solar substance, and he claimed that heat and light would affect the conduction of electricity by bodies. Biological and chemical processes were also connected in Hutton's scheme, and he contended that 'plants compose phlogistic matter in growing', for there was no evidence for decay without renewal: 'There would also appear to be in the system of this globe, a reproductive power, by which the constitution of this world, necessarily decaying, is renewed'.[52] Nature was thus conceived as a system of processes and transformations, different modifications of the ethereal solar substance serving as sources for the regeneration of nature.

In Hutton's system light emanated from the sun and was contained in bodies as phlogiston, an idea probably acquired from Black and Macquer;[53] and the operations of nature were maintained by the circulation and conservation of phlogiston. Hutton's conception of nature has analogies with Newton's theory of the 1670s, in which ether was an underlying principle maintaining the activity of nature by its circulation, though Hutton's theory reflected the transformation in the Newtonian concepts of ether and active principles in the eighteenth century. Hutton rejected Newton's dichotomy between active principles and passive matter. Rejecting Newton's atomism and theory of the passivity of material entities, Hutton distinguished between the phenomenal manifestations of substances, which he termed 'body', and the underlying substratum, or 'matter'. This substratum was defined in terms of its intrinsic activity: 'Matter may thus be considered as acting powers'.[54] In Hutton's dualistic theory the ethereal solar substance functioned as an inherently active substance endowed with repulsive force, balancing the inherent attractive force of gravitative matter. On the theological level this conception of nature contrasted with Newton's stress on divine sustenance by means of active principles, for activity was subsumed in the acting powers constituting material substances. Nature was thus a self-regenerating system of active powers, its self-sufficiency maintained by the inherent activity of material substances.

Ether functioned as an active substance immanent in the fabric of nature, conserving and recruiting activity by its transformations and circulation: 'Nature is a perpetual circulatory worker'.

Hutton's emphasis on the role of phlogiston suggests that he considered chemical principles applicable to the powers that characterised the invisible realm of matter. Devoid of solidity, extension, and inertia, Huttonian matter could not be characterised by the passive laws of motion.[55] As John Playfair expressed it, in Hutton's system 'the chemist . . . is flattered more than anyone else with the hopes of discovering in what the essence of matter consists; and Nature, while she keeps the astronomer and mechanician at a great distance, seems to admit him to a more familiar converse, and to a more intimate acquaintance with her secrets'.[56] Hutton thus suggested a cosmology determined by irreducibly chemical principles.

Adam Walker's textbook *A system of familiar philosophy* (1799) provided a full exposition of the unified ether theory. Walker asserted the identity of fire, light, electricity, and phlogiston as 'modifications of one and the same principle', the ethereal repelling substance. The operations of nature were determined by a 'balance' of the two 'powers' of attraction and repulsion, which were 'opposing or antagonistic principles . . . in a state of unceasing warfare'. Electricity was the 'genuine principle of light and fire', and the emanation of the 'ethereal matter' of electricity from the sun activated the universe, for electricity was 'the soul of the material world'.[57]

The stress on the unity of natural phenomena thus led to enunciations of a dualistic world view, in which the transformations of an ethereal, active repellent substance balanced the attractive power of gravitating matter, providing a coherent scheme for the systematisation of a diversity of phenomena and the interactions between natural agents. In his *Mathematical and philosophical dictionary* (1795–6), a reliable guide to contemporary attitudes, Charles Hutton reported that 'there was such a strong affinity between the elements of fire, light and electricity, that we may not only assert their identity upon the most probable grounds, but lay it down as a position against which at present no argument of any weight has an existence'.[58]

The influence of unified ether theories

Between 1798 and 1806 there were several new developments in British natural philosophy that initiated the decline of the imponderable fluid theories. In 1798, Rumford rejected the imponderable fluid theory of heat, arguing that the theory could not explain the generation of heat by friction. The invention of the battery and the discovery of electrolysis in 1800 led to Humphry Davy's theory of electrochemistry; in a full statement of his theory in 1806, Davy explained electricity by the forces of chemical affinity, abandoning the theory of the electric fluid. Thomas Young's advocacy of an undulatory theory of light in papers written between 1799 and 1804 led him to reject the concept of light as an ethereal elastic fluid analogous to fire, as well as Newton's 'emission' theory of light as the projection of 'rays' of discrete corpuscles.[59] Although these developments illustrate the decline of imponderable fluid theories in the early nineteenth century, the conceptions of nature explicated by Rumford, Davy, and Young demonstrate the continued influence of the theory of unified ethereal substances on the development of natural philosophy.

The impact of the unified ether on the ideas of Rumford, Young, and Davy can be seen by a brief survey of their theories of nature. Rumford rejected the theory of heat as an imponderable fluid in favour of a theory in which the effects of heat were held to be the result of the interaction between the motion of the particles of ordinary matter and the vibrations of an ambient ethereal medium. Supposing a dualism of ordinary matter and ether, he asserted that the vibrations of the 'atmospheres composed of aether' surrounding the particles of ordinary matter were communicated to the surrounding ether and finally to other particles of ordinary matter.[60] This theory of the transmission of heat was reminiscent of Gowin Knight's account of the propagation of light based on a dualistic theory of matter and ether. Rumford echoed concepts characteristic of eighteenth-century ether theories in maintaining that '*motion is an essential quality of matter*'; matter was inherently active and hence 'rest is nowhere to be found in the universe'.[61]

The theory of ethereal atmospheres was also adopted by Young in his early discussions of the mode of action of ether. He supposed a universal ethereal substance that 'may possibly be the ground work of all the phenomena of nature', speculating that 'light, heat, cohesion and repulsion' may 'depend on some modification of the actions of the medium [the ether]', which was 'connected with the electric fluid'. Although he abandoned this programme of a unified natural philosophy grounded on the concept of ether, probably because he could not satisfactorily explain cohesion and repulsion through the theory of ethereal atmospheres, the theory of the unified ether shaped the development of Young's natural philosophy, and the concept of a luminiferous ether as the vehicle of light and radiant heat became the central feature of his theory of optics (Cantor, 1970).

In an early essay, Davy argued that 'the electric fluid is probably light in a condensed state . . . its chemical activity upon bodies is similar to that of light'; he claimed that 'the different species [of matter] are continually changing into each other'. Davy referred to James Hutton, and this commitment to the unity and interconversion of natural powers was characteristic of Hutton's unified ether theory. This conception of nature was of fundamental importance for Davy's formulation of his electrical theory of chemical affinity. Davy argued that the electrical and chemical powers were so interconnected that it was likely that the electrical and chemical forces were 'identical'. Although he abandoned the theory of the electric fluid, his emphasis on the unity of natural powers echoed the unified ether theory. Davy continued to maintain a dualistic natural philosophy in which an 'etherial matter' endowed with repulsive force was conceived as the *'antagonist* power to the attraction of cohesion',[62] a concept characteristic of the unified ether theory.

Davy's theory of nature exemplifies the manner in which the unified ether theory continued to shape the commitments of natural philosophers, even though the supposition of imponderable fluids – to explain the phenomena of heat, repulsion, and electricity – was increasingly being called into question. The unified ether theory emphasised the balance of forces, the unity and interconversion of natural phenomena, and the self-sufficiency of nature, contending that the activity of nature was maintained by forces of attraction and repulsion. These ideas, ultimately divorced from imponderable fluid theories, had a continued and significant influence on the development of British natural philosophy in the nineteenth century. The concepts of the balance of forces and the self-sufficiency of nature were developed by Faraday and Joule in the 1830s and 1840s into the theory of the convertibility and indestructibility of natural powers, one of the strands that, transformed, became explicated as the 'conservation of energy' by about 1850.[63] The theory of the unified ether thus had an enduring impact on the history of natural philosophy.

Notes

1 Heimann (1973); P. M. Heimann, 'Voluntarism and immanence: conceptions of nature in eighteenth-century thought', *Journal of the History of Ideas 39* (1978), 271–83.
2 R. McCormmach, 'John Michell and Henry Cavendish: weighing the stars', *British Journal for the History of Science 4* (1968), 150; R. McCormmach, 'Henry Cavendish: a study of rational empiricism in eighteenth-century natural philosophy', *Isis 60* (1969), 293–306.
3 Schofield (1970), 15, asserts a rigid dichotomy between 'mechanists' who sought 'the causation for all phenomena of nature . . . [in] the primary particles of an indifferentiable matter . . . and the forces of attraction and repulsion between them', and 'materialists' who believed that 'the causes of phenomena inhere in unique substances'. These categories are derived from a distinction between force and ether theories in Newton's natural philosophy. Cf. P. M. Heimann, 'Newtonian natural philosophy and the scientific revolution', *History of Science 11* (1973), 1–7.
4 I. Newton, *Opticks; or, a treatise on the reflections, refractions, inflections and colours of light,* 4th ed., 1730 (reprinted, London, 1952), 347–70.

5 I. Newton, 'An hypothesis explaining the properties of light', in T. Birch, *The history of the Royal Society at London*, 4 vols. (London, 1756–7), 3:247–305; reprinted in *Isaac Newton's papers and letters on natural philosophy*, ed. I. B. Cohen (Cambridge, 1958), 177–235; text in I. Newton, *The Correspondence of Isaac Newton*, ed. H. W. Turnbull, J. F. Scott, A. R. Hall, and L. Tilling, 7 vols. (Cambridge, 1959–77), 1:362–86.

6 Newton to Oldenburg, 25 Jan. 1675/6, in R. Boyle, *The Works of the Honourable Robert Boyle*, ed. T. Birch, 5 vols. (London, 1744), 1:74; reprinted in Cohen, *Papers and letters*, 254. See Newton, *Correspondence*, 1:413–14.

7 Newton to Boyle, 28 Feb. 1678/9, in Boyle, *Works*, 1:70–3; reprinted in Cohen, *Papers and letters*, 250–3; text in Newton, *Correspondence*, 2:288–95. See also I. Newton, 'De aere et aethere', in *Unpublished scientific papers of Isaac Newton*, eds. A. R. Hall and M. B. Hall (Cambridge, 1962), 214–20. On the date of this manuscript of the 1670s cf. Hall and Hall, *Unpublished papers*, 187; Westfall (1971), 373, 409–10.

8 Newton, *Opticks*, 352.

9 I. Newton, *Mathematical principles of natural philosophy*, trans. A. Motte, rev. F. Cajori (London, 1934), 546. Hereafter *Principia*. On Newton's rejection of Cartesian vortices see Whiteside (1964, 1970).

10 Newton, *Opticks*, 362. On Newton's concept of pressure see A. E. Shapiro, 'Light, pressure and rectilinear propagation: Descartes' celestial optics and Newton's hydrostatics', *Studies in History and Philosophy of Science 5* (1974), 273–6.

11 A classic statement of the interpretation of Newton's ether as a 'mechanical' medium is in Maxwell's 1878 article 'Ether', in J. C. Maxwell, *The Scientific Papers of James Clerk Maxwell*, 2 vols. (Cambridge, 1890), 1:763–75. Commentators who adopt this interpretation take different attitudes to the significance of the ether. Rosenfeld (1965) views Newton's 'mechanical' ether as his preferred position; whereas Hall and Hall (1967) view the 'mechanical' ether as unimportant in Newton's natural philosophy, as it contradicts the methodology of *Principia*. Both these accounts begin from a mistaken premise: that Newton's ether was a quasi-Cartesian 'mechanical' medium.

12 Newton, *Opticks*, 351–2. Cf. Bechler (1974).

13 *Four letters from Sir Isaac Newton to Doctor Bentley containing some arguments in proof of a Deity* (London, 1756), 25; reprinted in Cohen, *Papers and letters*, 302.

14 Newton, *Opticks*, 401.

15 Draft letter from Newton to the editor of *Memoirs of Literature* (written some time after 5 May 1712) in reply to a letter of Leibniz's, in Newton, *Correspondence*, 5:298–300.

16 Newton, *Opticks*, 401. Cf. McGuire (1968), 187–208.

17 In drafts he referred to a 'subtile Aether or Aetherial elastic spirit' (quoted in Guerlac [1967], 48); this was reflected in his allusion to a 'subtle spirit' in the *General Scholium* to the 2nd ed. of *Principia* (1713). This was qualified as an 'electric and elastic' spirit in a marginal note in Newton's own copy of *Principia* and in Motte's 1729 English translation. See Newton, *Principia*, 547. On Newton's use of the phrase 'electric and elastic' see A. R. Hall and M. B. Hall, 'Newton's electric spirit: four oddities', *Isis 50* (1959), 473–6; A. Koyré and I. B. Cohen, 'Newton's "electric and elastic spirit"', *Isis 51* (1960), 337. In a draft reply to Leibniz (see n. 15), Newton argued that gravity could be explained 'by a power seated in a substance in wch bodies move & flote without resistance & wch has therefore no *vis inertiae* but acts by other laws than those that are mechanical'. Newton, *Correspondence*, 5:300.

18 Newton, *Opticks*, 380, 401. Cf. McGuire (1968), 164–74.

19 Newton, *Correspondence*, 1:364–6.

20 B. J. T. Dobbs, *The foundations of Newton's alchemy or 'the hunting of the Greene Lyon'* (Cambridge, 1975), 204–6. The tentative suggestion by Walker (1972) that Newton's ether is to be identified with the neo-Platonic *spiritus mundi* is mistaken. Cf. McGuire (1977), 107–9.
21 Newton, *Opticks,* 401. Cf. McGuire (1967), 85; Heimann (1973), 8–10.
22 Quoted in R. Kargon, *Atomism in England from Hariot to Newton* (Oxford, 1966), 138.
23 *Acta eruditorum* (1720), 185–8.
24 R. Greene, *The principles of the philosophy of the expansive and contractive forces; or, an inquiry into the principles of the modern philosophy: that is, into the several chief rational sciences, which are extant* (Cambridge, 1727), 1–2; H. Pemberton, *A view of Sir Isaac Newton's philosophy* (London, 1728), 377. See also B. Worster, *A compendious and methodical account of the principles of natural philosophy,* 2nd ed. (London, 1730), 28.
25 B. Martin, *Philosophia Britannica; or, a new and comprehensive system of the Newtonian philosophy,* 2 vols. (Reading, 1747), quoted in Thackray (1970), 135.
26 Newton, *Opticks,* 351, 352, 396.
27 S. Hales, *Vegetable staticks; or, an account of some statical experiments on the sap in vegetables* (London, 1727), 178.
28 [H. Boerhaave] *A new method of chemistry: including the history, theory and practice of the art: translated from the original Latin of Dr. Boerhaave's Elementa chemiae,* trans. P. Shaw, 2 vols. (London, 1741), 1:220, 236.
29 Ibid., 1:208, 223, 359, 362.
30 Ibid., 246–7.
31 J. Rowning, *A compendious system of natural philosophy* (London, 1737–43), iii.
32 Peter Collinson to Cadwallader Colden, March 1745, quoted in Cohen (1956), 435.
33 I. B. Cohen (ed.), *Benjamin Franklin's experiments* (Cambridge, Mass., 1941), 213–14.
34 Ibid., 233, 210.
35 G. Knight, *An attempt to demonstrate, that all the phaenomena in nature may be explained by two simple active principles, attraction and repulsion* (London, 1754), 4–5.
36 Ibid., 10, 66–7. Home (1977a), 262, considers Knight a Cartesian. This view fails to acknowledge the distinctive Newtonian framework of Knight's theory: cf. Heimann and McGuire, (1971), 296–9.
37 C. Colden, *The principles of action in matter, the gravitation of bodies and the motion of the planets explained from those principles* (London, 1751), 27, 28, 73. Cf. Schofield (1970), 130–3; Heimann and McGuire (1971), 303.
38 C. Maclaurin, *An account of Sir Isaac Newton's philosophical discoveries* (London, 1748), 381–9. See Heimann, 'Voluntarism and immanence', 275; R. Lovett, *The subtil medium prov'd* (London, 1756), n.p., preface.
39 Heimann (1973). For a survey of eighteenth-century chemical writings on this topic see Ziemacki's discussion (1974) of phlogiston and ethereal agents in eighteenth-century British chemistry. This work supplements the study of 'Newtonian' chemistry by Thackray (1970), who ignores the development of imponderable fluid theories in chemistry.
40 P. Shaw, *Chemical lectures publickly read at London* (London, 1734), 146.
41 Boerhaave, *Chemistry,* 155–7, 173.
42 A. L. Donovan, *Philosophical chemistry in the Scottish Englightenment: the doctrines of William Cullen and Joseph Black* (Edinburgh, 1975), 141–51.
43 Quoted in D. McKie, 'On some Ms. copies of Black's chemical lectures', *Annals of Science 21* (1965), 223.
44 Quoted, respectively, in Donovan, *Philosophical chemistry,* 229, and in T. Coch-

rane, *Notes from Doctor Black's lectures on chemistry 1767/8*, ed. D. McKie (Wilmslow, Cheshire, 1966), 83. On Black and theories of heat, see D. McKie and N. H. de V. Heathcote, *The discovery of specific and latent heats* (London, 1935); Fox (1971), 6–67; Donovan, *Philosophical chemistry*, 222–77; Talbot (1967).

45 J. R. Partington and D. McKie, 'Historical studies in the phlogiston theory: III, Light and heat in combustion', *Annals of Science 3* (1938), 338–71.

46 P. D. Leslie, *A philosophical inquiry into the causes of animal heat: with incidental observations on several physiological and chymical questions connected with the subject* (London, 1778), 104, 119, 124.

47 B. Higgins, *A philosophical essay concerning light* (London, 1776), 13.

48 Ziemacki (1974) distinguishes between theories in which phlogiston was considered to be a universal active medium and theories in which phlogiston was identified with an ordinary ponderable substance. A parallel development occurred in France in the 1760s: Rouelle, Macquer, and Venel equated phlogiston, Stahl's inflammable principle, with Boerhaave's fire, thus transforming both concepts. For Boerhaave fire was a physical instrument, the source of chemical change while itself remaining unchanged; for Stahl phlogiston was one of the chemical 'principles' that formed the basis of chemical substances. In identifying Stahl's phlogiston with Boerhaave's fire, Rouelle emphasised that fire was a chemical constituent of bodies, like phlogiston. See M. Fichman, 'French Stahlism and chemical studies of air, 1750–1770', *Ambix 18* (1971), 94–122; H. Metzger, *Newton, Stahl, Boerhaave et la doctrine chimique* (Paris, 1930), 209–45; R. Rappaport, 'Rouelle and Stahl – the phlogistic revolution in France', *Chymia 7* (1961), 73–102; Love (1974).

49 J. Hutton, *Dissertations on different subjects in natural philosophy* (Edinburgh, 1792), 246. On Hutton's natural philosophy cf. Heimann and McGuire (1971), 281–95.

50 Hutton, *Dissertations*, 263, 505.

51 Ibid., 175, 517, 519.

52 Ibid., 214, 218.

53 McKie, 'Black's chemical lectures', 215; P. J. Macquer, *Dictionnaire de chymie*, 2nd ed. 4 vols. (Paris, 1778), 3:144.

54 Hutton, *Dissertations*, 501.

55 Ibid., 257. See Heimann (1973), 19–21, on Hutton's theory of chemistry.

56 J. Playfair, 'Biographical account of James Hutton', in J. Playfair, *The works of John Playfair*, 4 vols. (Edinburgh, 1822), 4:83.

57 A. Walker, *A system of familiar philosophy*, rev. ed., 2 vols. (London, 1802), 1:6, 14, 18, and 2:1, 74.

58 C. Hutton, *A mathematical and philosophical dictionary*, 2 vols. (London, 1795–6), 1:473.

59 Fox (1971), 99–103; P. M. Heimann, 'Conversion of forces and the conservation of energy', *Centaurus 18* (1974), 152–3; Cantor (1970).

60 B. Rumford, *The complete works of Count Rumford*, 4 vols. (Boston, 1870–5), 2:172.

61 Ibid., 2:104. Goldfarb (1977) argues convincingly that Rumford's work has been wrongly interpreted as a precursor of the kinetic theory of heat, and stresses the role of the ether in Rumford's natural philosophy. However, in pointing to Boerhaave as Rumford's probable source, Goldfarb ignores the tradition of natural philosophy discussed in the present chapter. Another possible source of Rumford's ether is Lambert's *Pyrometrie* (Berlin, 1779), sections of which Rumford translated for Fourier. See Fox (1979).

62 H. Davy, *Collected works of Humphry Davy*, 9 vols. (London 1839), 2:28, 35; 4:44, 56, and 5:40.

63 Heimann, 'Conversion of forces', 147–61.

2

Ether and the science of chemistry: 1740–1790

J. R. R. CHRISTIE
Division of History and Philosophy of Science, Department of Philosophy, University of Leeds, Leeds LS2 9JT, England

The Scottish cultural context

Ether poses methodological problems for the historian of eighteenth-century science. The subtle fluid appeared in a proliferating variety of guises and performed in a bewildering number of roles, so that the most effective tactic seems to be to define a fairly narrow perspective and write ether's history from that point of view. Thus it becomes possible to write isolated sectors of the subject, in case studies that examine ether's role with regard to theoretical developments within a particular science; to epistemology, methodology, and theology; or to natural philosophy in general. This technique tends to reduce a Protean subject to manageable units of analysis, and has that to recommend it. However, its logic has problematic implications in the long run, for it will tend only to produce a cumulative display of ether's profusion, doubtless with numerous roles and functions added to those already instanced. It would be perfectly possible, for example, to write of ether's association with moral theory in the eighteenth century, or even to write its political history.[1] Simply to provide a coherent summary of existing work would be a formidable enough task. To hope that all these essentially analytical perspectives, reflecting as they do the interests and skills of many diverse specialists, may one day prove susceptible of unifying synthetic interpretation is pious in the extreme.

Caught between the Scylla of analytic possibility and the Charybdis of plausible synthetic explanation, ether's history urgently requires a canvass of alternative modes of approach. One such approach that begins to avoid the current dilemma is to examine in detail the uses ether was put to, and the pressures that shaped its cognitive development, within a particular cultural context. Immediately, the historian is thereby freed to see ether not in narrow vertical slices (methodology, epistemology, theory . . .), but in relation to its position within a culture-specific, overall cognitive structure. The isolated slices now appear as organically related parts of a cultural system. Further, if the social dimensions of that system be capably laid out, then ether can be seen as responsive not merely to narrowly conceived cognitive inputs, such as limitations of experimental technique or ideas of what could pass as a legitimate theory, but also to the larger economic and ideological forces that moved and shaped the culture. This approach thus gives full and integrated play to ether's multifarious nature while connecting it with those factors generally recognized by historians as responsible for the formation of the world in which men, even scientists, perceive, think, and act.

The arena chosen for this novel pursuit of ether is the mid-eighteenth-century culture of the High Enlightenment, in particular its Scottish manifestation. Two things render this choice appropriate. The social bases of the culture have recently been explored with a degree of success; so there is a reasonably solid body of opinion from which to work.[2] Also, ether was on the rise in this period, partly within medical science and physics, partly within certain epistemological writings, but above all in chemistry, which by 1755 had become a leading sector in scientific culture in its experimental success, growth in theory, institutional development, and social patronage.[3] Ether's place in the socially and cognitively expanding science of chemistry therefore offers the most likely scene of successful investigation. Further, the direction that studies of Lavoisier and the chemical revolution in France have now taken reveal the central place of ethereal concepts in the origin and continued prosecution of Lavoisier's chemical theory.[4] This material will provide the opportunity to close with an examination of French chemistry, in order to determine whether French chemistry's use of ether can be compared with the Scottish, and seen as an extension of it.

Eighteenth-century Scotland's most committed ethereal scientist was the chemist and physiologist William Cullen. Ether featured strongly in his expositions of the theoretical bases of both disciplines, and in the case of chemistry it did so from the time he first started teaching the subject at the University of Glasgow in 1747.[5] During his time as a teacher and practitioner of chemistry at Glasgow, then at the University of Edinburgh from 1756 to 1766, Cullen continued to deploy, and significantly to develop, the concept of ether. But Cullen was not Scotland's first etherealist. George Martine, whose critical experimental investigations of current opinion regarding rates of heating and cooling were to prove important for Joseph Black, had given sympathetic mention to Newton's ether in *An examination of the Newtonian argument for the emptiness of space and of the resistance of subtile fluids,* published in 1740. So too had Colin Maclaurin, in *An account of Sir Isaac Newton's philosophical discoveries,* posthumously published in 1748.

More importantly for this analysis, another sector of intellectual culture was already proving significantly favourable to ethereal conceptions, in ways both direct and indirect. This was the investigation of human nature and cognition put forward by David Hume in *A treatise of human nature* (1739) and *An enquiry concerning human understanding* (1748), and by Adam Smith in *The principles which lead and direct philosophical enquiries, illustrated by the history of astronomy.* This last was not published until 1795, after Smith's death, but it was first composed and written in the 1750s,[6] and was arguably a development of material presented in his professorial lectures at Glasgow, where he was accustomed to discuss the 'general powers of the human mind'.[7]

In his treatment of human nature and knowledge, Hume made no major and direct use of ether in the manner of David Hartley,[8] though he did make occasional use of an active spirit hypothesis.[9] Indeed, Hume generally was by no means averse to hypotheses, a matter that appears to create a difficulty when juxtaposed with his strictures on the impossibility of knowing 'the ultimate springs and principles' of natural phenomena,[10] and with the Newtonian severity of many of his methodological prescriptions.[11] James Noxon suggests part of the solution to this awkward juxtaposition, commenting that Hume did not 'distinguish in so many words between groundless speculative hypotheses and verifiable working hypotheses. The distinction is, however, implicit in his usage'.[12]

This reconciliatory point can be extended by emphasising the psychologising elements in Hume's analysis of causation. Although this analysis reduced necessary connection, philosophically considered, to constant conjunction, Hume stressed the determining power of associative patterns of belief in producing the idea of necessity,[13] and elsewhere insisted on the status of psychological necessity attained by such beliefs. Without the regular patterns of causal belief induced by association, perceptual and imaginative coherence and stability would be lost, and life would become, so to speak, phenomenally unliveable.[14] Coupled with Hume's own occasional use of an active spirit hypothesis and Noxon's argument, these psychologising elements warrant the conclusion that Hume's methodology and epistemology by no means ruled out ethereal speculation.

This is still some distance from the claim that Hume might actually have encouraged such speculation. However, there is evidence suggesting just such encouragement. Hume's critical opposition to the participation of religious and theological tenets in the rational discourse of moral and natural philosophy is well known. With regard to natural philosophy, aspects of this opposition were formulated some time before the classic assault on divine teleology in nature contained in the *Dialogues concerning natural religion*. In particular, he had attacked the argument of Cartesian occasionalism for the supreme deity 'as the immediate cause of every alteration in matter'.[15] For Hume this was an exercise in tautology. He had shown that we have no philosophically intelligible idea of causal power; thus the referral of such power to the deity was a further retreat into unintelligibility.[16] Nor did Hume confine his attack to Cartesian occasionalism. In *An enquiry*, the Newtonian variant received similar treatment, though Newton himself was exempted for, from this essay's point of view, the most significant of reasons. Newton, not intending 'to rob second causes of all force or energy', on the contrary 'had recourse to an etherial active fluid to explain his universal attraction; though he was so cautious and modest to allow, that it was a mere hypothesis, not to be insisted on, without more experiments'.[17] Ethereal speculation was seen by Hume as legitimate, a secular and therefore a preferable alternative to any reliance on a realm of divine causation in nature.

Hume's philosophical analysis of natural science was continued by his intellectual colleague and friend Adam Smith. In the spirit of 'the true old Humean philosophy',[18] Smith developed less an epistemology of science than a psychology of theory invention. He located the motivations of the scientific enquirer in two human passions, surprise and wonder, the former produced by an unexpected occurrence, the latter either by a singular novel object or by a succession of familiar objects in unfamiliar sequence.[19] These passions were functions of a more fundamental psychological mechanism, the association of ideas, and occurred when the expectations created by association about the regular behaviour of the external world were frustrated.[20] The way in which the imagination then grappled with these frustrations constituted the central concern of philosophical enquiry, philosophy being defined as 'the science of the connecting principles of nature . . . it represents the invisible chains which bind together all these disjointed appearances'.[21] Faced with these intervals in its expectations, the imagination literally pictures something to unite the disjointed appearances, thus enabling the passage of thought to resume a 'smooth, natural and easy' course.[22] The supposition of this linking chain of intermediate and imperceptible events succeeding each other 'in a train similar to that in which the imagination is accustomed to move'[23] is the only recourse open to the imagination in such circumstances. Smith offered here a psychologising account of causal inference, revealing both its analogical strategy and its necessary use of imperceptibles. The main example he used to illustrate the process was ethereal,[24] and he clearly regarded such speculative hypotheses as a naturally determined, integral, and central part of scientific enquiry. He was clear, too, in his denial that the imagination's 'invisible chains' had any native ontological status; they were associative chains, the product of mind, a point emphasised with direct reference to the Newtonian system of universal gravitation.[25]

Hume's analysis of causation and induction had tended to move science into a world of custom and belief, properly controlled by accurate experience and reflection. The net effect of Smith's work was to confirm this move, and extend it, as overtly as possible, into the province of passion and imagination. As regards ether, their treatment allows the proposition that classical empiricism not only need not have frowned upon ether, but indeed could generate powerful arguments in its favour.

The salience of this proposition lies not only in the interest and force of the arguments themselves, and hence in the likelihood of their influence. The arguments were part of a general cognitive field, a large philosophical endeavour which at that time was assuming considerable cultural and educational importance. One significant feature of this cognitive field's structure concerned the relation of epistemology and psychology, the science of human nature, to natural science. Hume had stated this programmatically in the introduction to *A treatise,* claiming that *'Mathematics, Natural Philosophy* and *Natural Religion,* are in some measure dependent upon the science of MAN; since they lie under the cognizance of men, and are judged of by their powers and faculties'.[26] The relation of the science of man to natural science was therefore one of priority. Natural science was the client, awaiting the benefits and strictures delivered by its patron. This structural relation gained a general acceptance in Scottish culture, evidenced not only in the work of individuals such as Hume or Reid, but also in its embodiment in university curricula.[27]

From its basic engagement with questions of human nature and cognition, the philosophy of Hume and Smith moved to moral theory, politics, and economics.[28] There are of course very significant differences between the conclusions that each of them reached, but certain crucial aspects of their work possessed a unifying similarity. Moral theory, for example, was approached from the point of view of psychological naturalism and of social determinism. Moral judgement and action originated in human sentiment and tended to conform to shared, communal norms. Economic theory, developed particularly by Smith, focused upon and attempted to reduce to systematic understanding those forces, such as division of labour, responsible for economic expansion and the creation of wealth.

Hume's and Smith's understanding of these topics cannot easily be separated from their earlier investigations. Hume had originally intended his science of man to provide foundations for what followed. Smith, though less overt in his intentions, is now coming to be understood as one whose philosophy of cognition and morals bears strongly upon his better-known work in economic theory.[29] Setting aside the issue of how far they managed to fulfil intentions to construct methodologically and substantively coherent intellectual systems, this is simply to say that there is adequate warrant for treating their work not as sequences of separate and distinct investigations, but as unified enterprises whose parts potentially possess manifold connective relations. Apparently restricted, technical discussions of the psychological origin and epistemological standing of scientific knowledge cannot with safety be isolated from systems of thought whose foundations they supported.

Broadly speaking, Hume and Smith were tackling a series of issues raised in acute form by the commercial growth of their nation, and the impact of that growth upon politics, social development, and human conduct – the problems of a modernity long sought by the Scots and now being achieved, not without anxiety about the implications of the achievement. Indeed, the relevance of their work was considerable, and this in turn depended upon perception of their modes of analysis and substantive doctrines as relevant to the ideological needs of the rising generation of Scotland's landed and professional elite.[30] Elite interest and participation in the philosophical culture provided by Hume and Smith can be persuasively documented through the elite's membership in and activities within societies such as the Philosophical and Select Societies of Edinburgh,[31] groups whose programmes and concerns directly reflected some of the philosophers' central preoccupations,[32] and whose avowed central aim was to improve Scottish life by stimulating progress in knowledge, agriculture, and manufacture.[33]

The substantial social relations of Scottish philosophy can, then, be adequately demonstrated. It is possible now to begin to inquire how this cultural complex affected ether. One aspect of this is already visible, for Hume's antioccasionalist support of ether may be taken directly as an expression of the modernising secularisation of culture whose main spokesman he was, and for which he incurred the considerable hostility of some elements in the Church of Scotland.[34] The question now to be posed is whether ether can be seen as having continued to respond to the shift towards secularisation, and also to those other features of philosophical culture so far instanced: the cognitive priority of epistemology, moral determinism, and economic advance.

As earlier suggested, the rising chemical science of William Cullen is the best site on which to examine expectation of ether's sensitivity to the concerns outlined. Expectation is heightened immediately by comprehension of Cullen's involvement with the culture of his day. As a university teacher he played an important role in the establishment and expansion of modern scientific education so earnestly desired by the Scots.[35] He was also a zealous improver, undertaking work in the technology of agriculture and manufacture. His playing the improvement card was important in the advancement of his career, because it gained for him the patronage of Lord Kames and the Duke of Argyle.[36] It established him ideologically as a patriotic citizen actively concerned with the promotion of his community's material prosperity. Cullen was also a prominent member of the Philosophical and Select Societies, contributing papers on chemical theory and on the steps to be taken to improve the teaching and practice of his science.[37] His adopted roles as reforming educator and civically minded scientist show therefore the considerable strength of his involvement with contemporary culture, through the fostering of modernity.

There are also strong circumstantial indications that certain of his intellectual loyalties were to Hume and Smith. He was a colleague of Smith's at Glasgow. This of course need not indicate intellectual loyalty; indeed, it may cynically provoke the opposite assumption. However, their joint sympathy for the mind and person of David Hume was revealed when Smith moved from the chair of logic to that of moral philosophy. Hume was a candidate for the vacated chair, and both Smith and Cullen wanted him as a colleague at Glasgow.[38] Although his candidacy failed, defeated by the church, Hume expressed copious thanks to Cullen for his support and friendship.[39] This occasion of Hume's attempt to enter academia lends credence to the view that Cullen should be seen as philosophically allied to Hume and Smith, particularly when one considers that Cullen's commitment was maintained in the face of by no means negligible clerical hostility.

That commitment was clearly expressed in Cullen's efforts to redesign the philosophical basis of chemical science. He accepted first of all the priority of epistemological analysis. He told his students:

> Logic is a very necessary part of introductory learning. By Logic I mean the analysis of the human mind, such as may be found in Mr. Locke's excellent treatise of the understanding. This is not only necessary in chemistry, but also in every other science where there is danger of error. I cannot but lament that students of medicine in the university are not obliged to go thro certain preparatory branches of learning.[40]

Accepting the cognitive priority of 'Logic', what precise form should it take? Locke was adduced as the *type* of thing Cullen recommended, but the exact form revealed a fundamentally Humean connection. 'In recommending this Study of Logic, we would, if we could venture, recommend it in a particular form. Not an obstinate disbelief in every *thing* and every *fact,* but that kind of Scepticism which the poet calls "The slow consenting Academic Doubt" '.[41] Cullen surely had his friend, the arch-sceptic of the age, in mind. Further discrimination is necessary here, however. Cullen was recalling Hume not at the end of book 1 of *A treatise,* teetering rhetorically on the brink of the abyss,[42] but in the essay 'Of the academical or sceptical philosophy', which concludes *An enquiry*. There academic scepticism is supported as a moderating corrective against Cartesian scepticism, held to be *'antecedent* to all study and philosophy',[43] and against thoroughgoing Pyrrhonism, which teaches men 'the absolute fallaciousness of their mental faculties, or their unfitness to reach any fixed determination in all those curious subjects of speculation, about which they are commonly employed.'[44] Cullen used the mitigated scepticism advocated by Hume to warn against acceptance of the 'concurrent testimony of a great number of authors . . . all facts which are said to be universal, are also to be suspected. General principles . . . [are] always to be received with diffidence'.[45] It was also used to point to our liability 'to mistakes in assigning causes for Phenomena'.[46] Cullen's sceptical mode of empirical epistemology exhibited congruence with Hume's. Now Hume's sceptical empiricism, in its methodological expression of temperate inductivism, has been seen to have coexisted with a psychologising stress on theory and hypotheses, a position developed by Smith. Cullen's congruence followed. The general principles mentioned by Cullen, though to be received with diffidence, were also 'certainly very necessary'.[47] Cullen's general attitude to speculative theory was positive. In his history of chemistry he emphasised the science's appeal to the liberal and speculative mind,[48] and later said, with reference to fire:

> You must in this and other subjects indulge me in giving much Theory. For tho' no body would recommend a wantonness of theory less than myself; yet I must be an advocate for its utility under proper restriction. It is a most powerful means of exciting us to experiments, and consequently the knowledge of facts.[49]

Theorising was empirically and heuristically justified, existing in subtle balance with its sceptical control. The stress on psychological satisfaction seen in Cullen's assertion that chemistry could gratify the liberal and speculative mind was also used in his extended discussion of causation in chemistry: 'But when several phenomena are referred to one common cause, we are satisfied'.[50] This recalls Smith's judgement of the satisfying effect brought about by the unifying power of a hypothesis.[51] Two further Smithian points emerge. First, with reference to Cullen's willingness to reason analogically in matters of theory, he can be seen adverting to the 'atmosphere of excited electricity which determines bodies once got within its sphere of attraction' when introducing his theory of chemical attraction by ether.[52] Second, although spending much time on this and other ether-based speculations, he insisted, like Smith, on their conjectural rather than ontological status.[53] Altogether then, Cullen's philosophical chemistry can be described as conforming with some precision to the work of Hume and Smith. The conformity may be seen in the general feature of epistemology's cognitive priority. It is seen, too, in Hume's, Smith's, and Cullen's substantive descriptions of scepticism's relationship to theory, the psychological aspect of speculation, and speculation's analogical derivation and conjectural status.

A plausible account of the intellectual coherence of Hume, Smith, and Cullen has now been established, and can lead into a demonstration of ether's relation to those expressions of modernity, secularization, political economy, and moral theory. It might be expected that the analytical task of showing the chemical ether's relation to cultural modernity is far more difficult and intricate than the effort of relating its supposedly closer cognitive connections of epistemology and method. This is not the case. The relationship to modernity is relatively immediate and easy to see. It is very directly illustrated in the case of secularization. Cullen was a secular as well as a sceptical and speculative chemist. In this he is to be contrasted, along with Hume, with the rest of his own and the preceding generation of Scottish philosophers and scientists. Thinkers such as Robert Wallace, Francis Hutcheson, Alexander Monro, Andrew Plummer, and Colin Maclaurin, all of them professors of science or philosophy at Glasgow or Edinburgh, embraced forms of reasoning that utilised First and Final Causes. Cullen explicitly rejected such reasoning. He was not a chemical theist, but that of course was natural enough, in that chemistry, constantly concerned with the *artificial* manipulation of bodies, allowed little opportunity for natural theological exposition in the manner of astronomy upon the *naturally* existing structure of the heavens, or of anatomy and optics upon the structure of the eye.[54] He had adequate opportunity, however, to introduce First Causes and the activity of God, as his predecessor Plummer had done with regard to chemical attraction.[55] But Cullen was no chemical occasionalist, and explicitly and curtly forbade such thinking in chemistry. 'Some say this attraction is the immediate act of the Creator. But this way of reasoning would soon put a stop to philosophical enquiries'.[56] In place of God as effective causal agent, Cullen put ether:

> Throughout all nature there seems to be an elastic repellent fluid, which is the cause of the phenomena we observe in nature; more particularly, of the various states of aggregation in different bodies . . . The attractive and repellant powers are constantly acting in op-

position to each other, and yet perhaps depend on the very same Aether acting by different circumstances.[57]

Ether now stood where once God stood in the structure of causation, as the immediate source of all active power. The pattern of Cullen's argument again remarkably evokes Hume, in the latter's invocation of Newton's ether in his sally against the occasionalists. Thus ether was used to develop its secularising potential, in chemistry as in epistemology, and the expression is perfectly direct.

Its connection from Cullen's chemistry to the economic basis of the culture and the perception of that base in political economy is less direct, but still strikingly close. Cullen did not construct arguments to show that ether promoted good theory that could then be applied to technological improvement. The relationship here is more subtle and more interesting and has to do with Cullen's conception of chemistry as 'philosophical' and with the precise socioeconomic status of the philosophical chemist. 'Philosophical chemistry' has to be defined very closely in this context, according to the exact construction placed upon it by Cullen. It did not mean epistemological discussion, nor did it necessarily mean systematisation. He described it as 'knowledge of those facts which must lead us to the knowledge of Causes, or the [philo]sophical part of Science'.[58] In other words philosophical chemistry was a science that gave causal explanations. The definition was comparable to Smith's description of philosophy as the 'science of the connecting principles of nature . . . representing the invisible chains which bind together all these disjointed appearances'.[59] Given that Cullen came to locate all chemical causality in ether, his ability to expound an ethereal chemistry was quite central to the image of philosophical chemistry he wished to purvey. That image was not only conceived in cognitive terms, however. It had important socioeconomic dimensions. As has been seen, Cullen believed in the 'liberal' appeal of chemistry – indeed, equated liberality with speculation: 'Or if more liberal still he aims at speculation . . .'[60] Chemistry as a liberal study was here contrasted with the traditional image of the 'chymists'; in Adam Smith's evocative phrase, 'those only who live about the furnace'.[61] Many years later, Cullen's achievement as a chemist was appraised by a successor in the Glasgow lectureship both as establishing philosophical chemistry and as 'taking chemistry out of the hands of the artists, the metallurgists and pharmacists, and exhibiting it as a liberal science, fit for the study of gentlemen'.[62] The liberal science, then, contrasted with chemistry as 'art', as practical technique and trade. The liberality was attained by rendering the science philosophical, an intellectual subject appropriate for attention from members of a leisured, gentlemanly elite who were not going to prosecute the practical side of the subject vocationally. The difference was essentially one of the labour process involved. In philosophical chemistry the labour was intellectual, meant to produce knowledge; in 'chymistry' it was technical and practical, meant to produce material goods. The different labour processes were attached, of course, to widely separate social categories. Yet Cullen, it has to be remembered, was an improver, and by no means wished for a separation of knowledge and practice. The image he sought was one that preserved the usefulness of knowledge while acknowledging the gulf between labour processes and between social categories. The terms of his perception here take us straight to the heartland of commercial expansion as it was then becoming to be understood in terms of the division of labour:

> It is the merchant that must inform us what arts there is a demand for and it is the merchant who must inform us of the marketable quality of manufactures . . . the merchant of more liberal education & extensive views must excite and direct the slow toiling industry of the artificer.[63]

It was the colleague of Adam Smith who spoke. The philosophical chemist concerned to improve technology did not set up his process and proceed to manufacture. He waited to be informed by the educated merchant about materially profitable objects of improvement, and he left to the merchant the stimulation and direction of the practical labourer. The labour process as it affected chemistry is divided in three parts. The labourer's task is directly productive, the merchant's directive and judgemental, and the philosopher's intellectual, to lend the aid of his special knowledge. In sociological terms, the division of labour adduced by Cullen mapped a social hierarchy with the philosophical improver now introduced at the top. Ether's relation to the economic base and to current analytic perceptions of it is now apparent. Because it was the substantive theoretical foundation of chemistry, and hence central to the science's new philosophical image, it was amalgamated, via the process of the division of labour, with the new economic role and social status of the philosophical chemist.

The final area of expectation, that of moral understanding, fails of any illustration in Cullen's chemistry, but is amply fulfilled in Cullen's medicine. This takes the argument beyond the strict bounds of this chapter while pursuing the trail of Cullen's thought. Fortunately, brief argument only is needed, as ether's role is once again obvious and direct. Hume's and Smith's description of moral conduct as originating in natural feeling rather than rational assessment, and their emphasis on its social function, could quite justifiably lead to their arguments' being seen as expressions of determinism, and this was in part the cause of the deeply disturbed reactions of some contemporaries, particularly those of the first generation of commonsense philosophers: Reid, Gregory, and Beattie.[64] Common sense was equally disturbed by another contemporary deterministic account of human nature, David Hartley's materialistic analysis, which relied upon the association of ideas and the nervous ether.[65] Cullen's theoretical medicine, like his chemistry, was dominated by ether, which underlay his concept of life and of disease, so it is hardly surprising to find commonsense opposition to Cullen in this sphere.[66] Although the terms of commonsense opposition to Hume, Smith, and Cullen had methodological expression, the radical concern was very much the preservation of active morality;[67] so in the field of morals the ether's materialistic associations raised implications of determinism for those contemporary moralists who sought to preserve a concept of morality based on active will.

Ether now stands fully displayed in the totality of its cognitive and cultural relations. Its modalities were those essential features of mid-eighteenth-century Scottish philosophical modernity, secularisation and determinism. Its mediation was the intensification of the division of labour, as that process created the new role of philosophical chemist. The analysis can fittingly end with the thought that Cullen himself might well have approved its spirit. With a sophistication regrettably rare among later chemical historians he commented in his history of chemistry on the subject's connection with the 'political and religious state of nations'.[68]

Ether theories of the Scottish chemists

Cullen's chemistry was of course far more than the impression embedded upon it by its cultural matrix. The matrix undoubtedly stimulated the appearance and shaped the acceptance and use of the science's central theory, but the theory also had to function as a theory. To achieve a complete portrait, it is necessary to turn now from ether's external relations and examine its domestic history.

The sources initially drawn upon by Cullen for his chemical ether appear to have been obvious ones in Newton's writings and in Bryan Robinson's accounts of Newton's ether. These, said Cullen, formed the 'most plausible scheme of chemical philosophy and will at least check the false theories of the corpuscularians'.[69] Robinson's speculations, and his aims, would seem to have been important for Cullen. In much of Robinson's work, the intention was to define ethereal activity in ways that made it more amenable to quantitative analysis, by reducing its species of action to more or less simple formulae, thus: 'If R. denote the Rarefaction of a Body by Heat, H the degree of Heat in the Body, and C the Strength of the Cohesion of its Parts; then, setting aside all external Compression, R will be as H/C'.[70] Such formulations might be dismissed as vacuous, a mere playing with letters that offered little real hope of assigning genuine quantitative parameters under experimental conditions. Yet the consistency of Robinson's attempt cannot be gainsaid, nor can the influence of its general desire for a quantified ethereal science. Indeed, the latter, it will be suggested, is of special importance in reaching an appreciation of Cullen's chemistry.

Some details of Robinson's notions concerning the interaction of heat and ordinary matter also deserve mention, bearing as they do a strong resemblance to the early stages of Cullen's thoughts on the same subject. For Robinson, heat was a vibratory motion of particles of ordinary matter, caused by vibrations of ether within ordinary bodies. After relating temperature to condensation of ether in the pulses of these vibrations, and describing how rarefaction of ordinary matter begins as its particles retreat from more to less dense pulsations, Robinson went on to reveal how this process terminated in fluidity:

> If the degree of Heat in a solid Body, to which the Expansive Force of the *Aether* between the Parts of Body is Proportional, be equal to the Force wherewith the Parts of the Body cohere; that Body, setting aside external Compression, will become fluid, and its Parts, by yielding to a Force impressed, be rendered capable of moving easily amongst themselves.[71]

An equilibrium of ether's elastic, expansive force and the cohesive force binding particles of ordinary matter causes fluidity. By the same token, once the degree of heat exceeds the cohesive force, a body's particles will recede from one another, attaining an aeriform state.[72]

Robinson held that ether's power was nonmechanical, was the power of 'Spirit', and that ordinary matter was inert.[73] Accordingly, he could not treat his force of cohesion as a property of ordinary matter. Dismissing cohesion as an effect of ether's gravitational action, he maintained instead that cohesion was an effect of light united with ordinary particles. This union caused increased density in the immediately neighbouring ether. No reasons for this were specified, but the effect of the increased ether density 'when the particles touch, [is to] press them on the sides opposite to the place of contact, and by that pressure make the particles to cohere with force'.[74]

As will be seen, Cullen by no means produced a straightforward rehearsal of Robinson's ideas when he came to develop his own theories of ether–matter interaction. However, once alerted by Cullen's recommendation of Robinson's works, it is reasonable to assume that features found there, particularly the impulse to quantification, the account of the states of matter, and the language of ether density, did inform Cullen's theoretical chemistry.

Out of these sources, Cullen developed two major theories – of chemical reactivity and the states of matter – which were applied to a variety of chemical problems. He also used ether for his theoretical understanding of fire, a topic to which he devoted much space. Underpinning and prompting these theoretical enquiries was a programme of experimentation.

A point requiring initial emphasis is that Cullen's thoughts on such matters underwent considerable development, as might be expected, between the late 1740s and early 1760s. Overall, this development was one that enhanced the role of ether in his chemistry. In the late 1740s, he explained chemical attraction by the relative balance of two sets of forces, the attraction of cohesion between particles of ordinary matter, and the repelling forces of the particulate ether. Given that different bodies contain different densities of ether:

> If the density of the one is not so much greater than that of the other as to overcome the attraction of its parts, the aether in each upon their approach will repel each other. If the density is greater, they will be attracted and join each other.[75]

As Robinson might have expressed it: Taking $A1$ and $R1$ to denote the attractive cohesion of the first body and the repelling ether contained in it, and $A2$ and $R2$ for the same in the second body, then attraction will occur either when $R1 > A2$, or when $R2 > A1$.

This same conception of the relative balance of attractive and repelling forces dependent on the relative density ratios of ordinary and ether particles was used in the account of matter's three states. Bodies whose cohesive attractions were greater than the repelling forces of contained ether were solid. When the two sets of forces were equal, bodies were fluid; they became elastic when the ether's repulsion was greater than the body's cohesive attraction.[76] The similarity of this scheme to Robinson's is obvious; but equally notable was Cullen's choice not to follow up Robinson's opaque speculations on the relation of light particles, ether density, and cohesion. Cohesion was treated as something distinct from, and acting in opposition to, ethereal repulsion.

At this early stage of ethereal speculation, Cullen faced one very considerable theoretical puzzle, and that was to account for the selective or preferential nature of chemical attraction. It was going to depend, obviously enough, on the relativities of ether's combination with chemically differing bodies,[77] but he was not able initially to offer a theoretical explanation of this. His own concept of philosophical chemistry as causal explanation coupled with his definition of the objects of chemical study as the particular qualities of bodies made the subject of election particularly pressing, and so far the general theory of reactivity was unable to cope with this most obvious feature of chemical particularity.

Cullen made progress with this puzzle by the early 1760s, if not before. An important stage in the sequence was a series of experiments he undertook in the first half of the 1750s, about which he read reports to the Glasgow Literary Society under the title 'On the generation of heat and cold by mixture'.[78] The experiments showed the opposite thermal effects of different chemical reactions. The crucial example was Glauber's salt, which in its deliquescent state with water generated heat, but in its crystalline state with water generated cold. As Cullen remarked, the chemical materials in each case were the same: the water and the acid and alkali composing the neutral salt; yet the thermal effect was opposite. He thought this could be explained by the different states of aggregation of the two forms of the salt, and the consequent different types of reaction they entered into with water: 'When a deliquescent salt is added to water, a part of this is united with it by way of mixture [i.e., combination], but when a crystalline salt is added to the same, it is united with it by solution only'.[79] The general law he formulated from his experimentation, despite the awkward counter-example of acids with volatile alkali, was that 'every mixture generates heat, and that every solution generates cold'.[80]

Cullen's explanation of this notable general phenomenon, to be found in his lectures, was in terms that related ether to state of aggregation. In mixture, condensation took place, releasing ether (identified with the matter of fire) and raising temperature. In solution, rarefaction occurred, absorbing ether and lowering temperature.[81] From this experimentation, then, Cullen was coming to see thermal effect as an empirical key to the discrimination of ether's particular union with bodies, and in a way that made conversions of state acutely important.

His biographer informs us that he went on to compile tables of these thermal effects; so the experimental prosecution continued.[82] So too did the theoretical development. The shift his theory now underwent was a major one, of ontological proportions. Instead of dealing in two causal parameters, the attractions of ordinary matter and the repulsions of ether, he located all activity in ether, leaving ordinary matter inert. The change is vividly seen in the theory of states of matter, now conceived by way of the relative density of ether internal and external to a body. It was solid when the external ether was greater than the internal, fluid when external and internal were equal, elastic when the internal was greater than the external.[83] The theory of attraction naturally underwent a similar shift. Employing the analogy of contemporary electrical theory, Cullen imagined that

> every body is surrounded by its own proper atmosphere of this fluid, which grows more dense as it recedes from the surface. This is analogous to the atmosphere of excited electricity which determines bodies once got within the sphere of its attraction to the surface of the electric body. It is to be observed, that bodies thus in contact with the excited body, remain, some longer, some a shorter time in contact with that body, until they have got an atmosphere of their own: then they are repelled till meeting with some other matter, they discharge their atmosphere, and are attracted and repelled as before.[84]

Attraction was now explained by the density gradient of an external ether atmosphere rather than by the relative densities of internal ether and ordinary matter. Cullen immediately applied the further detail of the electrical analogy to an additional elucidation of his distinction between mixture and solution. In mixture, bodies brought into contact have one common atmosphere, whereas in solution solute and solvent retain their own atmospheres and can thereby act separately upon other bodies.[85] How is it, though, that ether atmospheres determine bodies to a position where they have a common atmosphere and form a new chemical body? The most likely effect of the approach of two bodies each surrounded by ether is surely a repulsion, not an attraction. Cullen asked his audience to grant

> one postulatum, to viz [sic] that inert matter, in a certain contiguity of its parts, has a power to diminish the repelling power of the countervening aether betwixt its particles. This admitted, the attractive power may be the effect of repulsion, when two bodies are in such a close contiguity as to diminish the power of the intervening ether betwixt its particles.[86]

Ether between particles was perhaps being imagined as squeezed out or pushed aside by their contiguous approach. If not, if Cullen was thinking of some other power of inert matter, then the contradiction in terms showed the new ontology failing to sustain the theoretical exposition.

Moreover, the crucial issue of election remained. Cullen evidently believed that he had made some progress here, though his treatment was scattered and brief. When talking of the operations of chemistry, such as decomposition and combination, he asserted that combination depends on attraction, and attraction upon fluidity.[87] Shortly after, he claimed that 'fluidity is the only means of giving that contiguity which is necessary for the Attraction of Cohesion', but he then immediately canvassed electrical (i.e., ethereal) attraction as an *alternative* cause to fluidity.[88] Finally, having settled for the ether atmosphere variant just described, he nonetheless went on to add that 'the *elec-*

tive depends upon Fluidity'.[89] Clearly there was a pronounced degree of confusion as the crucial topic was approached. The most consistent gloss to be given Cullen's thought here would accept his insistence that contiguity was essential for a combination to take place, with the new particulate configuration possessing a single atmosphere. Fluidity might then be seen as enabling contiguity, because fluid particles, in the equally balanced distribution of their internal and external ether, were more easily determinable into contiguity by an approaching ether atmosphere.

Cullen himself did not face these acute problems of ether interface and balance; so, strictly from the evidence, the impression of confusion over election must stand. The new chemical ontology of ether, with its Smithian promise of unity and simplicity, broke down precisely where it needed to succeed. A final point should be made, however, which explains Cullen's desire to link fluidity and election. This desire related to the experimental side of his programme. If fluidity could have been established as the cause of election, then a potential route to a thoroughly quantified chemical science of elective attraction was opened up, by temperature measurement of the thermal effects of chemical reactions. Cullen's goal was therefore a large and ambitious one, and his means to it decidedly innovative. His degree of success in its pursuit is ultimately to be evaluated against the stature of his enterprise and the originality of the theoretical and experimental programme.

Cullen's reformulation of chemical theory in ethereal terms had a profound impact upon the direction and content of chemical science in the decades that followed. The work of Joseph Black, William Irvine, and Adair Crawford, to name only the better known of the Scottish chemical succession, was predominantly taken up with the question of heat in relation to the states of matter and change of state. However, the relationship of this later work to Cullen's cannot be assessed straightforwardly as a direct extension of Cullenian chemistry. Both Black and Crawford entered reservations about the ethereal concept of heat, and Irvine made some notable modifications to Black's doctrines of latent heat and heat capacity. Clearly then, there are problems in seeing Scottish chemistry simply as an extended working through of Cullen's basic ideas on ether–matter interaction.

Indeed, is there any point at all in maintaining Cullen's significance when faced with what seems to be a denial of his most fundamental concept by his most illustrious pupil, Joseph Black? Some historians, accepting Black's published views on the 'materialist' (i.e., ethereal) theory of heat, have wished to see his work as independent of any theoretical commitment on the nature of heat.[90] Others, however, wish to claim that Black was indeed a materialist, adhering to a notion of heat as subtle, self-repelling matter capable of entering into combination with matter by attraction, in a 'chemical' fashion.[91] This dilemma is capable of resolution. The answer to the question of Black's views on the nature of heat depends upon the timing of his utterances and the precise context of their composition. Once this is realised, it is possible to reassert the strong sense in which Black was Cullen's pupil.

The evidence for Black's theoretical neutrality concerning heat comes from the very end of his career, in John Robison's edition of Black's chemical lectures, where Black is made to say that the chemical theory 'will please the imagination, but does not advance knowledge. I therefore avoid such speculations . . .'[92] The avoidance, however, may be taken to be Robison's rather than Black's, for Robison had an avowed detestation of material fluid theories, and granted himself the most substantial editorial licence.[93] In fact, even in Robison's edition of the lectures, Black did not 'avoid such speculations', devoting space to a discussion of his pupil William Cleghorn's theory of heat,[94] a theory that, as Arthur Donovan has noticed, bore a strong resemblance to Cullen's first theory of the states of matter.[95] It is also possible to find Black considering heat as subtle matter in earlier sets of notes taken by students.[96] But the strongest evidence for seeing Black as a Cullenian etherealist comes not from any of his lectures, but from his notebooks, in the record of observations and queries that he followed in his formulation of the principle of latent heat. Some of these observations bore immediately upon Cullen's experiments on thermal effect in chemical reactions, particularly those where absorption and release of heat appeared as a consequence of change of state.[97] Others noted the necessity of heat for elasticity, and beyond that indicated an implicitly chemical conception of heat: 'Nitrous ether boils *in vacuo,* and grows vastly cold. Is not the cold produced merely by the evaporation of the elastic parts; or does it emit air, which requires heat for its elastic appearance?'[98] Elsewhere in his notebooks, Black argued that

> heat may be considered in nature as the great principle of chemical movement and life . . . we must consider it not as an accident in bodies, but as a separate and specific existence, not less so than light or electric matter; and though agreeing with these in some of its effects, in its nature possibly different from either.[99]

Although these notes contain no unambiguous formulation of heat as subtle matter, Black, like Cullen, considered heat as the cause of elasticity, as the fundamental active agent in chemical phenomena, and as comparable in some of its aspects to the matter of electricity. These features of Black's early work on heat surely indicate the large extent of his indebtedness to Cullen's ethereal chemistry at the moment of his most profound insight into the nature of heat's interaction with matter. Neither his later methodological caution in front of his students, at a period when commonsense philosophy's anti-ether and antihypothetical sentiments were gaining ground, nor Robison's editorial bias ought therefore to be allowed to obscure the close intellectual relationship of Cullen and Black before 1757.

Cullen, it should be emphasised, not only provided Black with interesting data and a general theoretical framework. His work provided the precise point of entry and exact form of problem that issued in Black's discovery of latency. Black's earliest notebook queries, before 1752, were on 'the nature of cooling mixtures, and the cold produced by liquefaction'.[100] The first entries quoted by Robison were on the reverse phenomenon, the heat released on state conversion from fluid to solid.[101] Black was concerned in the first instance not with the broad question of change of state, but with the narrower topic of fluidity, a point confirmed in Black's own reminiscences;[102] and fluidity, it may be recalled, lay at the obscure centre of Cullen's beliefs about elective attraction. Black may then have been initially engaged with the Cullenian problem of reactivity, but certainly not with the formulation of a general theory of change of state.

However that may have been in the early and mid-1750s, the development of Black's work on heat, and of the work of his pupils and successors, led Scottish chemistry away from Cullen's problem. He initiated a chemistry of ether–matter interaction that was of the highest importance for the origin of Black's work, but its extension involved the loss of at least one of his most cherished goals, the quantification of elective attraction. Black, Irvine, and Crawford became far more interested in the definition of and solution of problems to do with change of state, heat capacity, and animal heat. After Black's well-known discovery of latency and capacity, the most significant innovation in the Scottish chemists' study of heat was made by William Irvine. Irvine attempted an elegant and unifying reformulation of Black's principles by theorising that latent heat was a function of heat capacity: specifically, that the absorption and release of that heat necessary for change of state was the result of an enlargement or diminution of a body's capacity for heat.[103] Irvine's reformulation of Black's work had strong chemical implications that were made explicit in Adair Crawford's study of animal heat. Crawford held that air entering the lungs contained heat. Blood, returning from the body's extremities impregnated with phlogiston, was robbed of this phlogiston by the air, whose attraction for it was stronger. As this transference took place, the air emitted its heat, which was absorbed by the blood. The blood was able to absorb the heat because its capacity for heat had been enlarged by its loss of phlogiston.[104] In Crawford's work, capacity had become the key concept to understanding the absorption and emission of heat in animal chemistry. Further, Crawford was at pains to distance himself from what was by then the traditional way of conceptualising reactivity and heat. Attraction was only a metaphor,[105] heat not being capable of combining with matter chemically.[106] This point was made with reference to Lavoisier, who saw the heat evolved in combustion as the result of chemical decomposition. According to Crawford, this heat was the result of a change in capacity.[107] Irvine's idea of capacity was then the pivotal move in this final distancing of Scottish chemistry from its Cullenian origins.

Lavoisier as ether theorist

The foregoing description of Scottish chemical theory and its applications reveals a striking correspondence between certain of its aspects and important features of Lavoisier's chemistry as it developed after the mid-1760s. Recent studies persuasively advance the view, first, that his focus on gases, seen most famously in the case of oxygen, derived from a standing interest in the theory of the elastic state, which he early came to regard as the effect of the combination of ordinary matter with caloric ether, the matter of fire.[108] Second, they show that within his continued study of heat, prosecuted with Laplace, there was a notable effort to relate caloric ether to the understanding of chemical attraction.[109] This concordance of Scottish and French chemistry is highly significant for the history of ether. It shows in general terms the very considerable dominance that the concept achieved in chemistry in the science's most crucial formative years. More particularly, it illustrates the notable degree of unanimity imposed within the science by an overarching theory that allowed rigorous experimental prosecution.

An initial point to make before exploring this topic in more detail is that the Scots were not, so to put it, responsible for Lavoisier. They did not stimulate Lavoisier's interest in the elastic state, and the impact of their work when it reached Paris, though not negligible, was not radical.[110] The topic under discussion is therefore not a sequential development moving from Edinburgh to Paris. It is rather a question of outlining the way in which an analogous development created a similar conceptual set that therefore serves the historian to characterise the science in the broadest terms possible.

The fundamental reinterpretation wrought by Lavoisier scholars in the past decade has in turn evoked an interestingly problematic picture. Although the stages of his work on heat and elastic state theory have been brought clearly to light, and his attack on calcination and combustion thereby set in a novel context, the developmental process has not been characterised. The poles of that process can be seen in the following question: How did the geochemist of the 1760s with an interest in the German Eller's two-element theory of matter become the decidedly Newtonian caloric theorist of the 1780s? Though this question cannot be adequately answered within the confines of this chapter, it does focus clearly on the direction of Lavoisier's development. He came increasingly to employ the terminology of particles and attractive and repulsive forces, terminology that allows comparison with the Scots. The terminology of his comments on Eller's theory of the elements allows no such comparison.[111]

The overall process by which Lavoisier came to hold these views can be briefly summarised as follows. His quest for a theory of elastic state, achieved in some measure by 1772, focused his attention on the calcination of metals, because this phenomenon involved weight gain. Previously, Lavoisier had believed metals to lose air during calcination, because they afterwards no longer effervesced with acids.[112] The discovery of oxygen gave him the opportunity to formulate a new theory of acidity, which directed him to the chemical affinities of the oxygen principle.[113] The partnership with Laplace in the study of heat provided him with a theoretical terminology of particles, force structure, and balance from which he was able to manufacture a theory that enlarged and integrated his ideas on states of matter and chemical attraction.[114]

It is the latter stage that bears comparison with the Scots. By 1772, Lavoisier had firmly concluded that 'air is a particular fluid combined with the matter of fire',[115] but it was not until later, under Laplace's influence, that he overtly adopted terminology indicating that expansibility and elasticity were conceived in terms of a structure of forces. That he did adopt such terms by the early 1780s is seen in his discussion of the affinities of the oxygen principle, where he claimed that he had had this aim in view since the beginning of his work on oxygen and had been gathering materials for it for several years.[116] He offered some penetrating criticisms of the defects of contemporary affinity tables. They dealt only with simple affinities, whereas the chemist is normally encountering at least double or triple affinities. They did not take account of water's attraction in reactions performed the wet way. They ignored the crucial variables of degree of saturation and temperature.[117] This last point is especially relevant when the Scottish comparison is borne in mind. Lavoisier argued that all bodies are immersed in an all-penetrating igneous fluid that acts contrary to the force of attraction maintaining their aggregation. Their solid, liquid, or aeriform state is a consequence of the relative balances of the two sets of forces. The degree of heat in bodies must then affect the degree to which they are, or are not, combinable, because that degree of heat determines the relative firmness of each body's particulate union. Solid bodies such as lead and tin do not combine because the attraction between them is insufficient to overcome the attraction binding the particles of the separate substances. Once, however, the action of heat weakens the strength of their aggregation, the attractive force between them can operate, and a combination take place.[118] Lavoisier, very much under the sway of Laplace on this question of affinity, believed that this scheme held out great hopes for the progress of chemistry. It was not altogether chimerical to envisage a time when the chemist would precalculate the results of reactions, as the geometer now did for the motions of celestial bodies.[119] The fact that attraction varied with temperature offered the mode of measurement; according to Laplace, the balance of caloric and cohesion furnished a precise means for comparing chemical affinities.[120]

As the caloric ether assumed an ever-increasing role for Lavoisier, his theoretical chemistry came particularly to resemble the early version of William Cullen's. Their views on the states of matter had no significant differences, and on chemical attraction and its integration with states of matter the difference was slight. More generally, Lavoisier's chemical theory of heat can be closely compared with those of Black and Cleghorn. Certain discriminations require emphasis, however, particularly with regard to Cullen. Although both came to see temperature measurement as the key to understanding elective attraction, and to consider it an avenue for fundamental advance of their science, Cullen went on to abandon the attraction of cohesion for a unicausal chemical ether, whereas Lavoisier continued to think in dualistic terms of attractive cohesion and repelling caloric. Cullen, moreover, favoured ether density concepts that did not appear in Lavoisier, and his view of matter's liquid state did not incorporate the factor of external air pressure, as Lavoisier's did.

Yet despite these distinctions, the historian can surely have no doubt about ether's significance for the development of chemistry. It was so fundamental that the massive and transforming innovations brought about by Cullen and Lavoisier simply cannot be imagined without it. Although ether does not give one all there is to know about chemistry in this period, its position was central enough to propose the conclusion: no ether, no new chemistry.

Acknowledgments

I should like to thank the editors for their helpful comments on this chapter. Sections of it have been delivered as seminars at the University of Leeds and the University of Aberdeen. The points made by Peter Heimann and Nicholas Fisher on those occasions proved most useful.

Notes

1. Study of ether's role in moral theory might begin with David Hartley's *Observations on man, his frame, his duty, and his expectations*, 2 vols. (London, 1749). Its role in political ideology can be examined in the Revolutionary and Napoleonic periods. See J. B. Morrell, 'Professors Robison and Playfair, and the Theophobia Gallica: natural philosophy, religion and politics in Edinburgh, 1789-1815', *Notes and Records of the Royal Society of London* 26 (1971), 43-63.
2. N. T. Phillipson, 'Culture and society in the 18th century province: the case of Edinburgh and the Scottish Enlightenment', in *The university in society: studies in the history of higher education*, ed. L. Stone, 2 vols. (Princeton 1974), 2:407-48.
3. J. R. R. Christie, 'The rise and fall of Scottish science', in *The emergence of science in Western Europe*, ed. M. P. Crosland (London, 1975), 119-20.
4. Fox (1971); J. B. Gough, 'Nouvelle contribution à l'étude de l'évolution des idées de Lavoisier sur la nature de l'air et sur la calcination des métaux', *Archives Internationales d'Histoire des Sciences* 22 (1969), 267-75; H. Guerlac, 'Chemistry as a branch of physics: Laplace's collaboration with Lavoisier', *Historical Studies in the Physical Sciences* 7 (1976), 193-276; R. J. Morris, 'Lavoisier on fire and air: the memoir of July 1772', *Isis* 60 (1969), 374-80; R. Siegfried, 'Lavoisier's view of the gaseous state and its early application to pneumatic chemistry', ibid. 63 (1972), 59-72.
5. Cullen's chemistry lectures for 1748-9 show an already well-developed understanding of ether's uses for chemical theory. W. Cullen, Royal College of Physicians of Edinburgh (hereafter RCPE), MS C15.
6. It was certainly composed before the return of Halley's comet, which Smith referred to in the future tense. A. Smith, 'The principles which lead and direct philosophical enquiries, illustrated by the history of astronomy', in A. Smith, *The early writings of Adam Smith*, ed. J. R. Lindgren (New York, 1967), 106.
7. John Millar, quoted in *Lectures on rhetoric and belles lettres, delivered in the University of Glasgow by Adam Smith: reported by a student in 1762-3*, ed. J. M. Lothian (London, 1963), xvi.
8. Hartley, *Observations*, passim.
9. D. Hume, *A treatise of human nature*, in D. Hume, *Philosophical works of David Hume*, 4 vols. (Edinburgh, 1826), 1:88.
10. D. Hume, *Enquiries concerning the human understanding and concerning the principles of morals*, ed. L. A. Selby-Bigge, 2nd ed. (Oxford, 1902), 30.
11. Hume, *Treatise*, 8-10.
12. J. Noxon, *Hume's philosophical development* (Oxford, 1973), 91.
13. Hume, *Treatise*, 220.
14. D. Hume, *An abstract of a treatise of human nature*, eds. J. M. Keynes and P. Sraffa (Hamden, Conn., 1965), 32.
15. Hume, *Treatise*, 212.
16. Ibid.
17. Hume, *Enquiries*, 73.
18. The phrase is John Millar's. D. D. Raphael, 'The true old Humean philosophy and its influence on Adam Smith', in *David Hume: bicentenary papers*, ed. G. P. Morice (Edinburgh, 1977), 25.
19. Smith, *Early writings*, 32-9.
20. Ibid. 39-40.
21. Ibid., 45.
22. Ibid., 40.
23. Ibid., 40-1.

24 Ibid., 39, 41.
25 Ibid., 108.
26 Hume, *Treatise*, 'Introduction', 7.
27 This point is discussed by G. E. Davie, *The democratic intellect: Scotland and her universities in the nineteenth century* (Edinburgh, 1961), 3–25; and by Olson (1975), chap. 1.
28 A not untypical sequence for Scottish intellectuals of this period; compare this sequence with the careers of Thomas Reid and Dugald Stewart.
29 See J. R. Lindgren's excellent monograph, *The social philosophy of Adam Smith* (The Hague, 1973).
30 N. T. Phillipson, 'Towards a definition of the Scottish Enlightenment', in *City and society in the eighteenth century*, eds. P. Fritz and D. Williams (Toronto, 1973), 140–3.
31 R. Emerson, 'The social composition of enlightened Scotland: the Select Society of Edinburgh, 1754–1764', *Studies on Voltaire and the Eighteenth Century 114* (1973), 291–329.
32 Phillipson, 'Towards a definition', 139; Christie, 'Rise and fall of Scottish science', 118.
33 J. R. R. Christie, 'The origins and development of the Scottish scientific community, 1680–1760', *History of Science 12* (1974), 140.
34 E. C. Mossner, *The Life of David Hume* (Austin, Tex., 1954), 156–61.
35 J. Thomson, *An account of the life, letters and writings of William Cullen, M.D.*, 2 vols. (Edinburgh, 1859), 1:40–5.
36 Christie, 'Rise and fall of Scottish science', 120.
37 W. Cullen, 'An essay towards ascertaining the different species of salts', published by L. Dobbin, 'A Cullen chemical manuscript of 1753', *Annals of Science 1* (1936), 138–56; W. Cullen, 'Of the cold produced by evaporating fluids, and of some other means of producing cold', *Essays and Observations, Physical and Literary 2* (1756), 145–56.
38 Thomson, *Cullen*, 606.
39 Ibid., 72–3.
40 W. Cullen, RCPE, MS C10 (chemistry lectures of the early 1760s), 27.
41 Ibid.
42 Hume, *Treatise*, 340.
43 Hume, *Enquiries*, 149.
44 Ibid., 150.
45 Cullen, RCPE, MS C10, 30.
46 Ibid.
47 Ibid.
48 Ibid., MS C11 (history of chemistry), 3.
49 Ibid., MS C10, 26.
50 Ibid., MS C15, 1, 79.
51 Smith, *Early writings*, 45.
52 Cullen, RCPE, MS C10, 83.
53 Ibid., MS C15, 1, 79.
54 I owe this point to Dr. Nicholas Fisher of the University of Aberdeen.
55 A. Plummer, 'Experiments on neutral salts, compounded of different acid liquors, and alcaline salts, fixt and volatile', *Essays and Observations, Physical and Literary 1* (1754), 337.
56 Cullen, RCPE, MS C10, 68.
57 Ibid., 83.
58 Ibid., 22–3.

59 Smith, *Early writings*, 45.
60 Cullen, RCPE, MS C11, 3.
61 Smith, *Early writings*, 46.
62 J. Robison (ed.), *Lectures on the elements of chemistry, by Joseph Black, M.D.*, 2 vols. (Edinburgh, 1803), vol. 1, Preface, xxii.
63 W. Cullen, Glasgow University Library, Cullen MSS, no. 7.
64 Phillipson, 'Towards a definition', 144–6.
65 Hartley, *Observations*, passim.
66 Cullen's opponent was John Gregory, an early member of the Aberdeen Philosophical Society and latterly a professorial colleague of Cullen's at Edinburgh.
67 Olson (1975), 28–31.
68 Cullen, RCPE, MS C11, 1.
69 Ibid., MS C10, 87.
70 B. Robinson, *A dissertation on Sir Isaac Newton's aether* (Dublin, 1743), 95.
71 Ibid., 96–8.
72 Ibid., 100.
73 Ibid., 122.
74 Ibid., 114–16.
75 Cullen, RCPE, MS C15, 1:132.
76 Ibid., 130–1.
77 Ibid., 57.
78 Thomson, *Cullen*, 580–3.
79 Ibid., 581.
80 Ibid., 583.
81 Ibid., 53.
82 Ibid., 52.
83 Cullen, RCPE, MS C10, 112–13.
84 Ibid., 83.
85 Ibid., 84.
86 Ibid., 86–7.
87 Ibid., 68–9.
88 Ibid., 69.
89 Ibid., 84.
90 E. g., D. McKie and N. H. de V. Heathcote, *The discovery of specific and latent heats* (London, 1935), 27–8.
91 E. g., Schofield (1970), 185–90.
92 Robison, *Lectures*, 1:193.
93 See Robison's editorial correspondence, in *Partners in science: letters of James Watt and Joseph Black,* eds. E. Robinson and D. McKie (London, 1970), 320–77.
94 Robison, *Lectures*, 34.
95 A. Donovan, *Philosophical chemistry in the Scottish Enlightenment* (Edinburgh, 1975), chap. 9, n. 62, p. 308.
96 J. Black, Edinburgh University Library, MS Dc. 3 11, 4.
97 Robison, *Lectures*, 1:525–6.
98 Ibid.
99 A. Ferguson, 'Minutes of the life and character of Joseph Black, M.D.', *Transactions of the Royal Society of Edinbrugh* 5 (1805), 108.
100 Robison, *Lectures*, 1:xxiii.
101 Ibid., 525.
102 Ibid., 156.
103 W. Irvine, *Essays, chiefly on chemical subjects* (London, 1805), 88.
104 A. Crawford, *Experiments and observations on animal heat, and the inflammation of*

combustible bodies, being an attempt to resolve these phenomena into a general law of nature, 2nd ed. (London, 1788), 361–2.
105 Ibid., 363.
106 Ibid.
107 Ibid., 372.
108 See n. 4.
109 Guerlac, 'Chemistry as a branch of physics', 270–3.
110 Fox (1971), 29.
111 Siegfried, 'Lavoisier's view of the gaseous state', 61.
112 Ibid., 69.
113 Guerlac, 'Chemistry as a branch of physics', 270.
114 Ibid., 271.
115 Ibid.
116 A. Lavoisier, 'Mémoire sur l'affinité du principe oxygine', in A. Lavoisier, *Oeuvres de Lavoisier,* 6 vols. (Paris, 1862–93), 2:546.
117 Ibid., 547–8.
118 Ibid., 547.
119 Ibid., 550–1.
120 'Mémoire sur la chaleur', in Lavoisier, *Oeuvres de Lavoisier,* 2:317. Guerlac attributes authorship to Laplace.

3

Ether and physiology

ROGER K. FRENCH
Wellcome Unit for the History of Medicine, University of Cambridge, Cambridge CB2 3RH, England

Ethers have often been invoked to explain how matter, considered to be inert, could be endowed with motion. This problem was nowhere more acute than in the study of life; living things are self-moving, and the matter of living bodies was consequently held to be either innately mobile or moved by a special principle of life. Moreover, living things were seen not only as self-mobile, but as purposeful in their motions. Consideration of these problems highlighted two important questions that recurred throughout the history of physiology: What is the *source* of motion in the body? and what *controls* it? The theme of this chapter is that an examination of these questions will illuminate the role of subtle fluids in physiology.

Answers involving subtle fluids had been offered to these questions since antiquity. Purposeful, voluntary motions were universally held to arise from the rational soul of man, but this soul was also widely held to be immaterial and immortal, differing on both counts from the body. How then could the immaterial soul create motion in matter? One answer was to postulate a fluid so subtle that it somehow partook of the immateriality of the soul and of the materiality of the body. This answer took two rather different forms in the Galenic and Aristotelian traditions, both of which were subsequently of importance in the West.

For Galen, the 'animal spirit' was the agent of the rational soul, occupying the brain and nerves and communicating between the will and the voluntary muscles. The 'animal' or 'nervous' spirit remained the physicians' traditional subtle matter of physiology until at least the beginning of the eighteenth century, and was indispensable in explaining body–mind interaction in voluntary motions. The problem of interaction did not necessarily arise in the case of the involuntary actions, whether 'vital' or 'natural', for these did not involve the immaterial soul, but resulted from a 'faculty' or simply from 'nature'.[1]

Galen confessed that the term *faculty* was often used in ignorance of true causes, and the analogy he employed to illustrate the action of a faculty included the attraction exerted between similars: the motion of oil towards the wick of a lamp and the suction of air into bellows. Because there was no problem of interaction, Galen was not obliged to use subtle fluids to explain these motions, and his 'vital spirit' was a physiological agent of respiration, not motion. In no case did Galen face the question how the motion arose or how a very small motion of the barely material spirit in the nerves could produce great muscular power. His physiology was founded on the innate powers of living matter.[2]

By contrast, Aristotle in both his physiology and his physics drew a sharp distinction between the mover and the moved: Everything that moved was moved by something else and the ultimate source of motion was an unmoved mover. In animals the unmoved mover was the soul, and Aristotle here used the analogy of a mechanical puppet set in motion by pulling the strings. Because the unmoved mover was immaterial, Aristotle was faced with a problem of interaction when explaining the origin and control of bodily motion. He answered the problem by invoking a spirit (*pneuma,* the word used also by Galen) that mediated between the mover and the moved by supplying a force of expansion and contraction that, in his analogy, moved the strings of the puppet.[3]

The innate (not respired) spirit provided Aristotle with a physical answer to the question of the source of motion. However, in solving the second problem – the control of motion – Aristotle made the explanation psychological. The unmoved mover was the soul's perception of an external Good or Evil, and this aroused the motion of appetite. This in turn excited the motion of the animal in pursuing the Good or avoiding the Evil. The *rational* appetite was concerned with voluntary motion, but animals were compelled to act involuntarily by simple perception and appetition. Since the nonrational sentient soul of animals was a form of matter, the living creature carried with it its own causes of motion, but this living matter was not essentially mobile. It was the *pneuma* that in this way imposed motion upon matter that also carried the soul (*psyche*) from generation to generation and was responsible for the inception and development of the new individual. This *pneuma* was an active subtle fluid far removed from the elements of the mundane world but related to the ether of the heavens.

Subtle fluids in pre-Newtonian physiology

This common stock of traditional ideas provided the basis for a number of different seventeenth-century answers to the questions of biological

Ether and physiology

motion. The generation of animals was an irreducibly biological problem, and one for which subtle fluids continued to provide an answer. Harvey used Aristotle's notion of a quintessence, and similar ideas were used in the eighteenth century by Stephen Hales, Hermann Boerhaave, and Linnaeus. However, in the period between Harvey and Hales, science passed through a corpuscularian revolution: Matter was now believed to consist of particles, whether part of a universal plenum or atoms traversing a void. The doctrine of corpuscularity asserted that ultimate reality lay in particles and motion: What then of spirits, which had traditionally bridged the gap between materiality and immateriality?

Descartes banished them. The nervous spirits became, in accordance with the doctrine of a universal plenum, passive particles acting mechanically, subtle only in their smallness. Moreover, according to the Cartesian world picture the immaterial soul, which was a different category of existence from matter, nevertheless could produce voluntary motions in the body. The break with traditional ideas was decisive and lay at the root of much Continental (and particularly French) physiology in the later seventeenth and eighteenth centuries. However, in England physiology was already strongly influenced by Harvey, whose work could be seen as developing Aristotelian biology in a very effective way, with the result that Aristotelian and other traditional concepts did not suffer a sudden eclipse; certainly, corpuscularity forced a compromise with the old ideas, although only a few writers embraced Cartesian ideas. Thus the term *spirit* retained some suggestion of innate mobility and nonparticulate subtlety, and the word sometimes carried its theological connotations into physiology. For example, Francis Glisson[4] argued that spirits (demons and angels) were essentially alive, whereas the matter of man's body was 'accidentally' alive insofar as his rational, spiritual soul imposed spiritual life upon it. The growth of animals and plants was attributed to the action of a plastic force, the *archeus* of the 'chemists' (Van Helmont and others), which was a development of the primeval life of matter and which Glisson identified with the Hippocratic healing power of nature; the *archeus* was at least partly spiritual in nature. The notion of the Great Chain of Being also provided a reason for supposing the existence of spiritual forces at work in the physical world, and George Cheyne, early in the eighteenth century, held that there was, above man in the chain, *'an active, self-motive, self-Determining Principle'* that supplied motion to the inert matter of the universe.[5] Robert Whytt of Edinburgh believed in a similar chain of 'intelligences' above man, and he also believed that the inert matter of the body was moved by an extended but immaterial soul which, like that of Glisson and later animists, was to be identified with the Hippocratic healing force of nature (French, 1969). A contrast to the situation in Britain was provided by the Continental mechanists (before about 1740) who, under the influence of Descartes, could not believe that a soul or spirit could directly move matter (except in the case of voluntary motion). Thus, Friedrich Hoffmann of Halle[6] believed in a chain of spiritual entities but was obliged to accept ether to account for animal motions. His theory is discussed in the section of this chapter entitled 'Cartesian subtle matter'.

The absence of a Cartesian physiological revolution in Britain meant the survival not only of active spirits, but also of the soul to which they were closely related. A number of late seventeenth-century writers tried to dress up more or less traditional ideas in corpuscular clothing; thus, Thomas Willis retained a lower *anima,* controlling the grosser physiological activities, in addition to a rational *animus*. The *anima* was akin to fire, located in the blood, and found also in animals; and the *animus* was like light and was distributed through the brain and nerves. Willis also claimed that the blood was the source of the spirits, whence they were 'alembicated' by the heat of the body. These spirits were particulate but *active,* and were more closely related to the chemical spirits of the Oxford physiologists who had been working on respiration than to the mechanical spirits of Descartes. For Willis muscular motion was the result of the effervescent meeting of different kinds of spirit. However, Willis's scheme was entirely qualitative, and the activity of the spirits and the two souls – the lower of which was material – combined to obscure the questions of the origin and control of motion.[7]

Other English physiologists of the later seventeenth century reacted against Cartesian and corpuscular mechanism and developed their subject along Harveian or even Aristotelian lines. Thus, Glisson strongly objected to the view that matter was inert, claiming that to a certain extent all matter was alive and capable of motion, an idea fundamental to the problems of the origin and control of bodily motion. All matter, he argued, had 'primeval life' that possessed the fundamental properties of perception, appetition, and motion.[8] We have already seen that each of these, *perceptio, appetitus,* and *motus,* was a faculty of the Aristotelian soul, study of which was a regular part of the university course in Glisson's time. The idea was, moreover, elaborated in moral philosophy, through the problems of volition and appetition, in the school pneumatology of the early eighteenth century[9] and in didactic texts on natural law of the same period.[10] The argument common to all was that *perceptio* perceived Good or Evil, *appetitus* generated a desire to follow the first or avoid the second, and *motus* was the subsequent action. *Perceptio* was unconscious (except in the case of man's rational faculty), and *appetitus* and *motus* were involuntary; that is, they operated in the area that Descartes had relegated to machinery, and the notion of 'unconscious perception' remained for that reason problematic for the Continental mechanists.

Yet the idea of unconscious perception was used by the arch-mechanist of traditional historical interpretation, G. A. Borelli, who employed it precisely because he saw and tackled the problems of physiological motion that we are now considering. Since he and others used it we must begin to change our minds about what is meant by the traditional term *mechanist*. We must first distinguish between French and Italian ideas of mechanism in the late seventeenth century. Borelli's mechanism was inspired by the mathematical success of the physical science of his compatriot Galileo rather than by the beast–machine of French physiology after Descartes. Borelli successfully treated the gross movements of animals by considering their parts as levers and pulleys, but his theory was inadequate to explain the beating of the heart, the most notable of all the involuntary motions.[11] He argued that the heartbeat began when the presence of blood acted as an unpleasant stimulus to the embryonic ventricle: The stimulus was perceived as Evil by the soul, the *appetitus* of which desired to avoid the stimulus, and the *motus* of which initiated the first contraction of the ventricle to expel the offending blood. The returning blood caused the cycle to be repeated and the motion became in a sense habitual. To express its reliance on the soul, Borelli called this motion 'animastic'; we might call it 'tissue animism'.

Subtle matter in eighteenth-century physiology

There were several different kinds of subtle matter used in medical theory in this period to solve the problems of motion: first, the two surviving classical accounts described in the opening section of this chapter, which provided different models of imposed and innate motion (animal generation by subtle matter can be included in this group); second, electricity; third, heat considered as a fluid and the closely related ideas of 'light' and 'fire' as fluids; fourth, Cartesian subtle matter; and fifth, the Newtonian ether.

In taking these categories in turn we find that the first, traditional accounts of subtle matter, must include the dispute about the existence of nervous spirits. A subtle matter flowing down from the brain through the nerves was an answer to the problem of the control of motion, but there was some scepticism about the supposed cavity of the nerves. Normal aqueous fluids were too gross[12] to have the necessary features of subtlety and rapidity of motion, but, on the other hand, little else could be observed in nerves. As a result it was often assumed that the nerve juice was a mixture consisting of a coarse, passive, and nutritive part and a subtle, active part comparable to the macrocosmic ethers of the various theories of the time. Thus, G. D. Santorini held that the nerve juice contained an active ether composed of very small particles with great rapidity of motion.[13] In this way the nerve juice was said to be analogous to semen, in which the glutinous matter was contrasted to the active ethereal subtle fluid that was responsible for generation. The same parallel, between the action of ether in the nerves and that in animal generation, was made by Hoffmann[14] and was almost certainly a revival of the classical idea in an attempt to establish a comprehensive biological theory; many others made similar attempts.[15]

Electricity was often regarded as a fluid after the Leyden jar became well known, and it was undeniably active and subtle. Benjamin Franklin and others considered the electrical fluid particulate and able to lie quiescent in bodies as a lurking vital fire. By the middle of the century it was realised that electricity could stimulate muscular contraction, and it was not long before electricity was used in Edinburgh to treat a palsy by the administration of some six hundred severe shocks in three days (French, 1969:47). Haller stimulated the nerve of a muscle with electricity and obtained a contraction but with characteristic caution did not identify electricity with the subtle matter of the nerves. Even when electric fish were examined in detail and the copious supply of nerves to the electric organs of the torpedo were noted, there was no general agreement that the active agent of the nerves was electricity.

In posing again our questions about the control and origin of physiological motion, we can see that electricity regarded as a fluid had the qualities of subtlety and rapidity necessary for the control of muscular motion, if not for its source. Again, the motion of the heart can be seen as a critical case, because it was distant from the powers of the soul, and because it was a motion that could only with difficulty be reduced to mechanism. These factors encouraged the German worker Staehelin[16] to suggest that the nerve fluid was electrical in nature and that it reached the heart discontinuously, each drop stimulating the muscle of the heart to contract, while relaxation occurred as the fluid left the heart and entered the blood. But as we have seen, Staehelin's views were exceptional. The real problem was whether the stimuli of various kinds (including electricity) aroused the innate powers of the muscle, or whether they called into action a separate principle of motion. In the later part of the century, Felix Fontana and Luigi Galvani proposed that this principle was to be identified with the electrical fluid itself (Ritterbush, 1964:50, 51), but broadly speaking, eighteenth-century physiologists did not consider the body an electrical machine. The old problem of muscular motion, the problem of the source of physiological motion, continued to be discussed in its own terms. If there was a common notion of biological electricity, it was that organisms collected and concentrated the dispersed electrical fluid as a principle of vitality.

The notion of fire as a subtle fluid played a very much larger part in eighteenth-century physiological speculation. Fire had been the most active and subtle of the four elements of the Greeks, second only to the ether of the heavens, and the word came to express a number of different ideas: electrical fire; the corpuscular and chemically active fire of late seventeenth-century physiology; the subtle fluid of heat, caloric; and the principle of combustion, phlogiston. It is not always possible to distinguish sharply among these ideas, and they have in common the notion that heat was active and could move matter, for example, that of muscles. Often these ideas became intermixed: For example, if heat, as a secondary quality, was simply the result of particles in motion, then perhaps the particles involved were simply those of an elementary matter of fire; if heat was a fluid, could it not flow about the body, its particles, like those of light, moving with great speed in the nerves? What is quite clear is that the general acceptance of corpuscularity during the later seventeenth century greatly lessened the differences among fires, spirits, auras, ethers, heat, and light of an earlier age. All these terms, in a new corpuscular guise, were used in physiology to explain the action of the nerves, from Willis and his contemporaries in the 1680s to John Tabor and his contemporaries in the second decade of the eighteenth century.[17] Tabor located the soul within the brain and held, like Willis, that it radiated itself to the nerves and sense organs like light. In both cases the soul seems to have been conceived as subtle matter, the material *anima* of seventeenth-century binary animism.

Tabor's animism was perhaps a reaction against the growing influence of Boerhaave, whose mechanistic physiology was part of a medical scheme that was to become paramount before the middle of the century. Boerhaave gave an important role to 'fire' both in his chemistry and his physiology, and this drew attention to the problems of animal heat and motion. Most of the roles attributed to fire in animal motion were qualitative, but heat, particularly when conceived as a fluid, could also have its quantitative aspects. The intensity of the body's heat had traditional medical importance in judging the degree of fever, but now, in the eighteenth century, those who thought of heat as a fluid could also discuss its volume. Joseph Black's quantitative work on heat was known to William Cullen, who, as a medical man, was aware that biological heat was a special case, because of its 'innate' nature and association with life. Cullen would also have known of the seventeenth-century idea that respiration and combustion were very similar, both consuming air and generating heat, and he claimed that the production of animal heat by respiration was governed by the activity of an ethereal fluid in the nerves.

Cullen's notion of an ethereal nerve fluid was soon extended. Interest in the subject of animal heat, whether or not fluid, was high in Edinburgh in the 1770s, as shown by the large number of student dissertations on the topic. One student incautiously identified Cullen's ethereal nerve fluid with Newton's universal ether *and* with the electrical fluid. The student was rebuked at surprising length by William Smellie in the first *Encyclopaedia Britannica*.[18] The search for nonmechanical explanations of life led also to John Hunter's view of life as 'animal fire', freed from food as the result of respiration. This material basis of life was diffused through the body, upon which it imposed its motion. Possibly, Hunter thought of respiration in terms of a transfer of phlogiston, which is certainly what Priestley had in mind when explaining the relationships among respiration, combustion, and the vital activities of plants. Priestley also held that phlogiston was the cause of muscular motion. Another way of explaining muscular motion by fire-related subtle fluids was that employed by Buffon, who supposed that at a fundamental and universal level there were two kinds of matter, a *matière vive* and *matière brute*. *Matière vive* was essentially active, responsible for the phenomena of light and heat, and composing the ultimate 'organic molecules' that formed living bodies; it corresponded to the Newtonian force of repulsion. *Matière brute* was inert and passive, and corresponded to Newtonian attraction.[19] So here the Newtonian repulsive force – the hint is probably taken from Newton's idea that fermentative actions supply motion to the world – is made an innate property of a certain kind of matter, which becomes as active and as subtle as ether. An alternative was to emphasise the material activity of Newton's ether, and Buffon's countryman Lamarck assumed that an elementary, ubiquitous, and ethereal 'fire' produced sensible heat when activated, excited the motions of animals and plants, and was the source of motion in the macrocosmic world.[20]

Cartesian subtle matter

Another answer to the problems of the control and origin of motion that were allied to the problem of dualism could be found within Descartes' own system. Of the different grades of original matter described by Descartes, the finest could, with one change, serve the purposes of other subtle fluids: It was necessary only to attribute activity to it, and the Cartesian tradition in physiology on the Continent could be extended into the period when British physiology was dominated by Newtonian mechanism in the form of the hydraulic-machine model put forward by Archibald Pitcairne at the beginning of the eighteenth century. Thus, without departing very far from the Cartesian tradition, Johannes Gottsched, discussing muscular motion in 1694, denied

Ether and physiology 119

appetite and will in the natural motions of animals and claimed that all involuntary motions were due to the activity of an ether, which he also called *aer* and *aura*,[21] universal and innately active. In the body it moved the particles of blood in a fermentative motion and was distilled from the blood to form the active part of a dual-nature nerve juice characteristic of the time.

Gottsched explained muscular motion by the assumption that ether in the muscle fibre expanded with heat across the width of the fibre, thus reducing its length. This is a minor variant of older ideas about the inflation of muscles, and Gottsched employed the usual seventeenth-century speculative microgeometry of muscle structure to explain contraction. His explanation did not meet the criticism that muscles were known to decrease rather than increase in volume in contraction. Moreover, his explanation was 'kinematic' in failing to relate the quantity of motion in the ether to that in the muscle. Yet it is clear that Gottsched did face the problem of the quantity of motion in the body; although he did not entirely distinguish between the various quantitative aspects of motion such as the force, the distance moved, the velocity, and the amount of moving matter, yet he had a general notion that the muscles somehow produce a great deal 'more' motion than they get from the nerves. Indeed, as he observed, it was the common view that there was some mechanical contrivance in the muscles that multiplied the inflowing *virtus* of the nerves. Aristotle had said that motion resulted from the moving force overcoming the resistance, so that, for example, a man might fail to move a boat beached in sand. Yet, armed with a lever, he would perhaps succeed; it appeared that the lever was multiplying the moving force. Indeed, some seventeenth-century physiologists thought that muscles acted as levers or pulleys in this way, multiplying the force of the nervous spirits. Gottsched held that the microgeometry of the muscle could *not* multiply forces, and also that the force developed by the muscle was actually greater than appeared during its motion because of the disadvantageous position of the insertion of the muscle into the bone in relation to the position of the fulcrum and weight. It was obvious to him that the additional force of the muscle was derived from a separate principle, ether.

So Cartesian subtle matter could also be used to explain the motion of an inertly material Cartesian body. Those physiological writers who had not perceived the quantitative problem of the origin of motion in the body remained kinematic in outlook and felt no necessity to invoke ethereal explanations. The Spaniard Martin Martinez, writing in 1723 on the motion of the heart,[22] considered that a very small force or motion (*vis* or *commotio*) of the nerves could raise great weights by the machinery of the muscle, analogous to multiple pulleys. It followed that there was no need of an active principle, and Martinez denied spirits of all sorts.

Another reason for denying spirits and subtle fluids was the belief that the immaterial soul could directly produce motion in matter, and that no problem of dualistic control or origin of motion existed. George Ernst Stahl began to put forward such arguments through his students' dissertations as early as 1685:[23] Spirits were either material or immaterial, not both, and so might be acted upon either by the body or by the soul. But to rely on spirits as an intermediary between the material and immaterial was simply to put the problem of dualism one stage further back. It was preferable, he argued, to say that the immaterial soul acted upon matter by means of motion; motion implied direction, and direction, purpose; since no mere machine was purposeful, the soul regained its traditional importance in physiology.

Opposition to Stahl's theory came from his own university, Halle, in the person of Hoffmann, whose beliefs have been hinted at earlier in this chapter. Hoffmann devoted a separate tract to the subject, and from it we can see that he was fundamentally a Cartesian mechanist, but held a number of ideas from different sources that somewhat complicate his system. Spirits, including those above man in the Chain of Being, were essentially active and, being immaterial, were unable to move matter, except for the mind of man in creating voluntary motion; matter was essentially inert except for the simple motion of the particles of fluids. However, this motion was too simple to account for life, which Hoffmann explained by invoking an ether to impose motion on the not-quite-inert matter of the body. Life was no more than the sum of these imposed motions and was 'mechanical'. Ether itself seems to have been derived by Hoffmann partly from the work of the English respiratory physiologists of the previous century.[24]

Newtonian ether

The effect of Newtonianism on English physiology was to introduce in an acute form the problems of physiological motion that form the basis of our discussion and that had partly been obscured by seventeenth-century chemical physiology and innate Glissonian powers of matter. Newtonianism posed the question that the Cartesians had faced: If matter is inert, and the bodily machine is material, what moves the machine? The Newtonians thought about motion in general in quantitative terms, and when thinking about animal motion in particular, did much to make physiology a quantitative, 'dynamic' science. In general, they believed that there was a constant loss of motion in the universe in collisions between moving bodies, and that there had to be sources of new motion like ether, gravity, and elasticity; in the particular case of animal motion, they could also use an explanation that Newton had suggested in query 31 to the *Opticks* (1717) in which these general sources of motion were discussed, that is, the fermentative motion that supplied movement to the heart and blood.

The early Newtonians, like Archibald Pitcairne, considered the body a hydraulic machine, no doubt in Newtonianising the seventeenth-century tradition of work on the circulating fluids and glandular secretion, but at the same time there was an alternative tradition, dating back at least as far as Glisson, of examining the solid parts, particularly the muscles and nerves. Glisson's notion of the innate mobility of muscle is generally held to have influenced Haller's important work on the *vis insita* or irritability, and between the two men much effort had gone into attempts to establish a microgeometry of muscle structure and, alternatively, the structure of the ultimate fibre. Clearly, the solid parts were more gross than the moving fluids of the body, and where ether was used as an explanation of motion it was normally thought to act upon the solids. It was partly for this reason that Pitcairne did not employ ether as an explanation, and partly for the reason that his Newtonianism was derived from the *Principia* and not from the *Opticks*. Once more the question was faced, but more acutely than in the physical sciences: Was matter essentially passive or active?

The English and Scottish answer in the Newtonian period was generally that the inertness of matter required an external moving force. This view fitted well the prevailing natural theology, and Andrew Baxter in 1733 relied heavily on physiological evidence to assert the immateriality of the soul and the existence of God. Ether itself, he claimed, was material and therefore inert, and had no role in physiology, because all motion was derived directly from God. 'Let us look into any book of anatomy, or natural philosophy . . . and we shall not find that the particular instances alledged are intended to shew whence the origin of motion is; or that *fine machinery may be a power to itself*'.[25]

Similar arguments are found in Browne Langrish's book on muscular motion, published in the same year as Baxter's.[26] Like other post-Pitcairne medical men, Langrish drew more heavily on the *Opticks* and its queries than on the *Principia,* and he added to Pitcairne's hydraulic model a discussion of the solid fibres. Even in the hydraulic machine, Langrish was aware of some of the dynamic problems of motion: 'In all hydraulic Engines there is an absolute Necessity for a perpetual supply of the moving force'. The ultimate origin of motion seems to have been, for Langrish, outside the realms of physics and physiology, and probably he thought, like Baxter, that the ultimate mover was God: The 'principal agent' of motion was too subtle ever to be demonstrated mathematically, and it was only the physical manifestation of motion that was performed 'mechanically'.

So we cannot expect a physiological answer to the problem of the origin of motion from Langrish, despite the fact that he was aware of the problem, and his description of muscular action is consequently kinematic in following only the pathways taken by motion in the body, and not its quantity. Taking his hint from the *Opticks* and the Newtonians, Langrish asserted that the constituent particles of matter were polarised like magnets and that attraction depended on unlike poles facing one another. Crystallisation, chemical changes, and muscular motion were to be explained on this basis. The control of muscular motion by the nerves was explained by the nervous spirits' somehow adjusting the polarity of the particles of the muscle so as to achieve a greater or lesser attraction. This was the 'mechanism' of muscle action, yet, despite his claim that the resultant force of a machine could never be greater than that put into it, Langrish saw no discrepancy between the amount of motion exercised by the nervous spirits in rotating or otherwise adjusting the polarity of the muscle particles and the amount of motion developed by the contracting muscle. He consistently referred to the *qualities* of attraction and repulsion, and did not grasp Newton's concept of the 'quantity of matter' or its relationship to attractive forces.

Ether would have been an excellent answer to Langrish's unanswered problem of the hydraulic engine without a moving force, but he was prevented from using an ethereal explanation for two reasons. First, he disagreed with the well-known account recently published by Robinson (discussed later in this section) which accounted for muscular motion by an ethereal vibration of the nerves and muscular membranes, which were said to swell across their width and so contract along their length. Second, he was influenced by Glisson's nonmechanical account of muscular contraction without inflowing spirits and increased volume. By the time Langrish gave the Croonean lectures[27] in 1747 he probably felt that to use the term *ether* in relation to muscular motion no longer implied belief in Robinson's particular use of the idea, nor in an influx of Cartesian spirits. Moreover, he now felt free to use the term *ether* to describe the nervous spirits that imposed motion on matter.

One point in Langrish's criticism of Robinson illuminates very well the problems faced by British physiologists in attempting to cope with Continental mechanism. Langrish thought that the ganglia existed in order to prevent bodily vibrations from disturbing the nervous spirits that might otherwise produce false sensations or motions. Attention was given to the ganglia at the time because of a dilemma in physiology: If the nerves were indeed unbranched in their passage to and from the brain, then the soul must perceive everything that happens in the body. This agreed acceptably with the traditional faculties of the soul, but most Continental physiologists by then believed that the soul had the single faculty of conscious rationality. But it was clear that the soul was not conscious of the natural motions of the body (such as peristalsis of the intestines), and although Descartes had relegated these to machinery, it was felt in Britain that they involved some kind of vital activity. There was therefore a search for a vital centre outside the brain that organised nonconscious physiological motion. Perhaps the ganglia on the vagus and sympathetic nerves were such centres? Perhaps they were where the nerves branched? Perhaps they acted as filters to prevent voluntary commands from reaching the vital organs and sensations from moving in the reverse direction? The search for an extracranial organising centre played a significant part in the formulation of notions about autonomic functions of the sympathetic system and about the transmission of reflex actions across the spinal cord.

Bryan Robinson's *Treatise*,[28] although principally concerned with the hydraulics of the body–machine, borrowed from the *Opticks* the notion of ether vibrating in the solid nerves. Conducted to the muscles, this vibration caused a swelling in the membranes, which in turn contracted the fibres. The advantages of such a scheme in physiology were simply that there was no need to postulate invisibly small cavities in the nerves, and that the subtlety of ether accounted for soul–body interaction, that is, the control of voluntary motion. This problem was more obvious at the time than that concerned with the source and quantity of motion. Robinson's solution was employed by Richard Mead in his introduction to William Cowper's book on muscles[29] and by George Thomson in his treatise on bones.[30] Neither of these contributions went beyond Newton's discussion in the *Opticks;* medical men in general had but slight acquaintance with the *Principia* and the laws of motion, and for the greater part their physiology remained kinematic.

The most elaborate use of an explanation of nervous action resting on ethereal vibrations was that employed by David Hartley,[31] who related sensations in the soul to vibrations, analogous to those of heat and maintained by a Newtonian ether, in the medullary parts of the brain. The scheme provided a mechanical basis for associationalist psychology, but Hartley was anxious to declare that mental events themselves were not mechanical, and he made the pious declaration that his system could not be used to prove the materiality of the soul. He was left with the problem of interaction between body and soul, as Descartes had been, but, using Newton's principle of limited explanation, he professed himself unconcerned whether the problem was soluble in any of the ways in which the post-Cartesian philosophers had solved it. Nevertheless, he allowed himself to speculate that there might be an 'infinitesimal elementary body' transmitting the medullary changes to the soul. In physiological terms the idea was not very novel, for ether simply replaced the traditional nervous spirits. This modernisation of earlier ideas may have been the route by which Battie[32] arrived at a theory similar to Hartley's and which explained how 'pressure' was transmitted along the particles of the nerves to the brain, causing sensations in the soul.

Continental ethers in the period after Newton

The use of Newtonian ethers in physiology outside Newton's own circle of English and Scottish followers was much more limited, and it is not clear whether the ethereal fluids employed by eighteenth-century French and German writers were primarily Newtonian in inspiration. For example, F. Quesnai, writing in the 1740s,[33] gave ether a central role in his physiological scheme, but it was only partly Newtonian, for he drew freely from Descartes and the ancients. His ether was universal and particulate, penetrating gross bodies by way of their pores. It was the only active form of matter, was akin to fire, and was particularly concerned with the production of heat and cold and with the transmission of light. Its action was entirely by contact and was less in dense bodies because of their greater porosity. Physical properties like weight and elasticity depended upon the activity of the ether. Quesnai's description of the action of ether within the body makes it clear that he began to appreciate the quantitative problems of motion: The body was essentially inert, and even though it could produce subtle humours (the chemical spirits of the previous century), they were ultimately derived from food and were thus likewise passive; their motion was imposed from without, by ether. So great was the contrast with the essentially active ether that ether was conceived by Quesnai as a principle of life. This 'vital principle' resided chiefly in the nerves of the major part of the nerve juice (the minor and coarser part, like that described by Santorini, was nutritive) and so was ideally placed to control the motions of the body.

As the sole active form of matter, this vital principle also generated the motion of the body, but the life it gave to the body in this way was *mechanical* life. The idea of mechanical life was more comprehensible on the Continent (we have seen that it was used by Hoffmann) than it was in Britain, where the term *mechanism* suggested something that excluded life. The idea of mechanical life had arisen from Descartes' contention that animals were purely mechanical. This view was somewhat modified by the later mechanists, who, like Hoffmann, claimed that the parts of the animal machine may move with their own innate, if low-level, powers. Such a view made possible the unique biological properties conceived by such 'mechanists' as Haller, and we shall see below that a form of vitalism arose from such ideas. In a further modification of the original idea the term *mechanism* became interchangeable with *organic necessity,* an idea that stated that the motions of animals were not free but bound by necessity to the structure of the organic parts: This was not to deny that those parts may be alive in a fairly traditional and nonmechanistic way.[34]

Just as British physiological writers had difficulty with the notion of 'mechanical life', so their Continental counterparts found the concept of 'unconscious sensation' incomprehensible. We have seen that from Glisson onwards there survived a notion that the lower physiological activities were governed in some nonrational and perhaps automatic way by the soul, but on the Continent, so completely had Descartes banished the lower activities of the soul that 'soul' was equated with 'mind' or consciousness: It cannot be the soul that constructs the body or controls its every action, for we are not conscious of it doing so. Yet again the question arises: If the body is a machine, what moves the parts? Quesnai recognised this 'dynamic' problem, and tried to answer it. He saw that the problem arose because of the adoption of mechanism: 'On oublie la puissance motrice & directrice, qui est precisement l'Ame Vegetative'.[35] The very mechanism of the body emphasised to Quesnai that a moving force was needed, and this he found in his ethereal vital principle. Yet the problem of dynamics still remained, for the vital principle in the nerves had still to transmit enough motion to the muscles to explain the great force the muscles could produce. He was unable to suppose that such a great force could be channelled down the narrow passageway of the nerves and was obliged to conclude that the force of the muscles was derived from the activity of ether in the blood they contain. The source of motion, then, was localised ether, and its control was partly by nervous transmission by means of ether and partly by the organic necessity of the structure of the parts. Ultimately this organic necessity was referable to God, who had constructed the machine. The result was a series of nonpsychic biological properties very similar to those described by Haller. Although failing to provide an adequate dynamic answer to his problems, Quesnai had at least separated out the two questions of the source (and quantity) of motion and of its control, which had always been confused in kinematic physiology.

Animism, mechanism, and dynamic physiology

The question of the inertness or innate activity of matter was central not only to those who used ether as an explanation, but also to two groups long opposed in the traditional historical interpretation, the 'animists' and the 'mechanists'. These two groups will appear in a new light if we examine them in relation to the physiological problems that made the use of ether desirable: the problems of the origin, quantity, and control of motion.

It is characteristic of all mechanistic arguments that the body alone produced and guided its motions. Paradoxically, by following this article of faith, the mechanists to a greater or lesser extent abandoned the doctrine that matter is inert, for example, by postulating God-given natural forces that inhered uniquely in biological matter and that were responsible for motion and other activities. This was a qualitative notion, defended on a priori grounds, and it was the animists who developed a quantitative and penetrating attack on this position by showing that, however complex, biological matter remained essentially inert. The converse paradox was that for animists the body was a true machine, and could only be moved by an external force, the soul.

An early mechanist position, transferring some Newtonian physics into a Cartesian view of the body, is seen in the graduation thesis of J. Bajollet, 1735.[36] The body was conceived as a hydraulic machine, with its motions dependent on the arrangement and physical properties of its parts. There were no spirits and no chemical explanations. Each physiological motion was no more than the generation of momentum ('actual force') in proportion to the moving cause, and the result of action was 'impression' determined by the quantity of the resisting 'virtual force'. 'Function' was not a guided action and had no purpose. The result of two actions of equal but opposite momentum was either static equilibrium or constant motion about the point of equilibrium, that is, libration, which was the key to this severely reductionist physiology. 'The life of man', wrote Bajollet, 'is nothing other than a libration of the principle organs, or the heart and vessels with blood'. The valves of the heart turned this libration into a circuit of blood, and the continued balance of the librating forces represented health. Libration made the human machine move with perpetual motion: 'It is true that in the human machine there is this difference (from other machines) that it needs no restitution of weights or elasticity, at least at the sensible level, so that throughout life it is a specimen of perpetual motion'. 'Weights or elasticity' referred to the devices driving the clockwork, Bajollet's frequently used analogy, but nowhere in his dissertation is there any reference to forces being lost through friction; increased resistance of the smaller vessels in inflammation was simply a cause of an increased pulse in a system where the effects of opposing momenta alternately became causes. The whole system of alternations was so complex as to be quite outside the powers of mathematics to describe, and Bajollet's scheme remained as qualitative as it was intended to be quantitative in principle by reducing physiology to physics.

All of these points were to be bitterly disputed by the animists: the lack of quantitative reasoning; the omission, in a reductionist programme, of any explanation of the control of motions and purpose; the naive assumption of perpetual motion; and above all the refusal to acknowledge the need of a constant supply of a moving force.

The reaction against this extreme reduction, or identification of biology with physics, was soon seen, in Montpellier and elsewhere. Physiology was, after all, the study of man, and man was compounded of body and soul. Some signs of this reaction are seen in the 1738 thesis of N. Cambray.[37] The effects of the soul in the body were emphasised, including morbid changes not restricted to the voluntary organs. Unlike any machine, the soul could recognise Good and Evil, and this recognition became again a principle of motion. Cambray, like many others who followed him, consciously tried to place his own reasoning in the long tradition of European scholarship from which Cartesian and Newtonian mechanists (in physiology) were felt to have broken away.

It is probable that Cambray's thesis was inspired by Sauvages, who in the 1730s began to teach an animist doctrine in Montpellier, probably having been influenced by Stahl. Likewise, the results of Sauvages' teaching appeared in several other dissertations by his students.[38] The Stahlian position was modified in two ways: First, Sauvages was more interested in medical applications of animism, and second, the Stahlian soul was pushed towards the idea of Hippocratic 'nature' or its healing power. This medical interest forms a link among several groups of ideas we have already met; Sauvages said that all diseases were the body's attempt to reject intrusive morbific matter. This was a cardinal point in the war against the mechanists: no mechanical system could recognise morbific matter (indeed, no mechanical system could become 'ill'); and perception and reaction were closely related to the nonconscious perception, appetite, and motion, the pursuit of Good and the flight from Evil, that we have met in the neo-scholasticism of Glisson, the habituation theory of Borelli, and the general educational framework of the later seventeenth century. These issues were also closely involved with the stimulus and response theory, which in the hands of both Descartes and Bellini[39] was mechanical, automatic, but nevertheless purposeful in avoiding damage to the fabric of the body.

The same sets of ideas were reinforced in the new format in which the academic ideas of the eighteenth century were presented. We can see natural, unconscious appetite as a cause of animal motions in Hutcheson's pneumatology, which was taught in the Glasgow arts course; we may see something similar in another 'new' eighteenth-century subject, the law of nature and nations, and Otto's 1738 text on the subject used the same language as the physiologists: Moral actions were free, but other actions of the body and of the world at large were *mechanical*. The very inertness of matter, said Otto, removed any arbitrariness in the actions imposed on it, which were consciously determined, not free. The actions of the body were determined by the mechanical structure (i.e., organic necessity) of the solids and liquids, and death was a cessation of motions (i.e., the end of mechanical life). Yet it is clear that biological behaviour, lying between the moral actions of the rational soul and the necessary action of the mechanical body, did not fit any more neatly into enlightened treatises on natural law than it did into Cartesian mechanism. Once more, what moves, and what guides, the mechanism? Otto spoke of a force created originally alongside all matter (*vim omni materiae concreatam*) and inherent in it: He used the term *vis insita,* and in most respects this force acted as that later described by Haller. 'Nature', then, was this motive force guided by the structure of the body's solids and fluids. This was nature in the sense that Sauvages used it to recall the Hippocratic healing power, and it was nature in precisely the sense used by Galen at the opening of *Natural faculties:* an inherent, living force, not part of a soul, that determined the lower physiological activities.

Nature in this sense filled the gap in mechanical dualism, a gap that was obvious to those with experience of the biological world and that was being filled by other theorists with ether. To return to Sauvages,[40] we can see that it was Newtonian physics that caused his dissatisfaction with the mechanical account of the body. The most important point is that he rejected the notion of the body as a perpetual motion machine. To change the position or motion of a body, force was required; in any actual machine, part of the moving force was consumed in overcoming resistance, and only what remained produced useful work: No result was greater than its cause, no motions were multiplied. His model of a machine was the pulley, which mechanists like Martinez had held multiplied motions, but in which Sauvages balanced his equation of motion by distinguishing between force and velocity, so that $FC = PV$, where F is the force (weight) applied to the pulley, C is the velocity of the part it is applied to, P the suspended weight, and V the velocity at which it moves. From Newton, Sauvages accepted that motion was lost in collision (friction in the case of the circulating fluids) and that there must be a constant generation of new motion. The problem of dynamic physiology was squarely put by Sauvages, who said that in any actual hydraulic machine the ratio of ingoing to outgoing forces after loss by friction was 27:4. This made it quite impossible that the product of such a machine, that is, the animal spirits of the mechanists, should retain enough force to account on their own for muscular motion, including the beat of the heart, which in the traditional mechanist scheme was the source of all motions in the body, including that by which the animal spirits themselves were separated from the blood. As remarked previously, the heart was the critical case in these disputes for this reason, and the post-Harveian work on the circulation of the force of the heart by Borelli, Keill, Hales, Jurin, and others was closely and mathematically examined by the animists to prove the existence of a mover external to the hydraulic machine of the body.

Sauvages also denied 'libration' in the form of 'perfect elasticity'; again, he demonstrated the loss of motion in quantitative terms by measuring the decreasing distance of rebound in a bouncing ball, which he explained by assuming that motion was lost through internal friction and dissipated as sound and vibration of impact. The result of all this was to show that a pure hydraulic machine, conceived on Pitcairne's lines, needed a constant supply of a moving force. This was supplied by the essentially living and mobile soul, and there are suggestions in this dissertation that Sauvages conceived this force as being supplied along the nerves by the animal spirits travelling at nearly the speed of light, much as ether had been envisaged. However, as for the critical case of the still-beating isolated heart, Sauvages thought that the soul or vital principle might be distributed through the body and its principal organs. He had not clearly separated the rational *animus* from a nonconscious *anima* on traditional lines, and he stated that the joint principle was the chief characteristic of motion, whether physical or intellectual. This physical motion was direct horsepower, the ultimate answer to the problem of dynamic physiology. He consciously reintroduced one of the traditional faculties of the soul, but paradoxically his proof of the power of the soul made the body into a true hydraulic machine that he discussed quantitatively, whereas the mechanists to whom he was opposed avoided mathematical demonstrations. In line with this argument and opposed to that of Stahl, Sauvages said that the soul was bound by strict laws established by the Creator. He used an analogy between the soul and a clavichordist who may play what he pleases but is obliged to express himself according to the accepted scales. The analogy went further and explained how the player may play perfectly without directly attending to the motions of each finger, and even while his mind is occupied by another matter. Unconscious, vital actions were thus explained without involving the conscious rationality of the soul. Here the soul had prescribed laws for itself to enable these vital actions to proceed in the best manner without its perpetual attention. The 'best manner' was originally rational and wise, but became automatic through habit, and also involved the pursuit of Good and the flight from Evil. In medical terms this was the expulsion of morbid matter in disease, or in actions like coughing, which were voluntary (i.e., soul-directed) but automatic (i.e., habituated) upon the stimulus of particles entering the trachea. Finally, the musical analogy was extended to explain how the vibrations in the nerves of the brain were related to mental events: Such vibrations were 'isochronous' with the action of the *anima*. Although the substrate for these vibrations is not discussed, it is not difficult to see these ideas as a parallel to ethereal, Hartleyian psychology.

Vitalism and the disappearance of ether

In the animism of the eighteenth century the problems of the origin, quantity, and control of motion were squarely faced, and this constituted a major distinction between the animists and the mechanists. Against this background ether had an important role as a source of imposed motion and as a means of communication between body and soul.

The animist–mechanist dispute was not resolved but was transformed into a conflict between various forms of vitalism. In general vitalism avoided the most controversial features of animism and mechanism, particularly the unconscious perception, physiological activity, and even rationality of the immortal and immaterial soul. This was incomprehensible to those whose background was the Continental Cartesian[41] tradition that equated soul with the conscious mind. Most animists avoided the difficulty of unconscious sensation and rationality by attributing strict laws of union to the body–soul duality. Bound by these laws and sometimes located within the tissues, the soul was to a certain extent reduced to a vital force.

Vitalism also avoided the reductionism of the mechanists, a feature that the animists had successfully attacked, mathematically showing the weakness of the mechanical model. Avoiding the worst of both systems, vitalism also combined their best: that is to say, the features with the greatest explanatory value at the time. This came about partly as a result of experimental work, in which the problems of motion we have been discussing were conceived as the very important questions of irritability and sensibility. Experiments exploring the reaction of parts of the body to stimuli invariably involved the nerves, muscles, spinal cord, and brain, and although the competing animists and mechanists believed themselves to differ on fundamental grounds, their experimental results were similar. What emerged was the idea of biological properties, that is, qualities (principally of sensation and motion) that were unique to living systems and not to be derived from a mechanical model.

Hall[42] suggests that these explanations of life fall into three categories in which life was considered as immanent, inhering in a simple form in all matter; emergent, arising through the complexity and organisation of matter; or superadded to matter by a vital principle. In the first two of these categories our problems of motion were again obscured, and, as we have seen before, in such circumstances ether could play no explanatory role. But where motion and life were thought to be imposed the notion of subtle active fluids could again be of use.

Perhaps the most important concept of the later part of the eighteenth century was Haller's notion of *vis insita:* the innate ability of the muscle to contract. We cannot fit this into any of Hall's three categories because of Haller's studious avoidance of speculative explanations. The *vis insita* was a natural, God-given force, indefinitely greater than the stimulus that provoked it (as a small spark may explode any amount of gunpowder), and it did not necessarily arise from the complexity of the parts or reside in the ultimate particles of matter. Haller's limited explanation did not encourage a quantitative approach to the problems of motion, but it did enable him to dismiss all theories of subtle matter as fashionable and rash speculations.[43] This reliance on Newton's principle of limited explanation can be compared to the attack on hypothetical ethereal explanations made by Thomas Reid towards the end of the century.[44]

Only when discussing the motion of the heart did Haller imply that motion may depend upon complexity of structure (the unique branching of the muscle fibres of the heart), in which case the *vis insita* of the heart would be emergent. The difficulty with emergence was to decide at what stage of increasing complexity the nonliving becomes living, and one answer to this was to assume that at all stages matter had some form of life; that is, to posit the notion of immanence. Later writers like Maupertuis, writing in the middle of the eighteenth century, seemed to recall the doctrine of Glisson in claiming that the particles of matter have some form of 'desire, aversion and memory'.[45]

Although ether played no part in the immanent and emergent forms of vitalistic theory, it could be used where motion and life were considered to be imposed upon matter. This form of vitalism differed from animism in that the ultimate principle of life was no longer the immortal soul of Christian tradition, but some separate vital principle, and ether was less an agent of communication and more a vital principle itself. Again, it was the stubborn idea that matter, or at least some matter, was inert and passive that called upon a separate principle to explain life and motion. The *matière vive et brute* of Buffon is an example we have already met, and later in the century, Buffon's compatriot Bergier attacked d'Holbach's system of nature and defended theology on the basis of motion imposed upon inert matter.[46]

Much of what has now been said on vitalism can be illustrated by reference to the Montpellier followers of Sauvages. Theophile de Bordeu emphasised the biological properties of sensibility and irritability, of which each organ had its own kind, and which together constituted a principle of life quite distinct from the rational soul. Bordeu's student thesis also reveals his familiarity with the ideas of Glisson.[47] The views of Bordeu's successors at Montpellier had considerable influence in France. Menuret de Chambaud and Henri Fauquet published the new physiology in the *Encyclopédie,* and the 'organicism' of Bordeu is held to have been a model for Bichat's 'organic sensibility', one of the five fundamental vital forces described by him. P. J. Barthez adopted a Newtonian concept of the inertness of matter, and the vital principle to which he appealed as an answer to the problem of motion in the body was developed from Bordeu's account. In places, Barthez implied that the vital principle was a subtle fluid, but he always denied that it acted as an intermediary between body and soul.

Notes

1. See the opening chapters of Galen, *On the natural faculties,* trans. A. J. Brock (London, 1916).
2. A useful summary of Galen's physiology is given by M. T. May in her translation of Galen's *On the usefulness of the parts of the body,* 2 vols. (Ithaca, N.Y., 1968), 1:44–64.
3. *De motu animalium,* 701b (see also *De anima,* 432a), in Aristotle, *The works of Aristotle,* trans. J. A. Smith and W. D. Ross, 12 vols. (Oxford, 1908–52).
4. F. Glisson, *Tractatus de natura substantiae energetica* (London, 1672), 186 ff.
5. G. Cheyne, *An essay of health and long life,* 8th ed. (London, 1734), 144–55.
6. See his account of the differences between his system and that of Stahl: F. Hoffmann, *Opera omnia physico-medica* (Geneva, 1748–9), and *Supplementum pars secunda* (Geneva, 1749), separately paginated.
7. T. Willis, *The remaining medical works,* trans. S. P.[ordage] (London, 1681), 88.
8. Glisson, *Tractatus,* the *ad lectorem* and passim.
9. For example, F. Hutcheson, *Synopsis metaphysicae, ontologiam et pneumatologiam complectens,* 4th ed. (Glasgow, 1756).
10. For example, M. H. Otto, *Elementa juris naturae et gentium* (Halle, 1738).
11. J. A. Borelli, *De motu animalium,* 2nd ed. (Leiden, 1685), pt. 2, 109.
12. But see A. Stuart, 'Experiments to prove the existence of fluid in the nerves', *Philosophical Transactions 37* (1731–2), 324.
13. G. D. Santorini, *Opuscula medica* (pt. 2: *de nutritione animalium*), (Rotterdam, 1719), 117.
14. Hoffmann, *Opera* (the fifty-eighth and fifty-ninth 'difference'), 25.
15. A number of authors are listed by A. von Haller, *Elementa physiologiae corporis humani,* 8 vols. (Lausanne, 1756–66), 4:378.
16. Reported by Haller, *Elementa physiologiae,* 1:500, 501.
17. J. Tabor, *Exercitationes medicae* (London, 1724), 301.
18. See the article 'Aether', in *Encyclopaedia Britannica,* 3 vols. (Edinburgh, 1771). The Edinburgh student attacked in this article was G. R. Brown. Mendelsohn (1964), 111.
19. T. S. Hall, *Ideas of life and matter,* 2 vols. (Chicago, 1969), 2:116–7, 8–9.
20. J. Hodge, 'Lamarck's science of living bodies', *British Journal for the History of Science 5* (1971), 323–52.

21 J. Gottsched, *Dissertatio de motu musculorum* (Köningsberg, 1694), 383.
22 M. Martinez, *Observatio rara de corde in monstroso infantulo, ubi obiter, et novo de motu cordis, et sanguinis agitur* (Madrid, 1723).
23 G. E. Stahl (praeses), *Dissertatio physiologico-medica de sanguificatione in corpore semel formato* (Halle, 1704). This dissertation was disputed in April 1684.
24 It is unlikely that the Newtonian Langrish derived his idea of ether from Hoffmann, as Schofield (1970), 194, suggests.
25 A. Baxter, *An enquiry into the nature of the human soul*, 3rd ed., 2 vols. (London, 1745), 1:147.
26 B. Langrish, *A new essay on muscular motion: founded on experiments, observations, and the Newtonian philosophy* (London, 1735).
27 Royal Society of London, *Philosophical transactions from their commencement in 1665 to the year 1800, abridged by . . . C. Hutton, G. Shaw and R. Pearson*, 18 vols. (London, 1809), 10 (pt. 2):1199.
28 B. Robinson, *A treatise on the animal oeconomy* (Dublin, 1732).
29 W. Cowper, *Myotomia reformata* (London, 1724), lxiii.
30 G. Thomson, *The anatomy of the human bones* (London, 1735).
31 D. Hartley, *Observations on man, his frame, his duty and expectations*, 2 vols. (London, 1749), 1:511.
32 W. Battie, *A treatise on madness* (London, 1758), 25.
33 F. Quesnai, *Essai physique sur l'oeconomie animale*, 2nd ed., 3 vols. (Paris, 1747).
34 See Otto, *Elementa juris*, 10, 64, 92–7.
35 Quesnai, *Essai*, 3:120.
36 J. Bajollet, *Vitae ac mortis animalium conspectus medico-mechanicus* (Montpellier, 1735).
37 N. Cambray, *Dissertatio de vita corporis humani* (Montpellier, 1738). Cambray is recorded as the author, there being no professor as praeses.
38 Sauvages speaks of these dissertations as expressions of his own ideas. See R. K. French, 'Sauvages, Whytt and the motion of the heart: aspects of eighteenth century animism', *Clio Medica 7* (1972), 35–54.
39 R. Descartes, *Treatise of man*, trans. T. S. Hall (Cambridge, Mass., 1972), 35; L. Bellini, *De urinis, de sanguis missione et de febribus* (Frankfurt, 1698) (see *De sanguis missione*, 165: *De stimulis*).
40 The most complete statement of Sauvages' physiology is his *Physiologiae elementa* (Amsterdam, 1755). A terse essay is appended to his translation of Stephen Hales's *Haemastaticks*, in which he reveals his debt to the Newtonians. See E. [Etienne, i.e., Stephen] Hales, *Haemastatique, ou la statique des animaux* (Geneva, 1745).
41 The Cartesian tradition continued strong in France until challenged by Sauvages and his followers. The influence of Stahl was greatest in Germany, and was taken up with modification in Britain. The Low Countries retained the modified mechanism of Boerhaave, and the influence of the Italian mechanists prevented wide adoption of animism in Italy. See Haller, *Elementa physiologiae*, 1:bk. 4, and 4:bk. 2.
42 Hall, *Ideas of life and matter*, 2:pt. 5.
43 Haller, *Elementa physiologiae*, 8:378.
44 For Haller's work on irritability (and sensibility), which has received much attention from the historians, see his *A dissertation on the sensibility and irritability of the parts of animals* (London, 1755; reprinted with an introduction by O. Temkin, Baltimore, 1936). A more detailed discussion is given in Haller's *Elementa physiologiae*, 1:459–505, on the motion of the heart, and 4:514–63, on the contraction of muscles. The ultimate divine origin of the biological property of irritability is set out in the small tract *Ad Roberti Whyttii nuperum scriptum apologia* (Roche, 1764). Thomas Reid, the Scottish commonsense philosopher, wrote an unpublished paper

on muscular motion towards the end of his life (he died in 1796): '*Animal Spirits* . . . and vibrations in an elastick Ether which pervades all Bodies are all Hypotheses, and like all other hypotheses in Philosophy, labour under two defects. First they suppose the Existence of certain things of whose existence we have no Evidence, and Secondly when they are supposed to exist they do not account for the phenomena they are brought to explain . . . [Nervous transmission and muscular contractions cannot] be accounted for by any laws of Mechanism we know. It is something beyond Mechanism and of a superior Nature'. Transcript in the possession of Mr. G. Davidson of the Medical School, University of Aberdeen. See also T. Reid, *Essays on the powers of the human mind,* 3 vols. (Edinburgh, 1819), 1:121–38.

45 Hall, *Ideas of life and matter,* 2:25.

46 N. S. Bergier, *Examen du materialisme,* 2 vols. (Paris, 1772), 1:11.

47 I am indebted for this information to Dr. Elizabeth Haigh, who allowed me access to the typescript of her forthcoming book on French vitalism.

4

The theological significance of ethers

G. N. CANTOR

Division of History and Philosophy of Science, Department of Philosophy, University of Leeds, Leeds LS2 9JT, England

During the period covered by this volume, the discussion of ether theories was not limited solely to assessing their explanatory power but also covered their more general functions. In particular, many writers employed ether theories to help solve theological problems, and in this way ethers bridged scientific and theological contexts. The major problem that lay behind the theses discussed herein concerned the relationships between God and the physical universe and between mind (or spirit) and matter. Although comparatively few of those who subscribed to ether theories discussed their theological significance explicitly, the concept of ether was employed throughout the eighteenth and nineteenth centuries in the solution of this classic enigma of dualism.

For convenience the general problem of dualism will be subdivided. Five particular but related problem areas can be identified for which ethers were important, and these will be discussed in turn and illustrated with examples derived principally from British writers. The first section of this chapter concerns the role of ethers in natural theology, and the second explores their cosmological functions; ethers were sometimes related to the ultimate structure of matter and were equated with the protoplast. Third, ethers were employed to account for all motion and activity in the universe, thereby provoking theological responses ranging from assertions that ethers had biblical roots to claims that the role of God was being challenged by atheists who tried to account for all motion by a mechanical fluid. A related problem is discussed in the fourth section, which concerns the role of ethers as intermediaries either between God and the physical universe or, analogically, between mind and matter. Finally, the diverse relations between ether and spirit are analysed.

These five problem areas were widely recognised during the eighteenth and nineteenth centuries, and yet only a small proportion of writers utilised ethereal fluids in their solution. Even among these writers there was little consistency; they employed very different versions of ether theory, their choice reflecting both wider philosophical issues and also the current state of scientific knowledge. Moreover, those who used ethers for theological purposes extended across almost the whole spectrum of religious opinion; among those discussed here are fundamentalists, Christian mystics, High Anglicans, liberal Anglicans, dissenters, spiritualists, and atheists. Yet this diversity of uses to which ethereal fluids were put suggests that they may be viewed as a scientific resource that could fulfil any of a number of functions within a specific range of philosophical systems. There were, of course, many writers, such as Joseph Priestley, David Brewster, and Thomas Reid, who explicitly rejected ethers from their scientific, philosophical, and theological systems. With this division of opinion, questions concerning ethereal fluids often feature in controversies that spanned scientific and theological issues. The following discussion alludes to some of these controversies, but since the principal concern of this synoptic chapter is to explicate the different theological functions ethers have fulfilled, I will not pay great attention to the scientific, theological, and philosophical commitments of individual protagonists. However, the concluding section suggests how theological attitudes towards ether theories may be related to broader social and intellectual issues.

Ether and natural theology

Throughout the eighteenth century, and less pervasively in the nineteenth, it was widely considered that the universe had been designed by God and as such manifested signs of its wise, intelligent, and good Designer. This natural theological theme appeared frequently in the Boyle Lectures, the Bridgewater Treatises, and a variety of other works, the most famous of which was William Paley's *Natural theology* (1802). In these writings the functions of different parts of the universe were discussed within the total economy, and every part had its pre-ordained role. One traditional argument relating to the role of ether concerned the existence of what appeared to be empty spaces between the stars and planets. Some objected that the Creator did not allow void space, since this would have performed no function. Thus the universe had to be full and the apparent spaces were in fact filled with some form of ethereal fluid, which constituted a plenum.[1]

A more common claim concerning the significance of ether for natural theology was that it was the principal secondary cause by which God governed the normal operations of the universe. For example, for John Cook, an Essex doctor, the discovery of ether was the key referred to in the title of his book *Clavis naturae; or, the mystery of philosophy unvail'd* (1733). In this work he claimed that God employed ether as the immediate cause of all motion: '*AEther* is the *Rudder* of the Universe, or as the *Rod*, or whatever you will liken it to, in the *Hand* of the *Almighty*, by which he *naturally* rules and governs all *material* created *Beings* . . . Now how beautiful is this *Contrivance* in God'.[2] Similarly, Richard Lovett of the Cathedral Church, Worcester, who equated ether with fire and electricity, argued that

> though God alone is the Author and Preserver of all Things, and which he continually upholds with his immediate Hand; yet, the only instrumental Cause of our Being is this subtil Spirit . . . In a Word, this pure AEther or Fire, contain'd in Air, is the Cause of all Motion, animal and vegetable.[3]

A particularly detailed example of the use of ether within a natural theological framework dates from the early 1830s when William Whewell sought the theological significance of the recently revived wave theory of light. In November 1830, Whewell received a commission from the president of the Royal Society, Davies Gilbert, to write one of the Bridgewater Treatises. Two and a half years later his treatise, *Astronomy and general physics considered with reference to natural theology* (London, 1833), was published. In his section on light, Whewell indicated his commitment to the wave theory and to the existence of a luminiferous ether that produced a wide variety of optical phenomena. Ether, he claimed, was a providentially designed mechanism that enabled man to perceive the physical world through the sense of vision. He suggested furthermore, albeit in a rather speculative manner, that ether was the source of all activity and the cause of chemical, electrical, and vital phenomena. In his view ether constituted one of the three fundamental substances in the universe, the other two being solid matter and aeriform fluids. Whewell therefore argued that owing to the crucial role of ether in the economy of the universe, 'if the world had no ether, all must be dead and inert'. Furthermore, he claimed that if we contemplate the remarkable intricacy of the ethereal mechanism and its importance in the universe, we are forced to acknowledge that it was designed by a 'most wise and good God'.[4]

Whewell also composed a work entitled 'View of the modern theory of light'[5] that may have been a draft for the Bridgewater Treatise but was never published. It was, however, far more than an introductory overview of physical optics, since Whewell not only analysed the significance of the wave theory for natural theology but, more precisely, identified five aspects of the luminiferous ether that were of importance to theology; these were the ether's simplicity, scale, fullness, variety, and harmony.[6] His arguments, which will be discussed in turn, are somewhat naive, however, and tend to overlap.

1. Whewell was impressed by the way in which the luminiferous ether, obeying the simple laws of mechanics, produced a great diversity of optical phenomena. Thus the simplicity of the cause contrasted with the complexity of its effects, and this relationship between the two indicated that the Creator had designed ether in a highly efficient manner.
2. Divine power, Whewell considered, was manifested by the interstellar spaces, which involved distances many orders of magnitude greater than those usually encountered. Similarly, 'the range of density and elasticity which we find in the mechanism of light [i.e., the luminiferous ether] is upon a scale equally extraordinary, or still more so'. Since the ether's elasticity was considerably greater than that of terrestrial bodies and its density very much less than that of steam, these contrasts in scale indicated God's power in being able to create so great a diversity in the universe.
3. The mechanism of the luminiferous ether also indicated to Whewell 'how full the world is of [God's] contrivances' and thus of manifestations of intelligent design. He speculated further about the existence of other micro-mechanisms in space that might produce thermal, magnetic, and electrical phenomena. This argument appears to be a variation on the traditional principle of plenitude.
4. The fourth argument concerned the variety of ethereal fluids in the universe and their design in such a way that each fulfilled a specific function. By considering different examples of action at a distance, such as occur with light, sound, electricity, and heat, Whewell claimed that their means of propagation must be different. This argument for a diversity of ethers (which seems incompatible with argument [1]) indicated specific design and thus a choice of means on the part of the Creator. Furthermore, to Whewell this argument undermined the claims of materialists who considered that phenomena resulted from the necessary properties of matter.
5. Whewell's final point was that from empirical evidence the luminiferous ether appeared to interact with other subtle fluids, such as those responsible for chemical, vital, and electrical phenomena. The connections between these various classes of phenomena indicated an overall providential design and harmony in which the luminiferous ether played an integral part.

Whewell is generally portrayed as a liberal Anglican for whom natural theology, as opposed to revealed religion, was of central importance.[7] He disagreed with the Scottish Evangelical David Brewster on many issues, including theology,[8] Brewster insisting on a literal interpretation of the Bible as opposed to Whewell's latitudinarianism. Morse (1972) has even suggested that Brewster's commitment to an inductivist methodology that omitted unobservable entities (such as ether) from science stems from his Calvinist upbringing, which emphasised the manifest nature of God's works. Brewster's antipathy towards 'broad' natural theology and ether theories can be seen in his published response to Whewell's Bridgewater Treatise.

Brewster received a copy of the book from Whewell soon after it was published. His initial response was civil and complimentary; he wrote to Whewell that it 'has been my companion on two Journies to the Highlands, and I need scarcely say that I never derived more pleasure or instruction from any other book.'[9] However, writing anonymously in the *Edinburgh Review*,[10] he attacked, among other points, Whewell's claim that the luminiferous ether indicated God's design. Brewster did not accept the wave theory of light, and he resented the fact that Whewell had presented that theory as portraying the true nature of light. Moreover, he rejected unobservable fluids like the luminiferous ether that Whewell had identified with the wave theory. On the subject of natural theology, Brewster claimed that its legitimate scope involved only the 'Power, Wisdom, and Goodness of God as [actually] *manifested* in the CREATION'.[11] Thus he considered that it was legitimate to argue from the known structure of the eye to the existence and attributes of its 'Designer'. However, hypothetical entities, like the luminiferous ether, were not appropriate subjects for natural theology, since they could not be considered as true premises of the argument. Moreover, he feared that when the hypothesis was refuted by scientists, as would surely happen in this case, the apologetic argument would readily be exploited by atheists. Thus, Brewster rejected arguments from design employing ethereal fluids.

Ether and matter

Throughout the period under discussion natural philosophers frequently speculated about the structural relationship between ether and gross matter, and the role of God in establishing this relationship. Ether was usually considered a more basic substance than gross matter, and thus it was suggested that ether was the protoplast out of which God had formed matter. The most famous example of this use of ether occurs in a letter of Newton's to Henry Oldenburg published in 1757, although written more than eighty years earlier. Here, Newton speculated that the whole diversity of the physical world had been produced by a variety of 'aethereal spirits, or vapours' in a condensed state. Initially, God brought about this process but subsequently he delegated it to the 'power of nature'.[12] It has been suggested that Newton's concern with the ether as 'protoplast' was related to the alchemical tradition with which he was familiar (Dobbs, 1975).

Two centuries after this letter of Newton's was written, a similar view was articulated with respect to a contemporary scientific theory, the vortex theory of the atom (Silliman, 1963; Siegel, this volume). A number of eminent physicists, such as Lord Kelvin, Peter Guthrie Tait, and James Clerk Maxwell, advocated this theory in which vortical motions in homogenous, incompressible, and frictionless ether constituted the atoms of gross matter. Since these vortex atoms were stable and perfectly elastic they behaved in the same way as the particles postulated by the kinetic theory of gases. Furthermore, the vortex theory appeared to offer a means of explaining all chemical and electromagnetic phenomena within a unified world view in which ether was the fundamental substance. Writing in the ninth edition of the *Encyclopaedia Britannica,* Maxwell expressed the view that the vortex theory was a most important and promising scientific hypothesis.[13]

The vortex theory of the atom, together with other ideas drawn from contemporary science, was adopted by Balfour Stewart and Peter Guthrie Tait in their attempt to defend the coherence of science and religion in a popular book entitled *The unseen universe; or, physical speculations on a future state,* published in 1875. They suggested not only that gross matter was composed of ether, but that the particles of that ether were themselves composed of an even more subtle ether, and so on. Furthermore, the subtler higher-order ethers contained greater energy, so that any particular ether was able to form the next ether lower on the subtlety scale. From this thesis they also argued that the subtler the ether, the more stable and long-lasting, and the higher its energy content. In accordance with the purpose of the book – 'to show that the presumed incompatibility of Science and Religion does not exist'[14] – Stewart and Tait postulated that their scale of ethers rose towards God and tended to become identical with his attributes of omnipresence, omnipotence (having infinite energy), and existence for all time. Thus in their scheme they employed a recessive scale of protoplasmic ethers that bore a specific relationship to God. Thus, God's act of creating our world involved the localisation of energy in the formation of gross matter out of the lowest-order ether.[15]

Stewart and Tait's idea of a scale of ethers between God and matter may have been derived from the popular concept of the Great Chain of Being. A more direct source, however, and one that they cited, was a passage in Thomas Young's *A course of lectures on natural philosophy and the mechanical arts* (Cantor, 1970). Young, who came from a Quaker family but who became an Anglican primarily for careerist reasons, was committed to the existence not only of matter but also of such 'immaterial' substances as the electric fluid and

> either caloric or a universal ether; higher still perhaps are the causes of gravitation, and the immediate agents of attractions of all kinds ... It seems therefore natural to believe that the analogy may be continued still further, until it rises into existences absolutely immaterial and spiritual. We know not but that thousands of spiritual worlds may exist unseen for ever by human eyes.[16]

In contrast to Young, other writers, such as Oliver Lodge (see the section of this chapter entitled 'Ether as intermediary'), maintained that only one ether was required, which performed both physical and theological functions.

Ether and dynamics

Mechanical philosophers in the seventeenth century, such as Descartes, employed ethereal fluids filling space as the immediate cause of all motion and activity. Indeed, the idea of motion without a mover in contact with the moving body was deemed impossible and incomprehensible. It has often been contended that this theory of contact action was displaced early in the eighteenth century by Newtonian central forces acting at a distance. However, throughout the eighteenth century and well into the nineteenth there were a number of natural philosophers who rejected action at a distance and instead adopted a contact-action plenal ether. Many of these writers were theologically motivated and claimed the Bible as the source of their natural philosophy. Although they did not form a coherent group, a number of these authors were intellectually indebted to John Hutchinson (1674–1737), who had attacked Newton's natural theology and natural philosophy early in the eighteenth century (Metzger, 1938; Wilde, 1980). We shall take as our example of this genre the book by Samuel Pike entitled *Philosophia sacra; or, the principles of natural philosophy extracted from divine revelation* (1753). Pike (1717?–73) was connected with the dissenting ministry and, towards the end of his life, appears to have been associated with the Sandemanians – a sect that later numbered Faraday among its members.

Unlike Hutchinson, Pike found no conflict between Newton and the physical system revealed in the Bible.[17] He considered that Newton had discovered the law of gravitation but not the underlying cause, which was motion in the pure and subtle ether filling the heavens. In Pike's cosmology there was no void space; all actions resulted from impulsion derived from this ether, which underwent a cyclical, mechanical, and everlasting motion. The inert particles of ether were projected at high speed away from the sun and towards the circumscribing firmament, but as they approached it they slowed down, congealed into larger clusters, and returned to the sun. Here the clusters were broken down and the individual particles reemitted. On this theory gravitation was explained by the pressure produced on bodies by the moving clusters of ether particles. Likewise, the planetary motions and rotations were explained by the forces produced by both the individual particles flowing away from the sun and the clusters moving in the opposite direction. Similarly, Pike explained cohesion, electrical, and optical phenomena on mechanistic principles.

As Pike's title suggests, he attempted to 'extract' the whole of his natural philosophy from the biblical text. Indeed, he conceived his book as contributing to the reconciliation between science and what he understood to be a fully biblical theology. Adopting an ingenious but somewhat crude philological method, Pike reinterpreted several biblical passages and 'derived' from them his theory of a mechanical ether. For example, he translated the Hebrew word *SHa'HaKIM* (skies) as meaning 'the *strugglers* or *aethers in conflict*'.[18] From analysis of this kind he claimed that Scripture teaches us that all natural causes and effects are mechanical.

Central to Pike's theological concerns was the despiritualisation of the physical world. Neither gross matter nor the ether particles were endowed with intrinsic power. Instead, ethereal impulse not only solved the problem of action at a distance but also allowed the whole of the physical world (since the Creation) to remain ontologically distinct from God, in whom alone power resided. Yet, Pike argued, this mechanical universe was not independent of God. At the Creation, God formed both matter and ether and gave motion to the machine. Furthermore, he maintained the mechanism and could at will interfere with its running, for example, by performing miracles.[19]

It is, perhaps, paradoxical that while a fundamentalist writer like Pike employed a contact-action ether to explain phenomena, many natural theologians considered the attempt to explain gravitation by purely material causes tantamount to atheism. The spectre of materialism, traditionally associated with the names of Epicurus, Lucretius, and Hobbes, raised its head again at the turn of the century with the publication of the *Méchanique celeste* (1798–1825) of Laplace. For many this work became associated not only with atheism but also with the social upheaval of the French Revolution. Samuel Vince, professor of astronomy and experimental philosophy at Cambridge, was one of those perturbed by French atheism, in response to which he wrote his *Observations on the hypotheses which have been assumed to account for the cause of gravitation from mechanical principles* (1806). In his preface, Vince made clear his concern:

> Many of the most eminent Philosophers upon the Continent have been endeavouring to account for all the operations of nature upon merely mechanical principles, with a view to exclude the Deity from any concern in the government of the system, and thereby to lay a foundation for the introduction of Atheism.[20]

Vince surveyed various ether hypotheses and proceeded to 'demonstrate' that no hypothesis specifying the elasticity and density of an ethereal medium could account for the inverse square law of gravitation. He then argued that the failure of all ether hypotheses to explain both gravitation and the stability of the celestial system indicated that God did not act by material causes; instead, God must act directly on matter in accordance with the law of gravitation. Finally, Vince claimed that contemplation of this universal God-given law led to appreciation of the power, wisdom, and goodness of the Creator.

A number of other writers also contended that no quiescent or impelling fluid could account for gravitation. Alexander Crombie, for example, adopted the fairly standard distinction between the law of gravitation as a descriptive rule about the behaviour of bodies and the idea of gravity as a causal agent.[21] Like Vince and Thomas Reid (see the next section), Crombie accepted the former but not the latter. There was, however, a dissenting voice raised against Vince's pamphlet. Writing in the *Edinburgh Review* for 1808,[22] John Playfair, professor of mathematics at Edinburgh, pointed out that Vince's argument was mathematically false, since there were an infinite number of ether hypotheses from which the inverse square law could be deduced. Playfair, who was a minister in the Church of Scotland, accepted that it was perfectly legitimate to inquire into the cause of gravitation and that such research should not be constrained by ideological considerations. Moreover, he was 'convinced that the issue of this argument is quite immaterial to the truths of natural religion, which must rest on the same immoveable foundation, whether the physical cause of gravity be discovered, or not'.[23] Playfair, then, rejected the connection that Vince claimed existed between ether and atheism, and he appears to have viewed Vince's pamphlet merely as the reaction of a frightened man to political developments in France.

Consideration of the effect of a ubiquitous stationary ether on the progress of the world is related to this controversy. Unlike Pike's cosmology, which ensured that no physical cause could terminate the solar system, those natural philosophers who adopted a stationary ether were faced with the theological implications of the ether's effect on planetary motions and the ultimate decay of the world. As Schaffer (1977) has shown, Edmund Halley in 1693 recognised that ether's resistance would lead to the secular acceleration of the planets and result in their spiral trajectories into the sun. However, Schaffer claims, Halley did not draw any strong theological conclusions from this argument. Newton, by contrast, abandoned altogether the idea of a Cartesian ether, since he demonstrated its incompatibility with celestial motions. His own concept of the stationary ether – as a rare, tenuous, and highly elastic fluid – was reintroduced in the 1717 edition of the *Opticks,* where he confronted the problem of the effect it would produce on planetary motions. In solving, or rather sidestepping, this problem he showed that if ether were rarer and more elastic than air by factors of 700,000, its presence 'would scarce make any sensible alteration in the Motions of the Planets in ten thousand Years'.[24] Thus, for Newton, ether by itself might produce decay in the world, but the rate of decay would not be observable.

Two very different theological arguments intersect in considering the history of the solar system. Within a natural theological context the stability of the system would indicate God's intelligence in creating a perfect mechanism. Though this view was logically compatible with a creationist theology it might also be employed by the atheist, who would argue that the physical world, being self-sufficient, has existed from eternity. The other factor concerned the physical implications of the creationist doctrine. Although a steady-state solar system did not imply that it was divinely created, a decaying system (without a compensatory process such as that envisaged by Kant) unequivocally implied creation at some past time and the gradual death of the world at a future time. Some theological purchase could thus be obtained from both of these theories, but particularly the theory of decay. Ethers were occasionally employed as a cause of this decay, although there were many other agents that might produce a similar result. However, this role for ether took on new significance in the 1820s and 1830s when the theory of the stationary ether gained support from two empirically based arguments. One was the interpretation of the wave theory of light that required ether to be stationary with respect to the sun and stars in order to account for stellar aberration. The other arose from the secular acceleration of Encke's comet, which was widely interpreted as the effect of resistance by an ethereal fluid. It was the latter that particularly attracted the attention of William Whewell. In his Bridgewater Treatise of 1833 he extrapolated from this explanation of Encke's comet in order to predict the behaviour of the earth and other heavy bodies. He argued that although the resistance to the earth's motion would be slight, nevertheless it would inevitably lead to its destruction, albeit in the distant future. This argument provided further evidence for the 'universal law of decay', which, Whewell believed, dictated changes in both the earth and the cosmos – in accordance with his interpretation of the nebular hypothesis[25] – and also in human societies. For Whewell the decay of the world had a specific theological implication: that it must also have had a beginning. In turn, this necessarily implied the existence of God, the Creator, 'a First Cause which is not mechanical'.[26]

Ether as intermediary

An important theme in Western philosophy has been the doctrine of dualism in which two distinct substances, mind and matter, are postulated in order to distinguish God from the material universe and our souls (or minds) from our bodies. Although these two substances are considered to be incommensurable they must be able to interact with each other, for example, in the acts of perception and volition. However, this dualistic ontology poses the problem of how these two radically different substances – the one incorporeal, penetrable, and intelligent and the other extended, impenetrable, and passive – are able to affect one another. Dualists, in proposing a variety of solutions to this problem, have sometimes adopted an ethereal fluid as the intermediary between mind and matter. An attraction of this solution is that ethereal fluids, as sometimes defined, partake of some of the qualities of both mind and matter; for example, both activity and extension have simultaneously been attributed to ether.

Probably the most frequently cited and most influential example of the ether as intermediary appeared in queries 23 and 24 to Newton's *Opticks* (1717). Here, Newton proposed the scarcely original theory that an ether fills the nervous system. In his explanation of vision, Newton suggested that rays of light falling on the retina cause this medullary ether to vibrate, and this vibration is then transmitted along the optic nerve 'into the place of sensation'. Likewise, in explaining how we move parts of our bodies, Newton considered that the 'power of the Will' causes the ether in the brain to vibrate. This vibration in the medullary ether is then propagated along the capillaries of the nerves and produces movement through either contraction or dilation of the muscles.[27] Both of these explanations were based on the hypothesis that Newton's ether could interact with both matter and mind, an issue that is discussed in some of his unpublished writings (McGuire, 1968).

Among those who adopted Newton's physiological use of ether was David Hartley, whose *Observations on man* first appeared in 1749. Hartley's ether was, however, unequivocally a form of matter, and hence this theory raised the problem of how a material ether could interact with an immaterial soul. Aware that this was a sensitive issue, Hartley was at pains to assert that his theory neither challenged the doctrine of the soul's immortality nor attributed sensation to brute matter. He explained that he merely posited vibrations in the medullary ether as physiological correlates to our sensations, ideas, and motions.[28] Despite his statements on this issue, other authors attributed to Hartley the view that the mind was merely matter in motion. Among these detractors was the Scottish dualist Thomas Reid, who contrasted the humble way in which Newton proposed a few open-ended conjectures about ether with the atheistic implications of Hartley's fully fledged system. Reid complained that since we have no direct evidence for the existence of ether, let alone its vibrations, Hartley had employed the illegitimate method of hypothesis. In comparison with God's infinite intelligence, man could not comprehend the extent of God's works, and it was therefore presumptuous of Hartley to have framed a hypothesis: Hypotheses were the creation of 'human imagination, [and] not the work of God'. Reid's other theological offensive concerned causality. He claimed that science was not concerned with efficient causes but that the scientist (*pace* Newton) should instead seek lawlike relations between observables. Hartley's ether, which mediated between the sensory organs and the soul, functioned as a spurious efficient cause of sensation, since for Reid the immediate source of sensation 'must be resolved into the will of God'. Any causal explanation of the interaction between mind and matter was theologically unacceptable, because this interaction was known only to God.[29]

Our final example of the use of ether as intermediary takes us to the early part of this century. Sir Oliver Lodge, the eminent physicist, attempted to solve the mind–body problem in a manner similar to Newton's two centuries earlier. Lodge's interest in the problem grew out of a general concern for a reaffirmation of Christianity in the face of modern science and rampant atheism. Moreover, after the death of his son in the World War I, Lodge became progressively involved with the problem of life after death, which he related to his long-standing interest in spiritualism. More precisely, he connected the central Christian doctrine of the immortality of the soul with the idea that we also possess an immortal 'spiritual body'. After death, which is only the death of the physical body, a person's soul remains localised in the 'spiritual body'. The existence of this 'spiritual body' is asserted in 1 Corinthians 15:44: 'There is a natural body, and there is a spiritual body'. Furthermore, Lodge employed this theory in his account of a number of 'spiritual' phenomena, particularly communication with the dead. In elaborating his theory he gave prominence to the ubiquitous ether, which he also utilised in his physics. He considered the ether a physical substance differing from both gross matter and spirit (soul) but able to interact with each. Moreover, he maintained that the 'spiritual body', which is immortal, is composed of ether. Hence every person has an etheric or 'spiritual body' in which the spirit (or soul) is localised: 'Our spiritual and real home is in the ether of space'. In putting forward this theory,

Lodge offered a solution not only to the traditional mind–body problem, but also to the problems of psychical research. Thus, for example, the ethereal body carrying the soul of a departed person is able to be present at a seance and to affect material objects by means of localised energy derived from ether.[30]

Lodge's interest in spiritualism reflected the late nineteenth- and early twentieth-century concern with a range of psychic phenomena. A large number of eminent scientists – including Lodge, J. C. Adams, A. R. Wallace, Lord Rayleigh, J. J. Thomson, and W. Crookes – were concerned with or actively investigated these phenomena, and many joined the Society for Psychical Research founded in 1882. Several recent studies have shown that many scientists of the period held deep-seated religious views that they attempted to integrate with contemporary scientific theories. Frequently ethereal fluids performed this mediating role (Wilson, 1971; Heimann, 1972; Kottler, 1974; Turner, 1974; Wynne, 1977, 1979).

Ether and spirit

A common view, particularly during the eighteenth century, was that matter was inert and thus some superimposed active agent was required in order to account for all types of phenomena ranging from gravitation and cohesion to animal growth. Frequently an ethereal fluid was proposed as the source of this activity. However, when ether was considered a purely material substance, the problem arose of explaining its own activity. To overcome this difficulty, Bryan Robinson suggested that 'Spirit' supplied this activity in the form of a repulsive force between the particles of a Newtonian elastic ether.[31] Another writer, Abraham Tucker, employed a 'spiritual substance', presumably identical to the neo-Platonist theory of the mundane soul, to account for the activity of ether.[32] Others, including George Berkeley and Michael Ramsay, claimed that ether was intrinsically active.

A particularly important and influential discussion of the active powers in nature (Ritterbush, 1964) appeared in Bishop Berkeley's *Siris* (1744).[33] Here, Berkeley maintained that an ether, which he equated with fire and the substance of light, was the secondary cause responsible for animating the universe. Yet Berkeley's ether was not conceived in mechanistic terms but was instead an active animate entity. Thus, Berkeley claimed that ether is 'the vegetative soul or vital spirit of the world' and is infused with divinity. In propounding this view he drew in part on early eighteenth-century chemical philosophy and particularly on Boerhaave's discussion of fire as the active agent. More significantly, he cited at length the *prisca sapientia* tradition[34] in which spiritualized ethers had been propounded by the ancient Egyptians, Greeks, and Persians and also in the Hermetic corpus and the Bible.

A slightly earlier writer who drew extensively on the *prisca sapientia* tradition was the Chevalier Ramsay, a Scot who spent most of his life in France and who wrote a popular exposition of the ancient theology in the form of a fictional travelogue. The second book of Ramsay's *The travels of Cyrus* (1727) concerned Cyrus's visit to the Magi of Persia. The Magi could supposedly disengage themselves from matter and thus communicate directly with the spiritual realm. Zoroaster, the Archmagus, initiated Cyrus into the hidden mysteries of the physical universe. Here the French and English texts diverge. In the latter, Ramsay outlined Newton's theory of gravitation (which he considered compatible with his ether theory), whereas in the French text he explicated the role of a pure and invisible ethereal fluid. Here he maintained that ether, which he identified with the *primum mobile* of the Greeks, was infinitely divisible and formed numerous fluids that were responsible for all celestial and terrestrial activity, including the motions of stars and planets, gravitation, light, and the growth of plants. Despite Ramsay's reference to *ce système Cartesien,* it is clear that his ether bore little resemblance to that of Descartes. He denied that ether acted according to the laws of blind mechanical necessity. By contrast, he claimed that ether 'is the *body* of the Great [God] Oromazes, whose soul is truth' and who is the 'first principle of things. He diffuses himself everywhere'.[35] Ramsay allied this belief in a divine function of ether with Newton's suggestion in the final query of the *Opticks* that 'God, being present everywhere by His will, moves all bodies in His infinite, uniform *sensorium,* and so shapes and reshapes according to His pleasure all parts of the universe'.[36]

Not only did Ramsay distinguish his ether from the subtle fluids of the mechanical philosophers; he also explicitly connected it with the *prisca sapientia* tradition. Walker (1972) suggests the identification of Ramsay's pure ether with the neo-Platonist doctrine of the mundane spirit. Moreover, he conjectures that a connection existed between Ramsay and Newton, and this may help explain the revival of Newton's concern with ether late in the first decade of the century.

Whereas Ramsay and Berkeley drew heavily on the ancient theology, other writers explicitly related ether to their Trinitarian beliefs. One such author was the Dublin clergyman Richard Barton, who discussed the analogies between the natural and moral worlds in a published series of lectures entitled *The analogy of divine wisdom* (1750). One topic under discussion was the analogy between the infinite divine Spirit or HOLY GHOST, and the UNIVERSAL AETHER or elemental Fire . . . Because as the mechanic philosophers make the Aether the cause of attraction, muscular motion and other extraordinary phaenomena of matter: So is the HOLY GHOST the cause of all spiritual conduct, which is consonant to the divine Law.[37]

Barton identified fire with ether, and also with electricity, and he cited Berkeley's *Siris* in order to show not only the role of fire in the world economy, but also that fire was the spirit of the world. In explicating the analogy between ether and the Holy Ghost, Barton pointed out that the Bible likened the appearance of the Holy Ghost to fire (Acts 2:3) and that Christ was able to baptise with fire and the Holy Ghost (Matthew 3:11). Another analogical connection was that ether functioned as 'a principle instrument in the conservation of the material world', and similarly the divine spirit had conserved religion and virtue in the moral realm. Furthermore, both ether and the Holy Ghost were universally present, yet both were unequally diffused. Thus while some material objects contained an excess of fire, some men by their virtuous acts had become more endowed with divine spirit. Barton's more general claim was that although our knowledge of ether and its functions was far from complete, science had shown that the whole economy of nature depended on ether. Likewise, he conceived a parallel dependence of the moral world on the divine spirit. From these and other analogies, Barton attempted to show the essential coherence between natural philosophy and divine revelation.[38]

In contrast to Barton, John Hutchinson not only drew analogies between ether and each of the three persons in the Trinity but also found evidence for his Trinitarian beliefs in the functions of the contact-action ether that he claimed was the source of all observed motions. Hutchinson believed that parallels existed between the divine realm and the structure of the physical world, such that the former could be appreciated by an analysis of the latter. Moreover, the key to understanding both was the Bible. In Hutchinson's physical theory the material world was governed by a single ethereal substance that was capable of three modifications: fire, light, and spirit (air). The analogue of this physical situation involved the three persons of the Trinity within the single Godhead (Elohim). In Hutchinson's scheme fire was related to the Father, light to Christ, and spirit (air) to the Holy Ghost. Furthermore, these relationships were sanctioned by their functional similarities and by lexical relations in the original Hebrew.[39]

Although Hutchinson rejected those natural philosophies that attempted to 'spiritualise' matter and instead insisted that the whole physical universe consisted of matter in motion, he was strangely silent about the human soul, which appears to have played no part in his ontology. Paradoxically, his emphasis on man as a mechanism, albeit one designed by God, brought him close to the traditional kind of materialism that has often been associated with atheism. The programme to reduce mind (and the soul) to the action of material particles has been articulated on numerous occasions both within and beyond the time period covered in this chapter. A central problem facing adherents to this programme, and one quickly seized on by their antagonists, has been the explanation of how inert matter could produce those kinds of activity that dualists considered belonged only to mind. However, one response available to materialists that went some way to meet this objection employed one or more ethereal fluids. To illustrate this strategy we will examine the theory proposed in a biography of Lucretius, published in 1805, by the surgeon and classical scholar John Mason Good.

Good, who appears to have been a Unitarian, clearly believed in the existence of God and in the truth of the Scriptures. However, he maintained that Lucretius had been correct in claiming that whereas man's body is composed of gross matter his soul is made up of a more refined, subtle form of matter. Certainly, he admitted, the soul was different from gross matter, but this did not imply either the soul's immateriality (which was not mentioned in the Bible) or the inertness of all matter. There was much evidence to the contrary: Matter manifested gravitation, and magnetic and chemical action; and more highly organised matter in animals and vegetables was subject to irritability. From animal sensation it was but a small step for Good to claim that the soul was material, but composed of a very subtle and volatile ethereal matter. In explicating this theory he pointed to Lucretius's doctrine that

 . . . the mind, in every act, we trace
 Most voluble, from seeds of subtlest size,
 Rotund and light, its mystic make must spring.[40]

Moreover, Good maintained that modern experimentally based physiology expounded a related doctrine involving the medullary ether and that there was no good reason why this theory, which explained the constitution of the nerves, should not be extended to include the mind. Thus he envisaged that the mind could modify the medullary ether.[41]

Even if Good was no atheist, his theory became associated with the materialist programme expounded by atheists.[42] Indeed, the belief that the mind or soul was composed of matter – probably a subtle form of matter – was expounded by such self-confessed atheists as Julien de la Mettrie and Baron d'Holbach (French, 1969).[43] In response to this challenge orthodox dualists recurred to the argument used by Samuel Clarke in his controversy with Henry Dodwell: Particles of matter irrespective of their subtlety or arrangement cannot have the property of consciousness. In Clarke's words: '*Consciousness* therefore cannot at all reside in the Substance of the *Brain,* or *Spirits,* or in any other *material System* as its *Subject;* but must be a Quality of some *Immaterial Substance'*.[44] Indeed, for many 'moderate' clergymen, gross matter and soul were sufficient and there was no need for recourse to ethereal fluids.

Concluding remarks

Having surveyed the theological functions that ethereal fluids fulfilled during the eighteenth and nineteenth centuries, we may now attempt to draw some general but tentative conclusions. One such conclusion concerns the temporal distribution of the writers cited. With the exception of Whewell's Bridgewater Treatise, all the works discussed herein that advocate ethereal fluids date from either before 1810 or after 1875. An explanation is required for this distribution (assuming that it does not result from an inadequate search of nineteenth-century literature). This earlier period corresponds roughly to the widespread interest among speculative natural philosophers in the unifying role of ethereal fluids (see Heimann, this volume). Moreover, the majority of our examples date from the mid-eighteenth century, the period when the ether theories of Newton and Boerhaave attracted most attention. After about 1830 a major scientific concern was the construction of mathematical and mechanical models of the luminiferous ether. However, the ether theories developed in this context were unlikely candidates for fulfilling four out of the five functions discussed in the section 'Ether and natural theology', although they still could be employed in the argument from design, which was their principal function in Whewell's unpublished manuscript. The late Victorian resurgence of interest is perhaps explained by the 'crisis of faith'[45] that affected many scientists, who then frequently attempted to reconcile science and religion through the somewhat simplistic strategies offered by spiritualism.

The following brief and very tentative comments, which relate to the earlier period, attempt to locate specific groups that tended to employ ethers for theological purposes. Initially, however, it may be helpful to reiterate the point that several different versions of ether theory have been discussed during the course of this chapter. Within this range two distinct types of ether may be identified; one class comprises 'animate' ethers, which were closely allied to spirit, and at the opposite end of the spectrum are mechanistic ethers, which were intended to 'despiritualise' nature. Those writers, such as neo-Platonists, mystics, and certain High Anglicans, who emphasised the immanence of God in an active universe and who rejected the notion of passive matter found the first class of ether theory acceptable but positively rejected the second. Among those discussed here, Berkeley, Ramsay, and Barton fall into this category, and it was also principally this group of writers who stressed the ancient roots of their theory. By contrast, contact-action ethers implied that God did not act directly on each particle of matter. This type of theory was most acceptable to those like Hutchinson and Pike for whom the Bible was the principal source of all spiritual and physical knowledge. This kind of ether was also acceptable to atheists, as was the related theory attributing all activity to the smallness or organisation of ether particles and denying the existence of immaterial souls. This last type of theory, together with others that emphasised ether as the 'active principle' in nature, appears to have appealed particularly to Low Churchmen such as Young and Good, and it may not be inappropriate to include Hartley and even Newton in this group. (Priestley, a Unitarian, likewise attempted a monistic synthesis, although not founded on ether theory.)

Missing from this discussion and underrepresented in the examples cited in this chapter is one significant cluster of theological writers – what might be called the 'moderate' or 'broad' church faction. The members of this group tended to be dualists who firmly distinguished mind from matter. This ontology denied the existence of active ethereal matter. Moreover, dualists usually considered God the immediate source of all activity in nature, so that no role was allowed for unobservable subtle media. It is no coincidence that most of those cited herein as objecting to ethereal fluids on theological grounds were dualists of the 'moderate' persuasion.

The theological functions of ether have too rarely been discussed by historians, who have usually concentrated on scientific theorising. Yet not only does this aspect of ether fail to exhaust its historical role, it also does not take us far in understanding why some writers were enthusiastic champions of ether while others vehemently refused to countenance its deployment or existence. The notion of an ether diffused through space involved far-reaching ontological and epistemological commitments (Laudan, this volume). Moreover, for those scientists deeply concerned with natural theology or the theology of nature, ether has helped solve such problems as the relation between God and the physical universe or the possibility of life after death. Whether this resource has been utilised, in preference to other strategies, has depended on individual theological and philosophical commitments and the position of ether theory in contemporary science.

The problematic relation between science and theology gives rise to two opposing theses. One posits the primacy of theology and conceives ether theorising as simply an extension or elaboration of an already held theology. The other suggests that ether theorising arises principally from the scientific arena and that scientists subsequently – often in old age – explore their extrascientific functions. Examples could be found to confirm each of these theses (Barton would be an example of the first and Lodge of the second), with the latter receiving more support from the nineteenth- than the eighteenth-century cases discussed in this chapter. At the same time many – both scientists and others – held strong predispositions for or against ether on theological grounds. In contrast to these two extreme positions, a more helpful general model would involve a dynamic interaction between scientific theorising and theology, with ether potentially providing an important element in this interaction.

One particular connection has recently received attention from Shapin (1980), who suggests that matter theory should be interpreted as reflecting the perceived social order: Thus the dualist symbolises in his matter theory – through the total separation of mind and matter – the distinction between classes in society, whereas Priestley is interpreted as having collapsed both mind and matter into a monistic theory analogous to his political egalitarianism. In his interesting but speculative scheme, Shapin concentrates on just one of ether's many functions, that of mediating between mind and matter. The social significance of this type of ether theory is not explained in detail by Shapin but perhaps reflects a stratified, socially mobile society. To what extent this thesis is helpful in understanding Lodge or indeed Newton – the socially mobile lad from Lincolnshire – is open to further research. However, this type of scheme, with significant modifications, has been utilised indirectly by Christie (this volume) in discussing the secular context of the work of Adam Smith, David Hume, and William Cullen. At the same time the evidence offered in the present chapter precludes the conclusion that ether was always associated with secularism, or indeed atheism.

While it may be difficult to identify *the* theological function of ether, ether theory may perhaps best be conceived as a resource that could help solve some recurrent problems in natural theology and the theology of nature. The present chapter has attempted to delineate five such functions employed by eighteenth- and nineteenth-century writers, and in this concluding section several general theses, founded on a limited sample of cases, are offered, albeit tentatively.

Acknowledgments

The author would like to thank John Hedley Brooke and Jonathan Hodge for their helpful comments.

Notes

1 E. Chambers, 'Aether', *Cyclopedia; or, an universal dictionary of arts and sciences,* 4th ed., 2 vols. (London, 1741), 1:n.p.
2 J. Cook, *Clavis naturae; or, the mystery of philosophy unvail'd* (London, 1733), 284–6.
3 R. Lovett, *The subtil medium prov'd; or, that wonderful power of nature so long ago conjectur'd by the most ancient and remarkable philosophers, which they call'd aether but oftener elementary fire, verify'd* (London, 1756), 64–5.
4 W. Whewell, *Astronomy and general physics considered with reference to natural theology* (London, 1833), 141.
5 W. Whewell, 'View of the modern theory of light', Whewell Papers, Trinity College, Cambridge: R.18.17^3. This appears to have been written in the early 1830s.
6 Ibid., 195–215.
7 W. B. Cannon, 'Scientists and broad churchmen: an early Victorian intellectual network', *Journal of British Studies 4* (1964), 65–88; J. H. Brooke, 'Natural theology and the plurality of worlds: observations on the Brewster–Whewell debate', *Annals of Science 34* (1977), 221–86; Cantor (1975); Morse (1972).
8 Brooke, 'Natural theology'; Cantor (1975); Morse (1972).
9 D. Brewster to W. Whewell, 10 June 1833, Whewell Papers: Add. MS a. 201^{82}. I am grateful to the Librarian of Trinity College, Cambridge, for permitting me to quote from this letter and also from Whewell, 'View'.
10 [D. Brewster], review of Whewell, *Astronomy, Edinburgh Review 58* (1834), 422–57.
11 Ibid., 428.
12 T. Birch, *The history of the Royal Society of London,* 4 vols. (London, 1756–7), 3:250. See also Heimann (1973 and this volume).
13 J. C. M[axwell], 'Atom', *Encyclopaedia Britannica,* 9th ed., vol. 3 (Edinburgh, 1875), 36–49.
14 [B. Stewart and P. G. Tait], *The unseen universe; or, physical speculations on a future state* (London, 1875), vii.
15 Ibid., 154–72.
16 T. Young, *A course of lectures on natural philosophy and the mechanical arts,* 2 vols. (London, 1807), 1:610–1.
17 S. Pike, *Philosophia sacra; or, the principles of natural philosophy extracted from Divine Revelation* (London, 1753), viii, 1–14.
18 Ibid., 15–33; G. N. Cantor, 'Revelation and the cyclical cosmos of John Hutchinson', in *Images of the earth: essays in the history of the environmental sciences,* ed. L. Jordanova and R. Porter (Chalfont St. Giles, 1979), 3–22.
19 Ibid., 77.
20 S. Vince, *Observations on the hypotheses which have been assumed to account for the cause of gravitation from mechanical principles* (Cambridge, 1806), 4.
21 A. Crombie, *Natural theology; or, essays on the existence of a Deity and of providence, on the immateriality of the soul and a future state,* 2 vols. (London, 1829), 1:106–7.
22 [J. Playfair], review of Vince, *Observations, Edinburgh Review 13* (1808), 101–16.
23 Ibid., 103. See also C. Maclaurin, *An account of Sir Isaac Newton's philosophical discoveries* (London, 1748), 389.
24 I. Newton, *Opticks,* 2nd ed. (London, 1717), 327.
25 Brooke, 'Natural theology'.
26 Whewell, *Astronomy,* 191–209. Cf. B. Powell, *The connexion of natural and Di-*

vine truth; or, the study of the inductive philosophy considered as subservient to theology (London, 1838), 164–6.
27 Birch, *History*, 252–4; Newton, *Opticks*, 328.
28 D. Hartley, *Observations on man, his frame, his duty and his expectations*, 5th ed., 2 vols. (Bath, 1810), 1:34, 525–6.
29 T. Reid, *The works of Thomas Reid, D.D., with notes and supplementary dissertations by Sir William Hamilton, Bart.*, 8th ed., 2 vols. (Edinburgh, 1895), 1:248–53. Related aspects of Reid's thought are discussed by Laudan (1970).
30 See, particularly, O. Lodge, *My philosophy representing my views on the many functions of the ether of space* (London, 1933).
31 B. Robinson, *A dissertation on the aether of Sir Isaac Newton* (Dublin, 1743), 122.
32 A. Tucker [E. Search, pseud.], *The light of nature pursued*, 2 vols. (London, 1768), 2:pt. 2, 81.
33 G. Berkeley, *Siris: a chain of philosophical reflections and inquiries concerning the virtues of tar-water, and divers other subjects connected together and arising from one another* (Dublin, 1744), 69–105.
34 This theme recurs in many of the works cited in this chapter and also in L. Dutens, *An inquiry into the origin of the discoveries attributed to the moderns: wherein it is demonstrated, that our most celebrated philosophers have, for the most part, taken what they advance from the works of the ancients; and that many of the important truths in religion were known to the pagan sages* (London, 1769), 181–92. See also Walker (1972); J. E. McGuire and P. O. Rattansi, 'Newton and the "Pipes of Pan" ', *Notes and Records of the Royal Society 21* (1966), 108–43.
35 A. M. Ramsay, *The new Cyropaedia; or, the travels of Cyrus: with a discourse on the theology and mythology of the ancients* (London, 1760), 73–85.
36 Walker (1972), 255–6. Cf. Newton, *Opticks*, 379.
37 R. Barton, *The analogy of divine wisdom, in the material, sensitive, moral, civil, and spiritual system of things, in eight parts* (Dublin, 1750), 62.
38 Ibid., 55–66.
39 J. Hutchinson, *Glory or gravity essential and mechanical*, in *The philosophical and theological works of the late truly learned John Hutchinson, esq.*, 12 vols. (London, 1748–9), 6:21–33.
40 Lucretius, *The nature of things: a didactic poem: translated from the Latin of Titus Lucretius Carus, accompanied with the original text, and illustrated with notes philological and explanatory* [by J. M. Good], 2 vols. (London, 1805), 1:405–7; J. M. Good, *The book of nature*, 3 vols. (London, 1826), 3:1–32.
41 Lucretius, *The nature of things*, 1:lxxxv–xciii.
42 J. Buchanan, *Faith in God and modern atheism compared, in their essential nature, theoretical grounds, and practical influence*, 2 vols. (Edinburgh, 1864), 2:70–134.
43 J. O. de la Mettrie, *Man a machine*, 2nd ed. (London, 1750); P. T. d'Holbach, *The system of nature: or, laws of the moral and physical world*, 2 vols. (Boston, 1889), 1:50, 62–3.
44 S. Clarke, 'A second defence of an argument made use of in a letter to Mr. Dodwell, to prove the immateriality and natural immortality of the soul', in S. Clarke, *The works of Samuel Clarke, D.D.*, 4 vols. (London, 1738), 3:799.
45 O. Chadwick, *The Victorian church*, 2 vols. (London, 1966–70), 2:112–50; A. Gauld, *The founders of psychical research* (London, 1968).

5

The medium and its message: a study of some philosophical controversies about ether

LARRY LAUDAN

*Center for Philosophy of Science, University of Pittsburgh, Pittsburgh, Pennsylvania 15260
USA*

It is by now a commonplace that the emergence of scientific theories has sometimes occasioned extensive philosophical discussion of the conceptual well-foundedness of the ideas on which such theories depend.[1] The eighteenth-century controversies about action at a distance and the nineteenth-century atomic debates are obvious cases in point. It has not, however, been widely appreciated that ether theories during their heyday produced a philosophical discussion at least as profound as, and probably more far-reaching than, those associated with atomism and noncontact action. My aim in this chapter is to explore briefly some of the philosophical aspects of 'the ether debates', with a view to documenting their impact, both on the fortunes of subtle fluid physics and on the nineteenth-century revision of the philosophy of empiricism.

Taking the long view, the ether debates erupted intermittently from Aristotle to Lorentz and Fitzgerald, but the special features of those debates that I shall discuss appear between 1745 and 1850. It is chiefly this period that exhibits a very striking interaction between ethereal physics and empiricist epistemology, an interaction that is the focus of this study. My claim is that during this period the character of this interaction shifted profoundly in ways that were to modify both science and philosophy.

The central theses of this chapter are these:
1. The epistemology prevalent in the second half of the eighteenth century was altogether incompatible with the various ether theories that emerged in the natural philosophy of that period.
2. Some of the early proponents of ethereal explanations chose to abandon or modify that prevailing epistemology so as to provide a philosophical justification for theorising about ether.
3. The modifications so introduced were unconvincing and inadequate,

leaving the scientific status of ether theories very unclear by the beginning of the nineteenth century.
4. The emergence of the optical ether in the early nineteenth century prompted a more radical critique of classical epistemology, a critique that produced some highly innovative and historically influential methodological ideas.

The first phase, 1740–1810

Our story should begin, as any account of Enlightenment epistemology must, by recalling the triumph of Newtonian mechanics and the trenchant inductivism associated with Newton's achievement. As numerous authors have shown, the half century following publication of the *Principia* was marked by a growing antipathy to hypotheses and speculations.[2] Induction and analogical reasoning were all the rage and Newton's doctrine of *verae causae* – adumbrated in dozens of eighteenth-century glosses on his first *regula philosophandi* – was thought to exclude any entity or process not strictly observable.[3] Whether we look to Berkeley and Hume in Britain, to 'sGravesande and Musschenbroek in the Netherlands, or to Condillac and D'Alembert in France, the refrain was similar: Speculative systems and hypotheses were otiose; scientific theories had to deal exclusively with entities that could be observed or measured. For half a century, many natural philosophers sought to develop theories satisfying those demanding strictures; 'moral philosophers' (e.g., Berkeley, Condillac, and Hume), for their part, explored the logical and epistemological ramifications of this new view of the nature of *scientia*.

However, long before epistemologists of science were able to digest these new challenges to the traditional demonstrative ideal of science, scientific developments themselves conspired to produce a significant shift. For, especially during the period from 1745 to 1770, many emerging theories within the sciences moved well beyond the inductive, observational bounds imposed by erstwhile Newtonians. Nowhere is this clearer than with respect to the development of mediumistic or ethereal explanations. In the 1740s alone, there were at least half a dozen major efforts to explain the behaviour of observable bodies by postulating a variety of invisible (and otherwise imperceptible) elastic fluids. In 1745, Bryan Robinson published his *Sir Isaac Newton's account of the aether*. A year later, Benjamin Wilson's *Essay towards an explication of the phaenomena of electricity, deduced from the aether of Sir Isaac Newton* appeared. Of greater moment, Benjamin Franklin developed his account of electricity as a subtle fluid; the Swiss physicist George LeSage articulated an ethereal explanation of gravity and chemical combination; and the highly controversial David Hartley embarked on a programme, culminating in his *Observations on man* (1749), to give a mechanistic theory of mind and perception, whose crucial ingredient was the transmission of vibrations in a subtle fluid or ether through the central nervous system. By the 1760s, the scientific literature abounded with ethereal explanations of heat, light, magnetism, and virtually every other physical process.

Two general points about these developments are especially relevant for our purposes. In the first place, by the 1770s, ethereal or subtle fluid explanations were very widespread among natural philosophers (with the exception, soon to be explained, of many Scottish scientists). Second, such explanations invariably violated the prevailing epistemological and methodological strictures of the age, strictures that, as already noted, would not countenance the use of theoretical or 'inferred' entities to explain natural processes. (After all, an entity that is regarded as in principle unobservable is scarcely consistent with an empiricist epistemology that restricts legitimate knowledge to what can be directly observed.)

Indeed, there was scarcely any domain of scientific theorising in the eighteenth century that left as much scope for speculative hypotheses about unseen agents as did ether theories. As Joseph Priestley remarked in his *History and present state of electricity:*

> Indeed, no other part of the whole compass of philosophy affords so fine a scene for ingenious speculation. Here the imagination may have full play, in conceiving of the manner in which an invisible agent produces an almost infinite variety of visible effects. As the agent is invisible, every philosopher is at liberty to make it whatever he pleases, and ascribe to it such properties and powers as are most convenient for his purpose.[4]

Not a tolerant epistemology at the best of times, classical empiricism (by which I mean the empiricism of Berkeley, Condillac, Hume, and Reid) left no scope for entities like ether. A few natural philosophers of the period failed to perceive the tension between the received epistemology and ethereal theorising. Leonhard Euler, for instance, could simultaneously maintain that the transmission of light depended upon vibrations in an imperceptible medium and insist, in his *Letters to a German princess,* that science should proceed by enumerative induction, eschewing all nonobservable entities. But most scientists and philosophers of the time saw the strain between the emergence of subtle fluid theorising on the one hand and the subscription to naive inductive empiricism on the other. Among this latter group some – such as Thomas Reid – were persuaded that epistemological doctrines took priority over physical theories and thus should be allowed to legislate fluid theories out of the scientific arena. Others, like Hartley and LeSage, saw that option as self-defeating and preferred instead to seek to develop new and more liberal versions of empiricist epistemology that could sanction subtle fluid theories. I want to examine both of these reactions.

The partisans of ether

Among the most persistent, not to say the most notorious, proponents of subtle fluid theories in the last half of the eighteenth century were David Hartley and George LeSage. Hartley foresaw many explanatory roles for the subtle, elastic fluid that he called ether: among them, explaining the transmission of heat, the production of gravity, electricity, and magnetism. Hartley's central concern, however, was to utilise ether, or rather vibrations within ether, to explain a large range of problems about perception, memory, habit, and other activities of the mind. On Hartley's view (which was a detailed elaboration of one of the more speculative conjectures of Newton's *Opticks*),[5] the brain and nervous system are filled with a highly subtle fluid that transmits vibrations from one point in the perceptual system to another. The vibrations in this ether, which are initiated by some external stimuli, subsequently cause the medullary matter composing the nerves and brain to vibrate in ways that are characteristic of the stimuli in question. In his *Observations on man* (1749), Hartley utilised the vibrations in the 'nervous' ether to explain a remarkably divergent range of phenomena, including 'sensible pleasure and pain' (pp. 34–44), sleep (45–55), the generation of simple and complex ideas (56–84), voluntary and involuntary muscular motions (85–114), the sensation of heat (118–25), ulcers (127), paralysis (132–4), taste (151–79), smell (180–90), sight (191–222), hearing (223–38), sexual desire (239–42), memory (374–82), and the passions (368–73). Indeed, most of the five-hundred-odd pages of part 1 of the *Observations on man* constitute a litany of phenomena that are explicable by Hartley's hypothesis of a vibratory ether.

As Hartley was perfectly aware, the whole structure of his argument was radically out of step with the prevailing inductivist temper of the age. Nowhere did Hartley attempt to 'deduce the ether from the phenomena'; nowhere could he use Baconian techniques of eliminative induction to establish the epistemic credentials of his enterprise. Neither could he point to any direct (i.e., noninferential) evidence for the existence of his ubiquitous subtle fluid; hence it is no *vera causa*. Rather, what Hartley had to settle for was a kind of post hoc confirmation. His arguments invariably have the structure of Peircean 'abductions':

Here is a phenomenon x
But if there were an ether, then x
___

(Probably) there is an ether

In short, straightforward hypothetico-deduction was the official methodology of the *Observations on man*.

I need hardly add that therein lay the trouble. For however tolerant later generations were to be about post hoc confirmations and the abductive schema, Hartley's contemporaries – as he knew full well – viewed hypothetical reasoning as inherently fallacious (after all, abduction *is* a form of the fallacy of affirming the consequent). Hartley's primary defence of his procedures involved a stress on the wide range of confirming instances that his theory could lay claim to. He suggested that its broad explanatory scope compensated for the unobservability of its explanatory agents and mitigated its failure to exhibit a traditional inductive warrant. As he remarks early in the *Observations on man:* 'Let us suppose the existence of the aether, with these its properties, to be destitute of all direct evidence, still, *if it serves to explain a great variety of phenomena, it will have an indirect evidence in its favour by this means*'.[6] Hartley likened the search for deep-structural theories like his own to the process of decoding a message. Just as the decypherer's task is to find a code that will render the encoded message intelligible, the natural philosopher's job is to find some hypothesis that will save the phenomena. The latter, like the former, must be content with indirect evidence: 'The decypherer judges himself to approach to the true key, in proportion as he advances in the explanation of the cypher; and this without any direct evidence at all'.[7] But Hartley must have known that this analogy, baldly stated, would not take him very far. If he was to establish that it was scientifically respectable to speak about unobservable entities for which there could be no direct evidence, then he would have to reorient the epistemological convictions of an age which took the view that, where hypotheses were concerned, indirect evidence was no evidence at all.

To that end, he composed a long section of the *Observations* dealing with 'Propositions and the nature of [rational] assent'. The unambiguous aim of that methodological excursus was to show that enumerative and eliminative induction are not the only routes to knowledge. He began the section on a then familiar note, to wit, that the methods of induction and analogy are the soundest methods of inquiry in natural philosophy. Indeed, he went so far as to give an associationist and vibrationist account of the mechanisms whereby repetitions of particular instances of a generalisation reinforce one another so as to habituate us to accept the generalisation of which they are instances.

Apart from the relatively strong assimilation of the method of induction to the calculus of probabilities, Hartley was here covering ground that was familiar territory to his contemporaries.

But Hartley went on to insist that the techniques of induction and analogy do not exhaust the methodological repertoire of the natural philosopher. There is, in addition, the *method of hypothesis*. Apropos of the Newtonian insistence that 'hypotheses have no place in experimental philosophy', Hartley was uncompromising: 'It is in vain', he explained, 'to bid an enquirer form no hypothesis. Every phenomenon will suggest something of this kind'.[8] Those who pretend to make no hypotheses are deceiving themselves and confuse their speculative hypotheses with 'genuine truths . . . from induction and analogy'.[9] Since the mind willy-nilly forms hypotheses when confronted by any phenomenon, it is far better to acknowledge tentative hypotheses than to acquiesce unwittingly in them: 'He that [explicitly] forms hypotheses from the first, and tries them by the facts, soon rejects the most unlikely ones; and, being freed from these, is better qualified for the examination of those that are probable'.[10]

Moreover, Hartley insisted, the examination and testing of hypotheses, even false ones, has the heuristic advantage of leading us quickly to the discovery of new facts about the world that we would be otherwise unlikely to discover:

> The frequent making of hypotheses, and arguing from them synthetically, according to the several variations and combinations of which they are capable, would suggest numerous phenomena, that otherwise escape notice, and lead to *experimenta crucis,* not only in respect of the hypothesis under consideration, but of many others.[11]

But even granting (as some of the most trenchant inductivists were willing to)[12] that hypotheses can be of heuristic value, Hartley was still confronted with the problem of explaining the circumstances under which it is legitimate, contra inductivism, to accept or believe a hypothesis involving unobservable entities. Unless it could be shown that there are some circumstances in which a hypothesis – not generated by inductive methods – warrants acceptance, Hartley's own programme for a hypothetical science of mind would be without foundation. Hartley himself propounded the conundrum:

> But in the theories of chemistry, of manual arts and trades, of medicine, and, in general, of the powers and mutual actions of the small parts of matter, the uncertainties and perplexities are as great, as in any part of science. For the small parts of matter, with their actions, are too minute to be the objects of sight; and we are neither possessed of a detail of the phenomena sufficiently copious and reg-

ular, whereon to ground an [inductive] investigation; nor of a
method of investigation subtle enough to arrive at the subtlety of
nature.[13]

It is disappointing that, after much fanfare, Hartley's defence for believing a speculative hypothesis that explains many phenomena took him no further than the early pages of *Observations on man* had done. Invoking again the cypher analogy, Hartley merely insisted that if a hypothesis is compatible with all the available evidence, then that hypothesis 'has all the same evidence in its favour, that it is possible the key of a cypher can have from its explaining that cypher'.[14] In a nutshell, Hartley's method of hypothesis boils down to the claim that a hypothesis warrants belief if it has a large number of known positive instances and no known negative instances. Confirmed explanatory scope thus functioned for Hartley as the decisive criterion for the acceptability of hypotheses.

Hartley immediately conceded that this criterion does not guarantee that the hypotheses it licenses will be true or even that they will stand up to further testing. They will possess none of the reliability (then) associated with the methods of induction and analogy. But, given the inevitability of hypotheses, what (he seems to ask) is the alternative? As he puts it, 'the best hypothesis which we can form, i.e. the hypothesis which is most conformable to all the phaenomena, will amount to no more than an uncertain conjecture; and yet still it ought to be preferred to all others, as being the best that we can form'.[15]

Not surprisingly, this epistemology carried little weight with most of Hartley's inductivist contemporaries. As they could point out, there were many rival systems of natural philosophy that – after suitable ad hoc modifications – could be reconciled with all the known phenomena. The physics of Descartes, the physiology of Galen, and the astronomy of Ptolemy would all satisfy Hartley's criterion. There was, in Hartley's approach to the epistemology of science, nothing that would discredit the strategy of saving a discarded hypothesis by cosmetic surgery or artificial adjustments to it.[16] As his critics pointed out, the great Newtonian epistemological innovation involved the insistence that 'saving the phenomena' or merely explaining the known data was an insufficient warrant for accepting a theory. Neither Newton nor his followers would quarrel with the view that it was a *necessary* condition of the acceptability of a theory that it must fit all the available data;[17] but they would not brook Hartley's transformation of this plausible *necessary* condition of theoretical adequacy into a *sufficient* condition for theory acceptance. Moreover, it was quite clear to any perceptive reader of Hartley's epistemological writings that they were meant to rationalise his ethereal neurophysiology.

Accepting the former meant acquiescing in the latter, which few were willing to do.

If Hartley chose to take on the inductivists somewhat obliquely, conceding to them that induction and analogy were sound modes of inference, expecting (but not receiving) in return an admission that the method of hypothesis, too, had its use, a more direct frontal attack on the prevailing epistemology came from another partisan of ether, Hartley's Swiss contemporary George LeSage. By his own account, LeSage discovered in 1747 the theory that was to make him alternatively acclaimed and notorious for well over a century. LeSage's approach involved postulating a medium surrounding all bodies. The corpuscles constituting this ether move in all directions and occasionally impact upon the particles constituting observable physical objects. The latter are 'semipermeable' to streams of ethereal particles; that is, most ethereal particles will pass completely through a macroscopic object (chiefly because the volume of its constituent particles is always a very small proportion of the space occupied by the body). Some, however, will collide with particles of the body; when they do, there will be appropriate transfers of momentum, with the ethereal particles rebounding and with the atoms of the body moving in the reverse direction. This kinematic ether was utilised by LeSage to explain a wide diversity of phenomena. In an article in *Mercure de France,* he utilised it to explain weight;[18] in his *Essai de chimie mécanique,*[19] it was invoked to explain many phenomena of chemical affinity; still more significantly, he used this approach in 1764[20] and again in 1784[21] to develop his famous kinematical model of gravitation.

The details of LeSage's ether model need not detain us here.[22] It was sympathetically, if critically, evaluated by Maxwell a century ago, and its mathematical and physical articulation was extensively explored by (among others) Preston, Kelvin, Croll, Farr, George Darwin, and Oliver about the same time. For our purposes a very brief summary of LeSage's model should suffice. LeSage explained gravity by assuming that there are streams of ethereal corpuscles flowing into the world from every direction. As already pointed out, ordinary bodies are highly porous and thus most of these *ultramondain* corpuscles move through a body with no interaction. A few, however, will impact with constituent corpuscles of the body. The result of such impacts is that some ethereal particles reverse their direction, while the body itself has a net force exerted on it by the collisions. In a one-body universe, there would be no resultant motion since there would be an equal number of collisions on all sides of the body. But if we introduce a second large object into this universe, each will act as a partial shield against the ultramundane corpuscles. This will lead to a pressure differential, with each body undergoing fewer collisions on one side than the other; as a result, each body will tend to move towards the other. In this way, the qualitative character of gravitational attraction is attained via contact action rather than action at a distance. The quantitative features of gravity (namely, its relation to the square of the distance and to the masses of the bodies) are explained in LeSage's full-blown *Traité de physique mécanique* (published posthumously by Pierre Prevost).[23] Although many of his predecessors (e.g., Fatio, Daniell Bernoulli) had sought mechanical explanations for gravitational attraction, LeSage's was the only one to emerge from the eighteenth century as a prima facie physically adequate model of gravitational interaction. As James Clerk Maxwell remarked of LeSage's theory: 'Here, then, seems to be a path leading towards an explanation of the law of gravitation, which, if it can be shown to be in other respects consistent with facts, may turn out to be a royal road into the very arcana of science'.[24]

But that reasonably flattering pronouncement by Maxwell is a far cry from the almost universal reaction of LeSage's contemporaries to his ethereal models. Immediately upon publication of his theory (and, since LeSage widely precirculated his ideas, in many cases before publication), it was subjected to a steady stream of abuse. Roger Boscovich called LeSage's system a 'purely arbitrary hypothesis', for which there is no direct proof.[25] The French astronomer Bailly, protesting against LeSage's model because it postulated hidden or unobservable entities, insisted that 'nous ne connoissons la nature que par son extérieur, nous ne pouvons la juger que par celles de ses lois qu'elle nous a manifestées'.[26] During the 1760s, Leonhard Euler wrote several encouraging letters to LeSage about the latter's work. Nonetheless, Euler made it clear that 'je sens encore une trés-grande repugnance pour vos corpuscles ultramondains, et j'aimerois toujours mieux d'avouer mon ignorance sur la cause de la gravité que de recourir à des hypotheses si etranges'.[27]

As these few passages suggest, not only was the reaction to LeSage's model largely negative; the grounds for criticism were generally *epistemological* rather than substantive. LeSage's critics were claiming that hypotheses about unobservable entities were no part of legitimate science. As early as 1755, LeSage complained that the common objection to his system of ultramundane corpuscles was that 'mon explication ne peut être qu'une hypothèse'.[28] In 1770, he worried to a correspondent that unless his system was presented very carefully, 'on le jugeroit sur l'etiquette comme une de ces hypothèses en l'air'.[29]

By 1772, LeSage became convinced that his theory was not getting the hearing it deserved because no one would evaluate it on its scientific merits. Instead, they dismissed it as a mere hypothesis. He claimed it to be an 'almost universal prejudice' of his age that any theory dealing with unobservable entities could not be regarded as genuinely scientific.[30] A decade later, LeSage had become so convinced that he could not get a fair hearing for his views that he withdrew publication of what was to have been his magnum opus. As he bitterly explained:

> Puisque vos physiciens sont si prévenus contre le possibilité d'établir solidment l'existence de mes agens imperceptibles, très-propres cepedant à rendre intelligibles les attractions, affinités, et expansibilités, que constituent à present toute la physique, je suspendrai encore quelque temps la publication des ouvrages que je préparois sur ses agêns.[31]

If the story ended here, it would amount to just one more case of a scientist whose works were suppressed because out of tune with prevailing epistemological fashion. But the LeSage case is more interesting and more important than that, because the difficulties he encountered with his physics prompted him to respond in kind, that is, to articulate a rival methodology to the dominant inductivist one. LeSage did this in a number of philosophical essays, which were designed to show that hypotheses in general and ones involving imperceptible entities in particular may have a sound epistemic rationale. LeSage dealt with this matter at greatest length in a much-quoted essay published posthumously by Prevost. Originally but unsuccessfully intended for publication as an article in the great *Encyclopédie,* the essay was titled 'On the method of hypothesis'.[32] In it, LeSage sought to show that enumerative induction and the method of analogy – the two dominant methods advocated by the ubiquitous Newtonians – were not as foolproof as their partisans claimed and that the method of hypothesis[33] was not so weak as its critics insisted.

As LeSage shrewdly recognized, the core presupposition of the method of enumerative induction is that a clear distinction can be made between what is observable and what is not. The traditional contrast between inductive methods and speculative ones was that the former stayed very close to sensory experience, whereas the latter moved a long way from it, thereby acquiring a much greater degree of uncertainty. LeSage would admit that there is possibly a distinction to be made between what is observable and what is not, but he insisted that rigid and exclusive adherence to the former would produce a very emaciated science. 'Those', he said, 'who disparage the method of hypothesis do not allow us to make conjectures, *except those which follow naturally and immediately from experience'.*[34] Although this view is 'repeated superstitiously', there is no 'precise idea' that can be attached to it. For 'what on earth is an immediate consequence deduced from the observation of a fact?

The existence of that fact and nothing more'.[35] But perhaps what was meant is that claims about the world that are closer to observation are better supported than those that are, as it were, several inferential steps away from sensory particulars. LeSage will have none of this. If, he says, the claims we make 'are hasty, what is gained if they are also immediate?' And if the claims we make are well evidenced, 'what does it matter whether [they are] immediate or as far removed from the phenomena as the last propositions of Euclid are from his axioms?'[36]

LeSage's point is that any form of theorising goes well beyond the available data; so there is no viable distinction to be made between theories that do and those that do not go beyond the evidence. Given that we must extrapolate from the known to the unknown, and that such a process is usually a risky business, what methodological rules can we use to ensure that such inferences are justified? The bulk of LeSage's essay is devoted to that task, and specifically to showing that the method of hypothesis is of greater moment in scientific reasoning than the rival methods of induction and analogy. As LeSage uses the term, 'the method of hypothesis' refers to any procedure that involves a comparison of the logical consequences of a theory with observations. The highest form of proof for a hypothesis would involve showing that all its consequences were true, that it exhibited 'exact correspondence with the phenomena'.[37] But, as he acknowledged, we are rarely if ever in a position to obtain such exhaustive evidence. Failing that, 'if the assumed cause [i.e., the hypothesis] is able to produce *all the presently known features* related to the principal effect, then it will have the highest degree of certitude that we can at present hope for in the circumstances'.[38] LeSage here added an important qualification. Before we accept a hypothesis that can explain all the available evidence, we must be sure that our evidence represents a large sample. Hypotheses that work well when the data are limited frequently break down as the data base is extended. LeSage insisted that our belief in a hypothesis should be a matter of degree: 'The greater the number of facts with which the [hypothesis] agrees, the more faith we should have in it'.[39] Pressing his analysis more closely, he stressed that mere agreement (i.e., logical consistency) between a hypothesis and the evidence is a very weak relation. If a hypothesis is to be solidly confirmed by a piece of evidence, then it must entail that evidential statement and none of its contraries. The greater the specificity with which the hypothesis correctly entails what we observe, the greater the extent to which it will be confirmed by those observations.[40]

The fact that hypotheses are not *generated* by an analysis of the evidence had been, for many of LeSage's contemporaries, a major strike against them. Inductivists, in particular, had stressed that the only legitimate theories are those that arise as generalisations from experience. LeSage replied that so long as we subject our hypotheses to a rigorous process of 'verification', it does not matter how they were generated initially. LeSage was as contemptuous as the inductivists of hypotheses that are not, or cannot be, verified. But he urged that we ought not to confuse the horrors of unverifiable hypotheses with the merits of verified ones. 'Thousands of times it has been said: the abuse of something, however universal, must never be taken as an argument against its legitimate usage'.[41] He maintained that Newton, in rejecting the method of hypothesis, 'never realized that one could utilize a method of research whose pitfalls he had recognized so well!'[42]

LeSage's next ploy consisted of a lengthy demonstration that, Newton's *hypotheses non fingo* notwithstanding, Newton's work in optics and mechanics was permeated by hypotheses and hypothetical reasoning.[43] Similarly, the works of Kepler, Copernicus, and Huygens rested on hypothetical modes of inference.[44]

He conceded that some hypotheses are spurious, singling out in particular the vortex theories of Descartes and his followers. 'The seventeenth century', he sagely observed, 'preferred to acknowledge every hypothesis, however implausible, while our century finds it more convenient to reject them all'.[45] Noting that epistemology, 'too, is subject to the rule of fashion and prejudice', LeSage maintained that there is a coherent middle way, which allows the use of hypotheses so long as strong empirical constraints are put on them. LeSage concluded his essay by pointing out the limited scope of analogical inference, insisting that it is 'inconclusive', 'arbitrary', 'impractical', and, in most cases, parasitic upon the method of hypothesis.[46]

Throughout this essay, as well as LeSage's other methodological writings, there are two levels of motivation. At one level, LeSage was genuinely concerned about epistemological issues in the abstract and, as a philosopher, felt it important to get as clear as he could about the logic of science. But lurking in the background (as in the case of Hartley's discussion) is LeSage the ether theorist, struggling desperately to get a fair hearing for a scientific theory that was being dismissed on what he regarded as flimsy methodological grounds. There is nothing untoward in all of this. We expect, after all, knowledge and the theory of knowledge to be closely intertwined. Nor are the excursions of LeSage and Hartley into epistemology merely self-serving apologiae. They are, of course, that; but they are substantially more than that as well. Chiefly, they are well-reasoned attempts to articulate and defend a hypothetico-deductive methodology in the face of inductivist criticisms.

There is another methodological theme common to the work of Hartley and LeSage that sets them apart from most of their contemporaries: a commitment to the progressive character of science. Throughout the earlier history of epistemology, the prevailing view was that putative scientific theories or doctrines were to be judged as true or false *simpliciter* and that the authentic methodology of science should be one that would produce true theories more or less immediately; provided, of course, that appropriate rules of inquiry were obeyed. Hartley and LeSage both protested against this all-or-nothing view of scientific theories. Acknowledging that their own theories might be false (because they were at best only probable), they stressed the approximative character of scientific inquiry. They insisted that theories, once promulgated, could be amended and improved to bring them into ever closer agreement with their objects. To illustrate the point, both likened the development of scientific theories to certain mathematical methods of approximation (in fact, both singled out the rule of false position and the Newtonian technique for approximating roots as examples). As Hartley remarked:

> Here a first position is obtained, which, though not accurate, approaches, however, to the truth. From this, applied to the equation, a second position is deduced, which approaches nearer to the truth than the first . . . Now this is indeed the way, in which all advances in science are carried on.[47]

LeSage took the process of long division as a paradigm case of successful approximation.[48] Each step in the process brings us another digit closer to finding the true quotient. The testing and correction of hypotheses against experiment is, he maintained, a suitable parallel.

This ploy proved very useful for LeSage and Hartley. Confronted with the inductivists' insistence that the method of hypothesis is inconclusive, they could grant the point; confronted by the claim that their specific ethereal models might be false, they could concede the possibility. But as they saw it, neither the fallibility of the method of hypothesis nor the falsity of some theories produced by it need force one to the conclusion that it has no role to play in science. On the contrary, once one admits that science is approximative and self-corrective, it becomes possible to envisage a theory that is both false and an important step forward.

Ultimately, in fact, the epistemological views of these two thinkers turned out to be significantly more influential than the ether models they developed (as we shall see). But before we turn to look at the later fortunes of both the method of hypothesis and the ether hypothesis, we need to examine the opposition more closely.

The early critics of ether and its epistemology

Although ethereal explanations of electricity, magnetism, gravity, and heat became increasingly frequent and respectable in the course of the second half of the eighteenth century, there were still many natural philosophers who refused to countenance them. This was particularly true in Scotland, where, the Scottish 'enlightenment' notwithstanding, many of the major philosophers or scientists were unwilling to acknowledge the importance of such theories. If ethers were generally regarded by the Scots as unsavoury, Hartley's nervous ether was reserved for special abuse. In the early decades of the nineteenth century the Scottish philosopher Thomas Brown noted, with some pride, that 'it is chiefly in the southern part of the island that the hypothesis of Dr. Hartley has met with followers'.[49] What Brown says about Hartley's theories can be duplicated for almost all the other major ethereal doctrines of the late eighteenth century; they rarely made it past Hadrian's Wall. Many of Scotland's major natural philosophers (including Robison, Leslie, Reid, and Hutton) rejected the imperceptible fluid hypotheses that were, by the 1770s, being widely discussed (and, in some cases, widely accepted) in England and throughout much of Continental Europe.

The primary reason for opposition to ether theories was the widespread acceptance among Scottish philosophers and scientists of a trenchant inductivism and empiricism, according to which speculative hypotheses and imperceptible entities were inconsistent with the search for reliable science. This linking of opposition to ether theories with an inductivist philosophy is neatly summed up in the 'Aether' article for the first edition of the *Encyclopaedia Britannica* (1771), which, in spite of its title, was a predominantly Scottish production. Barely was *ether* defined ('the name of an imaginary fluid, supposed by several authors . . . to be the cause . . . of every phenomenon in nature')[50] before the author of the article launched into a virulent attack on the method of hypothesis and a spirited defence of the inductivism of Newton and Bacon: 'Before the method of philosophising by induction was known, the hypotheses of philosophers were wild, fanciful, ridiculous. They had recourse to aether, occult qualities, and other imaginary causes'.[51] The article insisted that 'the way of conjecture . . . will never lead any man to truth'.[52] These passages from the *Britannica* reflect a thoroughgoing inductivism that influenced much Scottish writing on science of the period. Most of it sprang from the very influential work of Thomas Reid, leader of the so-called commonsense philosophers. As I have shown elsewhere, Reid's works are replete with abusive attacks on the method of hypothesis.[53]

What is important for our purposes is the explicit linkage between Reid's repudiation of Hartley's ethereal speculations and his attack upon the method that undergirds them. In his *Essays on the intellectual powers of man* (1785), Reid discussed Hartley's vibratory hypothesis at length. In the course of rejecting Hartley's 'hypotheses concerning the nerves and brain',[54] Reid embarked on the lengthiest methodological discussion in his opus. Reid quickly perceived that Hartley's *Observations on man* ran directly counter to the Newtonian inductivism that Reid himself espoused. He noted, significantly, that the epistemological chapter of the *Observations on man* was written to justify the methodology utilised in Hartley's research: 'Having first deviated from [Newton's] method in his practice, [Hartley] is brought at last to justify this deviation in theory'.[55] Reid claimed that Hartley was the only author he knew who rejected the principles of Newton's inductivism, and 'Dr. Hartley is the only author I have met with who reasons against them, and has taken pains to find out arguments in defense of the exploded method of hypothesis'.[56] It seems natural to infer, then, that most of Reid's methodological tirades against hypotheses were directed chiefly at Hartley, since he alone, in Reid's view, had criticised 'the true method of philosophizing'.

That 'true method', so far as Reid was concerned, was some form of *enumerative* induction.[57] Reid nowhere spelled out the rules of his form of induction (as even his followers had to concede),[58] but some of its features are clear. As he wrote in the *Intellectual powers of man:* 'The true method of philosophizing is this: From real facts, ascertained by observation and experiments, to collect by just induction the laws of Nature, and to apply the laws so discovered, to account for the phenomena of Nature'.[59] The nub of the issue concerns the kinds of inductive generalisations that we can perform on 'observations and experiments'. Reid was adamant on two issues at this stage: (1) that any entities postulated in a putative law should really exist 'and not be barely conjectured to exist without proof'; and (2) *all* the known deductive consequences of the law must be true.[60]

Condition (1), which Reid intended as a gloss on Newton's idea of *verae causae,* was extensively discussed, both in his published work and in his correspondence.[61] More often than not, Reid's first condition amounted to the rule that the scientist is allowed to postulate *only those entities that are observable*. Ethers and other imperceptible fluids are thus, by their very nature, disqualified from legitimate scientific status.

It is sorely tempting to side epistemologically with Hartley and LeSage against Reid. After all, the method of enumerative induction is not rich enough to build science; speculations and conjectures are inevitable. It does count in favour of a theory that it can explain a wide variety of phenomena, even if that theory postulates imperceptible entities. But to see the debate solely in these terms is misleading. In large part, the epistemological issue at stake between the inductivists and the hypotheticalists in the eighteenth century is simply this: Does a confirming instance of a theory automatically count as a ground for accepting or believing the theory? Both Hartley and LeSage insisted that any and every confirming instance provides evidence for the theory that entails it; accumulate enough such instances and the theory becomes credible. Thomas Reid, like such later philosophers of science as Whewell, Peirce, the Bayesians, and Popper, maintained that 'mere' confirming instances are not enough, that some additional demand must be met before we can legitimately say that true deductive consequences of a theory count towards warranted belief in that theory.

What motivates this conviction in Reid's case is a sound intuition that many unsavoury theories have some true consequences. If we were to regard every theory with some true consequences as well established, then we could never judge one theory to have stronger evidence than another, since 'there never was an hypothesis invented by an ingenious man which has not this [kind of] evidence in its favor. The vortices of DesCartes, the sylphs and gnomes of Mr. Pope, serve to account for a great variety of phenomena'.[62] Reid believed that a theory should not be regarded as confirmed or well established merely because it was sufficient to save the appearances. Of course, he did not quarrel with the view that legitimate theories must be compatible with the available evidence. But what Hartley and LeSage were willing to regard as a *sufficient* condition for an acceptable theory Reid viewed as a *necessary* but not sufficient condition.[63]

But in a classic case of babies and bath water, Reid's requirement went too far. His very narrow observational construal of the grounds for warrantedly asserting the existence of a thing left him completely unable to give an account of the success of the many deep-structural theories of his time. Because it demanded too much, his epistemology was altogether unable to come to grips with the contemporary theoretical sciences. But if Reid and the inductivists demanded too much, Hartley and LeSage ran the risk of requiring too little. After all, many vacuous explanations – witness the classic *virtus dormitivus* in Molière – have true deductive consequences. Neither Hartley nor LeSage provided any plausible criterion for distinguishing vacuously true theories from legitimate scientific ones. In the absence of such a distinction, they had no grounds for claiming that their theories should be accepted in lieu of a multitude of equally well-confirmed but vacuous ones.

In sum, by the late eighteenth century, neither the inductivists nor the hypothetico-deductivists had yet constructed a plausible epistemology of science; equally, the fortunes of ether theories were still unsettled, precisely because it was unclear whether there was genuine evidence for them. All this was to change profoundly in the next half century. Philosophers of science would articulate new and more detailed criteria for determining which deductive consequences of a theory were genuinely confirmatory, and ether theories, at least certain ether theories, would pass these criteria with flying colours.

The second phase, 1820–1840

The early nineteenth-century debate about imperceptible fluids and the methods for establishing their existence focused chiefly upon the wave theory of light (with its seemingly attendant commitment to a luminiferous ether). On the epistemological side, most of the interest centered around the emergence of a new methodological criterion for evaluating hypotheses. In brief, this criterion, which was nowhere prominent in the late eighteenth-century debates about the methodological credentials of subtle fluids, amounted to the claim that a hypothesis that successfully predicts future states of affairs (particularly if those states are 'surprising' ones), or that explains phenomena it was not designed to explain, acquires thereby a legitimacy not possessed by hypotheses that merely explain what is already known generally. The major figures in this part of the story are Herschel, Whewell, and Mill. To put the matter concisely, John Herschel accepted the new criterion, saw that it provided a rationale for the wave theory of light, but did not recognise how that criterion threatened the traditional inductivist programme to which he was committed. Whewell accepted the new criterion and used it both to defend the wave theory of light and to attack traditional inductive procedures. Finally, John Stuart Mill, perceiving that the new criterion undermined induction, repudiated the former and, along with it, the vibratory theory of light. It is this set of interconnections with which I shall be concerned in this section.

The wave theory of light was revived chiefly, of course, by Thomas Young and Augustin Fresnel in the early years of the nineteenth century. Through the first three decades of that century, opinion on the merits of the wave or the corpuscular theory of light was very divided among natural philosophers.[64] If the wave theory could provide a more convincing account of interference than the corpuscular theory, the latter seemed better able to explain problems of stellar abberation and double refraction. (Some participants in the ether debates, such as Comte, actually believed that the two theories were observationally equivalent and that it was a matter of complete indifference which theory one utilised.[65])

By the late 1820s, however, the balance of opinion was shifting perceptibly towards the undulatory theory. There were many factors responsible, but most important among them was the Fresnel–Poisson experiment. The idea for the experiment arose in response to an essay written in 1816 by Fresnel on the nature of diffraction. In this paper, Fresnel elaborated the wave theory of light and applied it to the explanation of diffraction phenomena. Poisson, a confirmed corpuscularian and member of a panel refereeing Fresnel's paper, observed that, according to the analysis of light that Fresnel was using, it would follow that the centre of the shadow of a circular disc would exhibit a bright spot. This predicted result was highly unlikely; it contradicted both the corpuscular theory and the scientists' intuitive sense of what was 'natural'. Indeed, the fact that the wave theory possessed this bizarre consequence was seen, prior to performing the experiment, as a kind of reductio ad absurdum of it. But when the appropriate tests were performed, the wave theory was vindicated by an exact concordance between what it predicted and the observed results.

There are two obvious methodological construals of the outcome of this experiment. On one interpretation, the experiment functions as a Baconian *experimentum crucis,* proving the falsity of the corpuscular theory and the truth of the wave theory. (Interestingly, although precisely this construal was given to the later Foucault experiments on the speed of light in water and air, this was not the dominant interpretation of this earlier result.) On another interpretation, widely adopted at the time, the disc experiment can be viewed as providing convincing evidence for the wave theory by virtue of its successful prediction of a surprising (i.e., unexpected on the background knowledge) observational effect. The logic that undergirds the former interpretation is the familiar logic of eliminative induction. But what provides the epistemic rationale for the latter interpretation is a set of developments within the methodology of hypothesis evaluation.

As we have seen, throughout the eighteenth century, proponents of the method of hypothesis pointed to post hoc (and, according to the critics of that method, presumably ad hoc) explanations of known phenomena as the chief vehicle whereby a hypothesis proves its mettle. But none of the early proponents of the method of hypothesis could show what distinguished arbitrary and vacuous hypotheses from genuine and worthy ones, since both classes possessed large sets of post hoc confirming instances.

During the 1820s and 1830s, proponents of the method of hypothesis articulated machinery for distinguishing 'artificial' hypotheses from legitimate ones. Specifically, they insisted that a proper hypothesis is one that not only explains what is already known, but also can be extended beyond the initial range of phenomena it was designed to explain. Particularly if the hypothesis can predict results that are unusual or surprising, the hypothesis loses its artificiality and becomes a legitimate contender for rational belief. What is involved in this modification of the method of hypothesis (a modification especially prominent in the work of Herschel and Whewell) is *nothing less than a redefinition of what constitutes evidence.* In stressing that a hypothesis must establish its credentials by going beyond its initial data base, Herschel and Whewell defined a promising *via media* between the earlier extremes of Hartley and LeSage, on the one hand, and Reid and the other inductivists, on the other. By claiming that nothing can be arbitrary about a hypothesis that is successfully tested against a body of evidence independent of the circumstances that the hypothesis was invented to explain in the first place, they thus defused the charge of arbitrariness that was traditionally directed against the method of hypothesis.

Given the familiarity of this idea to a modern reader, it is easy to underestimate how significant a shift in the history of methodology it represents.[66] Part of its importance lies in the stress it puts on testing claims against the unknown rather than the known. Before the method of hypothesis could be plausibly viewed as anything more than the logical fallacy which it had often been considered, something new was called for, something that would separate serious and legitimate hypotheses from bogus or specious ones. What suited the bill, at least so far as the 1830s were concerned, was what I shall call *the requirement of independent or collateral support*. In brief, this requirement amounted to the demand that before a hypothesis was credible, it had to explain (or predict) states of affairs significantly different from those that it was initially invented to explain. Evidence of this kind might come from one of two sources: either a surprising prediction of unknown effects or a successful explanation of phenomena that were already known but that did not serve as the original base for the formulation of the hypothesis.

This methodological requirement of independent support ought not to be confused with the earlier empiricist requirement that theories must involve *verae causae*. That earlier demand had nothing to do with the capacity of a theory to predict surprising results; it insisted, rather, that the entities postulated in a theory had to be directly observable or directly 'inferred from the phenomena'. In short, the methodological tradition of *verae causae* had rested upon a rigid distinction between directly observable entities and not-directly observable ones, endorsing the former and eschewing the latter. By contrast, what I am calling the requirement of independent support is indifferent to the question whether theoretical entities are observable. It focuses, rather, on the epistemic features of the sentences that can be deduced from a theory.

A theory like LeSage's gravorific ether, even if it had been entirely successful in its explanatory ambitions, did not seem to possess independent support in the sense defined. More to the point, no epistemologist in the eighteenth century would have been impressed if it had, for the notion of independent support in the sense under discussion here was very much a product of the early nineteenth century. By invoking this requirement, as we shall see, proponents of the optical ether were able to argue – as their eighteenth-century ethereal precursors were not – that there was some very impressive evidence available for a luminiferous ether, evidence that went well beyond the ability of that hypothesis merely 'to save the phenomena'. The Fresnel–Poisson circular disc experiment (a confirmed prediction), as well as the successful extension of the wave theory to polarisation, dipolarisation, and double refraction, betokened a degree of collateral support for the optical ether that earlier ethereal doctrines did not exhibit.

This set of issues was stressed with particular emphasis by both John Herschel and William Whewell. Herschel insisted that we cannot reasonably expect a theory to be a reliable predictor in the future unless it has also been so in the past. Unless we have seen that a theory enables 'us to extend our views beyond the circle of instances from which it is obtained', then 'we cannot rely on it'.[67] Before we accept any theory we must try 'extending its application to cases not originally contemplated . . . and pushing the application of [it] to extreme cases'.[68] Although Herschel frequently enunciated this requirement, and saw it as a vehicle for protecting us from 'the unrestrained exercise of imagination . . . arbitrary principles . . . [and] mere fanciful causes',[69] he did not discuss its rationale at any length.

What he did do, however, was to invoke this requirement repeatedly to show that the hypothesis of the optical ether was a sound one. Herschel pointed out that Young's wave theory, originally developed to explain reflection, refraction, and interference, was eventually applied with much success to the explanation of polarisation and double refraction.[70] He pointed out not only that (Fresnel's version of) the wave theory explained 'perhaps the greatest variety of facts that have ever yet been arranged under one general head'[71] but that Fresnel's theory also predicted 'a *fact* which had never been observed . . . and all opinion was against it'.[72] The confirmation of this surprising prediction did much, in Herschel's view, to establish the 'probability' of the undulatory hypothesis.

Like Herschel, Whewell found it necessary to supplement the weak demands of the traditional method of hypothesis by further constraints. He elaborated these in a lengthy section of the *Philosophy of the inductive sciences* (1840) devoted to 'Tests of hypotheses'. With LeSage and Hartley, Whewell insisted on the minimal condition: 'The hypotheses which we accept ought to explain phenomena we have observed'.[73] More precisely, he stipulated that every hypothesis had to be 'consistent with *all* the observed facts'.[74] But unlike Hartley and LeSage, who saw in such a requirement a sufficient condition for adequacy, Whewell argued that hypotheses 'ought to do more than this: our hypotheses ought [successfully] to *fortel* [sic] phenomena which have not yet been observed'.[75] At a minimum, these predictions should be borne out by tests of the hypothesis against phenomena 'of the same kind as those which the hypothesis was invented to explain'.[76]

The rationale for this rather more exacting requirement was precisely that it dissipates the air of arbitrariness surrounding a hypothesis whose only known instances are those used in its generation: 'Men cannot help believing that the laws laid down by discoverers must be in a great measure identical with the real laws of nature, when the discoverers thus determine effects beforehand in the same manner in which nature herself determines them when the occasion occurs'.[77]

Successful prediction of effects similar to those already known does much to increase our confidence in a hypothesis. 'But the evidence in favour of our induction is of a much higher and more forcible character when it enables us to explain and determine [i.e., predict] cases of a kind different from those which were contemplated in the formation of our hypothesis'.[78] Whewell's technical term for this particular mode of evidencing (which involves testing a hypothesis against types of processes different *in kind* from those it was devised to explain) is the *consilience of inductions*. In his view, this is the most impressive type of evidence that theories can possess. Whewell's stress on the special confirmatory value of successful predictions, and the contrast it marks with the earlier eighteenth-century discussions of the method of hypothesis, comes out very clearly in his unusual utilisation of the decypering analogy. We saw this analogy already in Hartley (and it was used before him by Descartes and Boyle, inter alia). In its pre-Whewellian form, the analogy had suggested that if a certain hypothetical assignment of letters to an encoded cypher produces an intelligible message, then this constitutes evidence that the hypothetical assignment is correct. In this version of the analogy, it is assumed that the entire cypher is known in advance to the decoder. In Whewell's version of the analogy, however, a portion of the cypher is initially concealed from the decoder and the test of his decoding is whether he can predict the character of the concealed cypher.[79] As Whewell's variation on this traditional analogy makes clear, his concern is to assign differential weights to the confirming instances of a theory.

Barely had Whewell defined this notion of consilience before he cited the wave theory of light *in extenso* as one of the few theories to have passed this demanding test. In quick succession, he ticked off the explanatory successes of the wave theory: reflection, refraction, colours of thin plates, polarisation, double refraction, dipolarisation, and circular polarisation. By contrast, the emission theory exhibited 'what we may consider the natural course of things in the career of a false theory'.[80] It could well enough explain 'the phenomena which it was at first contrived to meet; but every new class of facts requires a new supposition . . . as observation goes on, these incoherent appendages accumulate, till they overwhelm and upset the original framework'.[81]

So impressed was Whewell by the predictive successes of the wave theory that he used it as one of his two paradigm cases of exemplary theory development (Newtonian gravitational theory was the other), and his elaborate 'Inductive table of optics' culminated in the wave theory. Even in his earlier *History of the inductive sciences* (1837), Whewell saw the wave theory as the optical equivalent of Newtonian mechanics. After a lengthy discussion of what he then called 'the undulatory theory', he remarked:

> We have been desirous of showing that the *type* of this progress in the histories of the two great sciences, Physical Astronomy and Physical Optics, is the same. In both we have many *Laws of Phenomena* detected and accumulated by astute and inventive men; we have *Preludial* guesses which touch the true theory . . . finally, we have the *Epoch* when this true theory . . . is recommended by its fully explaining what it was first meant to explain, and confirmed by its explaining what it was *not* meant to explain.[82]

This passage neatly foreshadows Whewell's philosophy of science (with its emphasis on the independent support requirement) and the key role that the wave theory played in its formulation.

The Mill–Whewell debate

But what of the opposition? As we have seen, both Herschel and Whewell based their endorsement of the wave theory on its satisfaction of an innovative and highly controversial methodological demand (i.e., the requirement of independent support). If my account of the connection between views towards optics and this methodological rule is correct, we should expect that those who did not accept the Herschel–Whewell methodology would not share their enthusiasm for the undulatory theory. Confirmation for this expectation is ready at hand in the works of John Stuart Mill. Mill was, of course, the arch-foe of Whewell during the 1840s and 1850s. Their respective philosophies of science exhibited divergences at almost every major point. Not the least of these differences was the disagreement of the two men about the relative confirmational value of different instances of a theory. As we have seen, Whewell maintained that a theory was better confirmed by predictive instances or by explaining phenomena it was not originally devised to explain than it was by explaining phenomena it had been devised to explain. By contrast, Mill insisted that predictive successes, every bit as much as post hoc explanatory ones, were highly inconclusive; both, in Mill's view, if taken as grounds for asserting the theories that achieved them, were highly fallacious.

Significantly, this controversy emerges in Mill's *System of logic* (1843) with specific reference to the luminiferous ether. Mill's central point was that no number of confirming instances of a theory can establish it conclusively, chiefly 'because we cannot have, in the case of such an hypothesis (viz., the optical ether), the assurance that if the hypothesis be false it must lead to results at variance with true facts'.[83] The fact that the wave theory of light 'accounts for all the *known* phenomena' does not warrant the view that it is 'probably true'.[84] He then turned to consider Whewell's (and Herschel's) claim that it is chiefly the predictive successes of the wave theory that render it likely. 'It seems to be thought', Mill observed, 'that an hypothesis of the sort in question is entitled to a more favourable reception, if, besides accounting for all the facts previously known, it has led to the anticipation and prediction of others which experience afterwards verified'.[85]

Mill was scathing in his insistence that such a view flagrantly confounds the psychology of surprise with the methodology of support. Referring specifically to one of the predictions of the wave theory, he said:

> Such predictions and their fulfillment are, indeed, well calculated to impress the uninformed,[86] whose faith in science rests solely on similar coincidences between its prophecies and what comes to pass. But it is strange that any considerable stress should be laid upon such a coincidence by persons of scientific attainments. If the laws of the propagation of light accord with those of the vibrations of an elastic fluid in as many respects as is necessary to make the hypothesis afford a correct expression of all or most of the phenomena known at the time, it is nothing strange that they should accord with each other in one respect more.[87]

Mill was not taking exception to Whewell's psychological observation that many people are impressed by a theory that successfully makes surprising predictions. What he was calling for was a logical or epistemological account of why we should regard such instances as being of a privileged, probative character. Like Popper a century later, Whewell did not ultimately meet this challenge; he never showed why a theory's novel predictions should count for so much more, in terms of its epistemic appraisal, than its successes at explaining what it was devised to explain.

Nonetheless, Whewell tried to restate his case in his *Of induction, with especial reference to Mr. J. S. Mill's System of Logic* (1849). Whewell rightly summarised Mill's attack on him by saying that it amounted to the traditional inductivist charge that one ought not allow 'hypotheses to be established, merely in virtue of the accordance of their results with the phenomena'.[88]

Whewell reiterated that his was not merely the old-fashioned method of hypothesis, for he added the demands of independent support and consilience. As for Mill's charge that successful predictions should impress only the 'uninformed', Whewell insisted that 'most scientific thinkers . . . have allowed the coincidence of results predicted by theory with fact afterwards observed, to produce the strongest effects upon their conviction'.[89] (In the same passage he referred to 'the curiously felicitous proofs of the undulatory theory of light'.) But most of Whewell's reply consisted chiefly of pious hand waving rather than cogent arguments; he never satisfactorily met Mill's challenge to produce a plausible epistemological rationale for the requirement of independent support.

At another level, however, Mill had perhaps missed the point. It is one thing to stress, as he rightly did contra Whewell, that one or two surprising, but confirmed, predictions do not prove the theory that produces them. But if I am right, the motivation for introducing the predictive requirement was not to transform the method of hypothesis into a proof technique. Rather, the concern had been to find some way of reducing the arbitrariness and the ad hoc nature of hypotheses. So long as the only confirming instances that a hypothesis could claim were those used in its generation, there was no reason to expect that applications of it to further instances would be successful. After all, the hypothesis might simply have fastened on some noncausal or nonnomic accidents of the cases so far surveyed, and generalised these into a universal theory. But, as Herschel and Whewell observed, if the hypothesis can be successfully extended to cases or even to domains that were not used in its development, then it can no longer be claimed that the reliability of the hypothesis is limited to the phenomena that it was devised to explain.[90]

Conclusion and postscript

By the 1850s, the wave theory of light (and the associated hypothesis of a luminiferous ether) enjoyed a degree of acceptance among natural philosophers that none of the subtle fluid ethers of the eighteenth century had possessed. I have tried to show in this chapter that the different receptions afforded to the two sets of theories are to be explained primarily by the wave theory's possession of certain epistemic or methodological features not present in the earlier ethers. But the moral of the tale is not merely to be found in this difference; for even if earlier ethers had been predictive (in the sense indicated here), it is not clear that they would have been any more widely accepted than they were. What was needed was a shift not only in physics but in epistemology as well, so that independent support could be recognised as a decided epistemic virtue. That shift came (I have argued) in the 1830s, provoked in part by the wave theory itself, which served as an epistemological archetype for such philosophers as Herschel and Whewell. There is, of course, a circularity here. But, far from being of the vicious variety, it reflects the kind of mutual dependence between theory and praxis that has always characterised science and philosophy at their best.[91]

There remain many philosophers of science and theorists of scientific change who, though granting that substantive theories about the world do change, nonetheless adhere to the view that the canons of legitimate scientific inference are perennial and unchanging. (Included here are thinkers as diverse as Popper, Nagel, Carnap, Hesse, and Lakatos, among others.) The case we have before us stands as a vivid refutation of their claim that scientific standards of theory evaluation are immutable. It simply cannot be denied that, prior to the early nineteenth century, the ability of a theory to make successful, surprising predictions was no sine qua non for its acceptability; nor can it be denied that by the turn of the twentieth century, the requirement of predictivity was a commonplace in both scientific and philosophical circles.[92] Such a profound shift is irreconcilable with any philosophical dogma to the effect that scientific methodology has a fixed character.

The case before us poses an equally acute challenge to that group of relativists – associated with the work of Thomas Kuhn – who believe that new methods and new standards are paradigm-specific (and who argue that standards are retained or are rejected on the strength of the specific paradigms with which they are initially associated). Although the predictivity requirement may well have been ushered in by the wave theory of light (as I have argued here), it soon acquired a life of its own that dissociated its fortunes from the fate of the particular physical paradigm that had brought it to the fore.

Acknowledgments

The author is grateful to the National Science Foundation, which supported part of this research, and to the Librarian of the University of Geneva Library.

Notes

1 For a detailed discussion of the rationale for such interactions, see L. Laudan, *Progress and its problems* (London, 1977), chap. 2.
2 See, for instance, Laudan (1970); Cohen (1956).
3 That the prevailing eighteenth-century interpretation of Newton's first rule involves the demand that all theories restrict themselves to purely observable entities is slightly a matter of conjecture. What can be said with some confidence is that British and French glosses on Newton's *regulae* generally construe it in this way. (For a

typical and very influential discussion of Newton's first rule, see W. Hamilton (ed.) *The philosophical works of Thomas Reid*, 6th ed., 2 vols. Edinburgh, 1863), 1:57, 236, 261, 271–2.
4 J. Priestley, *History and present state of electricity*, 3rd ed., 2 vols. (London, 1775), 2:16.
5 See query 16 to Newton's *Opticks*, 2nd ed. (London, 1717).
6 D. Hartley, *Observations on man, his frame, his duty, and his expectations*, 2nd ed., 2 vols. (London, 1791), 1:15. A very similar methodological argument had been made two years earlier by Bryan Robinson: 'This *Aether* being a very general material Cause, without any Objection appearing against it from the Phaenomena, no Doubt can be made of its Existence: For by how much the more general any cause is, by so much the stronger is the Reason for allowing its Existence. The Aether is a much more general Cause then our Air: And on that Account, the Evidence from the Phaenomena, is much stronger in Favour of the Existence of the *Aether*, than it is in Favour of the Existence of the Air' [*A dissertation on the aether of Sir Isaac Newton* (London, 1747), preface, n.p.]
7 Hartley, *Observations*, 1:16. This decyphering analogy has a long prehistory among earlier proponents of the method of hypothesis. It can be found, for instance, in Descartes (*Oeuvres*, eds. C. Adam and P. Tannery, 12 vols. [Paris, 1897–1910], 10:323) and Boyle (Royal Society, Boyle papers, vol. 9, f. 63), among other seventeenth-century writers. As I shall show, it continued to be used by methodologists for well over a century after Hartley.
8 Hartley, *Observations*, 1:346.
9 Ibid.
10 Ibid.
11 Ibid., 347.
12 Many inductivists distinguished between what have subsequently been called the contexts of discovery and of justification. They were quite prepared to grant that hypothetical methods were useful for the former, but insisted that they had no role in the latter. As Thomas Reid succinctly put it, 'Let hypotheses . . . suggest experiments, or direct our inquiries; but let just induction alone govern our belief'. *Works*, 1:251.
13 Hartley, *Observations*, 1:364.
14 Ibid., 350.
15 Ibid., 341.
16 Indeed, Hartley even seems to endorse such a course of action. If, he says, our suppositions and hypotheses 'do not answer in some tolerable measure [to the real phenomena, we ought] to reject them at once; or, if they do, to add, expurge, correct, and improve, till we have brought the hypothesis as near as we can to an agreement with nature'. *Observations*, 1:345.
17 Indeed, Newton's first *regula philosophandi* insisted that theories must be 'sufficient to explain the appearances'.
18 [G. LeSage], 'Lettre à un academicien', *Mercure de France* (May 1756).
19 Published in Paris in 1758.
20 G. L. LeSage, 'Loi, qui comprend, malgré sa simplicité, toutes les attractions . . .', *Le Journal des Sçavans* (April 1764), 230–4. He wrote later papers on this same topic in the *Journal des Beaux-arts* (Nov. 1772) and (Feb. 1773) and in the *Journal de Physique* (Nov. 1773).
21 G. LeSage, 'Lucrèce Newtonien', in *Mémoires de l'Académie Royale des Sciences et Belles-lettres de Berlin* (Berlin, 1784), 1–28; reprinted in P. Prevost, *Notice de la vie et des écrits de George-Louis LeSage* (Geneva, 1805), 561–604.

22 Readers seeking details about LeSage's gravitational ether should consult either LeSage, 'Lucrèce', or Aronson (1964).
23 Cf. P. Prevost, *Deux traités de physique mécanique* (Geneva, 1818).
24 W. Niven (ed.), *Scientific papers of James Clerk Maxwell* (London, 1890), 2:474.
25 Prevost, *Notice,* 358.
26 Ibid., 300.
27 Ever an Enlightenment liberal, Euler added, 'Mais j'accorde très volontiers cette liberté à d'autres'. Euler to LeSage, 8 Sept. 1765, University of Geneva Library: MS Suppl. 512, f.314^r. (LeSage's transcription of Euler's letter is MS fr. 2063, f.141^r.)
28 Prevost, *Notice,* 464–5.
29 Ibid., 237.
30 He spoke of 'la prétendue impossibilité d'établir solidement un système, qui roule sur les objects essentiellement imperceptibles'. Ibid., 264).
31 Ibid., 242.
32 The full text of the essay was published in P. Prevost, *Essais de philosophie* (Geneva, 1804), 2:258 ff. The original can be found in the University of Geneva Library, Ms fr. 2019(2). Because there are some (largely minor) discrepancies between the printed version and the original, all my quotations shall be from the latter. I shall give in parentheses references to the appropriate section in Prevost's text.
33 By the term *method of hypothesis,* LeSage meant the view that science 'is conducted by the method of trial and error . . . by gropings followed by verification, by hypotheses which are then confirmed by their agreement with the phenomena'. University of Geneva Library, MS fr. 2019(2) (§5).
34 Ibid. (§26).
35 Ibid.
36 Ibid.
37 Ibid. (§7). Joseph Priestley would similarly insist that if a scientist can frame his theory so as to suit all the facts, 'then it has all the evidence of truth that the nature of things can admit'. Priestley, *History,* 16. For allusions to the very early history of this principle, see L. Laudan, 'Ex-Huming Hacking', *Erkenntnis 13* (1978), 417–35.
38 University of Geneva Library, MS fr. 2019(2) (§7) (my italics).
39 Ibid.(§15).
40 Cf. Ibid. LeSage even seemed to think, as did Hartley, that with a sufficiently large number of confirming instances, the hypothesis becomes virtually certain. As he wrote in his *Principes généraux de la téléologie:* 'Plus les phénomènes sont nombreux et plus la précision est grande; plus aussi ils jugent avec assurance qu'il ne sauroit y avoir d'autre hypothèse sur le même sujet, qui ait les mêmes avantages'. Prevost, *Notice,* 529–30.
41 University of Geneva Library, MS fr. 2019(2) (§18).
42 Ibid. (§19).
43 He claimed that 'almost all that the first two books of his *Principia* [i.e., Newton's] contain . . . is nothing more than a collection' of 'curious hypotheses'. Ibid. (§20).
44 Ibid. (§§23–5).
45 Ibid. (§29). Lest the wary reader may think LeSage was painting an exaggerated picture of the Newtonians' aversion to hypotheses, it is worth saying that his view is borne out by many of his contemporaries. Thus the *Encyclopédie* (in the article 'Hypothèse' observed that Newton 'et sur-tout ses disciples' were very opposed to hypotheses, regarding them as 'le poison de la raison et la peste de la philosophie'. Across the channel, the Newtonian Benjamin Martin noted in the 1750s: 'The Philosophers of the present Age hold [hypotheses] in vile Esteem, and will hardly admit the name in their Writings; they think that which depends upon bare Hypothesis and

Conjecture, unworthy the name of Philosophy'. *Philosophical Grammar*, 7th ed. (London, 1769), 19.
46 University of Geneva Library, MS fr. 2019(2) (§§30–8).
47 Hartley, *Observations*, 1:349.
48 University of Geneva Library, MS fr. 2019(2) (§5).
49 T. Brown, *Lectures on the philosophy of the human mind*, 20th ed. (London, 1860), 279.
50 *Encyclopaedia Britannica*, 3 vols. (Edinburgh, 1771), 1:31.
51 Ibid.
52 Ibid., 32.
53 Laudan (1970).
54 Reid, *Works,* 1:248–53.
55 Ibid., 250.
56 Ibid., 251.
57 Like Newton, in whose footsteps he sought to follow, Reid rejected *eliminative* induction, chiefly on the grounds that we cannot perform the exhaustive enumeration of possible hypotheses which that method requires. See, for instance, ibid., 250.
58 As Reid's successor Dugald Stewart observed: 'It were perhaps to be wished that [Reid] had taken a little more pains to illustrate the fundamental rules of that [inductive] logic the value of which he estimated so highly'. Ibid., 11.
59 Ibid., 271.
60 Ibid., 250.
61 See especially the exchange of letters between Reid and Lord Kames in ibid.
62 Ibid., 1:251.
63 That it is a necessary condition is made clear by Reid's condition (2), cited in the text.
64 I shall not discuss the first methodological debate that the wave theory provoked, namely, that between Young and Brougham. I skip over it for two reasons: (1) It has already been investigated at length by Cantor (1971); (2) it represents a more vituperative but less substantial replay of the earlier ether debates I have discussed, with Brougham playing Reid to Young's Hartley.
65 See especially A. Comte, *Cours de philosophie positive,* 2 vols. (Paris, 1924), 2:331–52. A general discussion of Comte's philosophy of science can be found in L. Laudan, 'Towards a reassessment of Comte's "méthode positive" ', *Philosophy of Science 38* (1971), 35–53.
66 Indeed, given the ubiquity of the requirement – or analogues of it – in contemporary philosophy of science, it is remarkable that its prehistory has not yet been explored. In an intriguing but false surmise, Karl Popper once observed that 'successful new prediction – of new effects – seems to be a late idea, for obvious [*sic*] reasons; perhaps it was first mentioned by some pragmatist'. *Conjectures and refutations* (London, 1965), 247.
67 For a useful discussion of related issues, see V. Kavaloski, 'The "vera causa" principle', unpublished doctoral dissertation, University of Chicago, 1974.
68 J. F. W. Herschel, *A preliminary discourse on the study of natural philosophy* (London, 1830), 167. See also ibid., 172, 203.
69 Ibid., 190.
70 Ibid., 259 ff.
71 Ibid., 32.
72 Ibid., 32–3.
73 W. Whewell, *The philosophy of the inductive sciences, founded upon their history,* 2 vols. (London, 1847), 2:62.
74 Ibid.

75 Ibid.
76 Ibid., 62–3.
77 Ibid., 64.
78 Ibid., 65. For a fuller discussion of these issues, see L. Laudan, 'William Whewell on the consilience of inductions', *Monist* 55 (1971), 368–91.
79 'If I copy a long series of letters, of which the last half dozen are concealed, and if I guess these aright, as is found to be the case when they are afterwards uncovered, this must be because I have made out the import of the inscription'. *Philosophy of discovery* (London, 1860), 274.
80 Whewell, *Philosophy of the inductive sciences*, 2:72.
81 Ibid.
82 W. Whewell, *History of the inductive sciences from the earliest to the present time*, 3rd ed., 3 vols. (London, 1857), 2:370.
83 J. S. Mill, *System of logic, ratiocinative and inductive*, 8th ed. (London, 1961), 328.
84 Ibid.
85 Ibid.
86 In early editions of the *System,* he referred here to the 'ignorant vulgar' rather than the 'uninformed'.
87 Mill, *System*, 328–9.
88 Whewell, *Discovery*, 270.
89 Ibid., 273.
90 The Herschel – Whewell requirement of independent support has shown up in a new guise in the work of E. Zahar, especially his 'Why did Einstein's programme supersede Lorentz's?' in *Method and appraisal in the physical sciences*, ed. C. Howson (Cambridge, 1976), 211–76. Zahar is no more successful than Herschel and Whewell were in providing a philosophical justification for the requirement. (For a latter-day 'Millean' critique of recent work in this area, see Laudan, *Progress*, 114–18.)
91 Cantor (1975). There are many other dimensions of the ether debates that deserve serious exploration. For instance, some natural philosophers regarded the mathematical formalisms of the wave theory as well established but refused to regard this as evidence for the existence of a luminiferous ether. What was at stake was whether a theory could be 'accepted' without its ontological presuppositions being taken seriously.

I have argued for the symbiotic character of the general relationship between science and the philosophy in Laudan, *Progress*, chap. 2, and in Laudan, 'The sources of modern methodology', in *Historical and philosophical dimensions of logic, methodology and philosophy of science*, eds. R. E. Butts and J. Hintikka (Dordrecht, 1977), 3–20.
92 A minor caveat is in order here. Several philosophers and scientists before the nineteenth century (e.g., Boyle, Huygens, Liebniz) had claimed that the ability of a theory to make surprising predictions was an epistemic advantage. But prior to the 1820s no *systematic* arguments had been made to the effect that such an ability was a sine qua non for an adequate theory.

6

The electrical field before Faraday

J. L. HEILBRON

Office for History of Science and Technology, University of California at Berkeley, Berkeley, California 94720 USA

'Aether, being no object of our sense, but the mere work of imagination, brought only on the stage for the sake of hypothesis, or to solve some phenomenon, real or imaginary; authors take the liberty to modify it as they please'.[1] We have ethers continuous and discontinuous, material and immaterial, subject to and free from the laws of ordinary mechanics; ethers filling the heavens, pervading the atmosphere, penetrating hard bodies; 'ethers cluttered by a great variety of concepts'.[2] They elude classification by the historian almost as effectively as they have escaped detection by the physicist.

I define ethers as subtle substances that mediate interactions between gross bodies. *Subtle* means very tenuous or rare, highly penetrating, and undetectable directly by sense. *Mediation* signifies the power to propagate action or the potential for action without moving as a whole.[3] Neither light particles nor air is ether: The former are subtle but not a medium; the latter is a medium, at least for sound, but not subtle.

The subset of ethers under discussion here are representations of the electric and electromagnetic fields discussed by physicists from about 1780 to 1830. *Field* in general signifies a region of space considered in respect to the potential behaviour of test bodies moved about in it;[4] the electricians of 1780 lacked the word but not the concept, which they called 'sphere of influence', *sphaera activitatis,* or *Wirkungskreis*. The electric field or *Wirkungskreis* has been represented geometrically, by lines the number and direction of which indicate potential velocities, orientations, and directions of motion; and physically, by a medium whose local pressures, stresses, or other circumstances cause the behaviour of the test bodies. Such a medium, if subtle, qualifies as an electric ether.

Strictly speaking, a field and its representation must act by local forces, by forces exerted between contiguous elements of space or ether, or, in the case of a discrete ether, between a particle and its nearest neighbours. One might therefore reject as a representation of a field any medium the elements of which act not only on nearest neighbours but also on further elements. Such close discrimination is doubtless desirable in an analysis of classical field theories. For my purpose, however, it would not be useful. The older writers seldom developed their theories in sufficient detail to allow an exact judgement of the range or mode of action of an ether particle. I admit all subtle media the function of which is to propagate electric and magnetic action between ponderable bodies appearing to interact directly over macroscopic distances.

Mediation by a classical field takes time. The older writers seldom stated explicitly that the ethers with which they modelled the field worked in time; instead they made analogies to the spread of light and references to the progressive displacements of ether elements. No more did they explicitly declare that the field had energy. But again it is plain that they held the proposition implicitly, for their ethers could be strained or tensed, and moved internally. The ethers we examine here played the same role as the 'medium or substance' that transmitted electrical action in Maxwell's theory, and that he commended to the attention of physicists in the last paragraph of his *Treatise*.[5]

Discomfort with postulating actions over macroscopic distances is endemic among physicists. It afflicted Aristotelians, corpuscularians, Newtonians, and electricians, among others, long before it bothered Faraday. Seeking its etiology in a particular philosophy or epistemology is not promising. Nonetheless some historians, considering the concept of the electric field as an artifact of discomfort with distance forces, pursue its origin in metaphysics. A more obvious and, in the event, a more fruitful approach is to read the old electricians, and to try to grasp the problems that might have recommended the field concept to them. It turns out that the prime mover towards field representations was not a fundamental principle of metaphysics, epistemology, or natural philosophy, but nagging problems in electrical theory.

Both the existence of these representations and their source in standard problems in electricity counter a common opinion about the origin of field theory and Faraday's part in it. Although my main purpose is not to correct this opinion, but to call attention to a neglected class of electrical theories, scholarly civility demands mention of the points at issue.

It is held that Faraday was the 'father of field theory'[6] and that the seeds for his fathering came from Kant and Boscovich. The first assertion concerns field theories that attempt to make do without electrical matter or fluid. It is the theory at which Maxwell aimed in following out Faraday's later ideas, a theory 'out of harmony' with the 'gross' conception of a 'molecule of electricity'.[7] This conception of the field rules out Lorentz's electrodynamics as well as pre-Faraday electric ethers.[8] In trying to do without accumulations or deficits of electrical matter, Faraday did take a fresh tack in his time, although one common in the eighteenth century; insofar as the result is *defined* to be field theory, Faraday must have been its father. But this is only to say that Faraday invented Faraday's theory.

The second assertion rests on poor historiography. About 1800, we are told, everyone subscribed to the 'dogma of action-at-a-distance physics'. To reach field theory, a 'metaphysical revolution' was required. Faraday alone made it, taking metaphysics from Kant and, especially, Boscovich; 'from his earliest productive years' he believed ponderable bodies to consist of seedless, multivalent, indefinitely extended, mutually interpenetrating 'atoms'.[9] By 1831, or perhaps a decade earlier, he had made a similar theory of electricity: Localised electrical fluids and magnetic poles had been exploded into the electromagnetic field.[10] This interpretation has been attacked on the pertinent ground that there is no good evidence for it.[11] The ground has been conceded without relaxing the interpretation. The very want of evidence has become a supportive argument. 'Hesitant to discuss his metaphysics', inhibited by 'a conspiracy of silence', Faraday did not write what he thought.[12] Here we leave metaphysics for metapsychics.[13]

The historiographical method that produced this interpretation is not uncharacteristic of recent writing in history of science. It appears to rest on the assumption that the innovative scientist acquires his basic and guiding conceptions from general philosophy and metaphysics. This preconception inspires a search for what is literally a philosopher's stone, a single key that unlocks the corpus, the philosophical thread, the leitmotiv, that runs through a life's work. The quest distracts the adept from the first task of the intellectual historian, reconstructing the public state of knowledge. Only after surveying the literature that his scientist may be presumed to have known can the historian identify what in the scientist's work might need explanation by intellectual biography. The explanation will invoke whatever then seems pertinent, including philosophical commitment. The alchemical historian, on the other hand, sublimes directly to the invention of the explicandum, and concocts pseudo-problems in place of historical ones. The alchemical analysts of Faraday have worked out answers to several pseudo-problems. It is pertinent that the latest of these adepts does not mention any of the ether theories with which we shall be particularly concerned, although he says that he has tried 'systematically to find the *problems* which led to the field theories'.[14]

I begin by outlining the consensus about electrical theory reached late in the eighteenth century. The consensus distinguished the roles of field and charge, which previous theories had conflated, and called forth speculations about ether mechanics. Next come accounts of ideas about electric ethers and galvanic currents from 1800 to 1820, when Ørsted's discovery augmented the load that both sorts of models had to carry. The third and last sections concern electromagnetic fields and ethers between 1820 and 1831, when Faraday found the inverse of Ørsted's effect. With that discovery, Faraday began his systematic work on electrodynamics, which led him far beyond the old speculations to the idiosyncratic field theory on which Maxwell built. The development of Faraday's ideas about fields must be reserved for a future essay.

Electrical theories of the late eighteenth and early nineteenth centuries

The physicists of the seventeenth century referred electrical attraction to the action of an effluvium or subtle vapour elicited from electric bodies by friction. The effluvium drew either by spearing chaff and rebounding with it or by mobilising the atmosphere, 'thinning' the near air to cause an inrush of that beyond.[15] The model does not qualify as an ether; it operates by projection and impact, not by mediation.

The discoveries of electrical repulsion and of sparking early in the eighteenth century made the model inadequate. Among the new mechanisms proposed was the 'atmosphere', an aura of electrical matter surrounding an electrified body. Pulsations of the atmosphere cause attractions and repulsions; ruptures of the electrical matter, which contains or is related to fire or light, create sparks and glows. In Franklin's version of the theory, the atmosphere stands quietly; positively electrified bodies repel one another via short-range forces between the particles of their respective atmospheres. Note that the atmosphere is the *charge:* The surplus of electrical matter that constitutes positive electricity spreads over and beyond the surface of the body it clothes. Such an atmosphere functions both as the source of the field and as an ether: As source it can be accumulated and conveyed; as ether it acts by pushes between neighbouring particles.

This model could not be extended to cover the air condenser, which can sustain a charge with its plates so close together that the grounded one lies within the atmosphere of the other. In this situation, according to Franklinist theory, the condenser should short internally; the atmosphere, as charge, should run to ground. Similar puzzles were presented by the dissectible condenser, or electrophorus, invented in 1775. Physicists escaped from their difficulty by dissociating the functions with which they had burdened the atmosphere. As source of the field it became a charge placed very close to the electrified surface. As the field, it lost its material connection with the charge, and ceased to be an ether: *Atmosphere* became synonymous with *Wirkungskreis* and *sphaera activitatis,* with the space in which electrical forces could be detected.

Consensus regarding the localisation of charge was established during the 1780s. The rules of the *Wirkungskreis* accounted for the phenomena: Accumulations of charge have atmospheres, or spheres of influence, within which the electrical matter normally contained in conductors segregates into plus and minus charges. 'Electrical motions' of ponderable bodies were usually represented as consequences of distance forces acting between the accumulated and segregated (or induced) charges, and between electrical and common matter.

The consensus included acknowledgment that no known experiment could decide whether electricity came in one fluid or two. In the former case one had to postulate a repulsion between particles of common matter or devise an ether that could otherwise account for repulsion between negatively charged bodies. In the latter case symmetry between positive and negative electrification was obtained at the cost of reasoning about a second electrical substance. One affected to choose on the basis of simplicity or convenience. The English mind tended to be singlist; the Continental, and especially the French, dualist. Although the singlists included authorities like Volta and van Marum, they probably made a minority among electricians by 1790, and certainly did so during most of the nineteenth century. The equal and contrary electrochemical powers of the poles of the pile, invented in 1800, strengthened the position of the dualists.[16]

The consensus of the 1780s referred to the localisation of charge, to the rules of the *Wirkungskreis,* and to peaceful coexistence between singlists and dualists. It did not bring agreement about the machinery that might account for the rules of electrical interaction. Many proposals were made, including some ether representations; the physicists of the eighteenth century were not content with the elementary scheme of distance forces with which historians credit them. Their disquiet arose not only from the usual and usually inconsequential scruple against action at a distance, but also, and primarily, from a non sequitur, or even a contradiction, in the rules of the *Wirkungskreis:* the connection between motions of electrified bodies and their supposed cause, attractions and repulsions between elements of electrical fluid(s).

Electrical matter placed inside a conductor spreads instantly to its surface. Similarly, the natural supply of electrical matter in a conducting body segregates immediately under the influence of an external charge. It appeared that no force exists between the electrical fluid(s) and the molecules of conductors, or at most a force very small compared with that between elements of the electrical matter. Two puzzles resulted: What retains a charge at the surface of a metal sphere? How do electrical attractions and repulsions develop the ponderomotive forces needed to account for electrical motions?

The usual answer to the first puzzle rested on accurate but misleading experiments. Rarified air does not begin to insulate until reduced to a pressure of about 10^{-3} atmospheres. The best air pumps of the eighteenth century did not reach pressures so low; not until after 1850, with the help of Geissler's mercury pump, did physicists find the region of increasing dielectric strength.[17] Throughout our period they reasoned that, since an electrified conductor would discharge immediately in a perfect vacuum, its electricity must be clamped upon it by the pressure of the surrounding air. As Biot explained, the surface layer of repellent charge exerts an outward force on each particle of the electric fluid proportional to the local depth of the layer (or to its surface density). These forces, when added together, produce a pressure proportional to the square of the thickness (or the density), and this pressure must be resisted by the weight of the adjacent atmosphere.[18] As Faraday remarked, Biot's theory assumes the existence of 'gross mechanical relations' between 'the ponderous air and the subtle . . . fluids of electricity'.[19]

These gross mechanical relations provided an answer to the second question as well: The ponderomotive forces responsible for electrical motions arise from imbalances in air pressure set up by charges. For example, surplus fluid given a pair of touching suspended cork balls accumulates opposite to the point of contact, and creates there a large pressure against the atmosphere. 'Evidently,' that is, in order to save the phenomena, this excess in pressure drives off some of the air, allowing the normal air between the balls to force them apart. Attractions occur via the pressure gradient called up by charge accumulations on the facing surfaces of the balls.[20] In general, the fluids distribute themselves as required by their distance forces, and the resultant accumulations cause motions by mobilising the air.[21]

The account just sketched is not an ether theory or even a medium theory, since the air acts not by propagating pressures but by immediate contact. It is easy, however, to construct a rudimentary ether theory by substituting for the 'gross mechanical relations' forces between the charges accumulated on bodies and the electrical fluid(s) naturally present in the air. Perhaps the earliest example of this approach is John Canton's representation of electrical atmospheres as 'alterations of the state of the electrical fluid contained in and belonging to the air'. In Figure 6.1, A, B, and C are neutral, positive, and negative, respectively. Since B's charge repels surrounding electrical matter

(Canton was a singlist), the air near B has less, and beyond B more, than its usual amount, up to a certain distance, represented by the outermost lines in the figure, where the state returns to normal. Since the air's electrical matter moves towards C, the gradient about C is opposite to that about B. An electrified body polarises the medium around it; a conductor suffers induction under the influence of the displaced electrical matter of the air *adjacent* to it.[22]

Canton's suggestion propagated through a string of British singlists, including Tiberius Cavallo and Thomas Milner, who tried to base a theory of electrical motions upon it.[23] Milner thought to obtain ponderomotive forces as interactions between the electricities of bodies and the polarised air contiguous to them: 'The air thus charged is the medium which enables one electrified body to act upon another at some distance'.[24] Cavallo explained how the trick is played: Neither of two touching electrified balls can sit at the centre of the system of polarised air it creates. Hence, on the assumption that 'there is something on the surface of bodies which prevents the sudden incorporation' of the contrary electricities of the balls and the air, the balls will move

Figure 6.1. Canton's representation of the electric field. (From Priestley, 1775, *The history and present state of electricity*, London.)

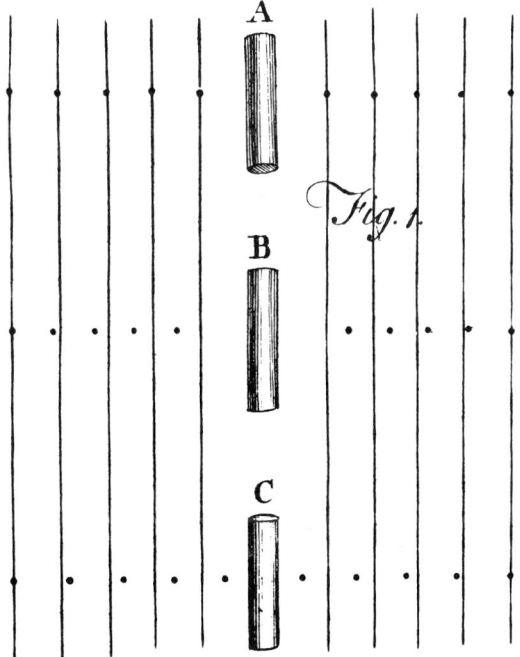

apart into the centres of their atmospheres under the attractive force assumed to act between electrical and common matter.[25]

The implausibility of Cavallo's special mechanism – the initial polarising of the air about 'centres' outside the balls – was granted by everyone, including himself. Other singlists tried to do better, inspired by a desire to avoid Aepinus's solution to the problems that recommended dualism: the infamous force of repulsion between particles of ponderable matter. Although they continued to use the analogy between air and insulators known to polarise, like glass and tourmaline,[26] they now emphasised the conductivity of the atmosphere.[27] The electrified balls polarise the air, but they also charge it by conduction. When the air between them has acquired a surplus of electricity of the same sign as theirs, they are pushed apart by the contiguous electrical matter, assisted, perhaps, by the polarisation of the remoter air.[28] Whatever the details, the general proposition that electrical motions and inductions over air gaps occur via electrical matter made active in the surrounding air was well represented in texts of the early eighteenth century.[29]

A notable form of the ether theory of electrostatic interactions was published in 1806 by Amedeo Avogadro, who began with Canton's ideas as mediated by Beccaria.[30] According to Beccaria, 'the electricity of a body resides within the superficial pores of it, and actuates the ambient air, not by diffusing itself into it, but by exciting either a tension or relaxation in the natural [electrical] fire inherent in it'. That he thought of these actuations as polarisations appears from Figure 6.2. Induction occurs via the tensed air. So do ponderomotive forces, if one allows the principle that bodies interacting electrically strive to move so as to minimize the tensions in the air.[31]

Avogadro goes forward from the proposition that no electricity can appear on any surface unless an equal and opposite electricity can develop on an-

Figure 6.2. Beccaria's representation of the electric field. (a) About a positive body, (b) about a negative one, (c) between positive bodies, (d) between negative ones, (e) between unlike ones. (From Beccaria, 1772, *Elettricismo artificiale*, Turin.)

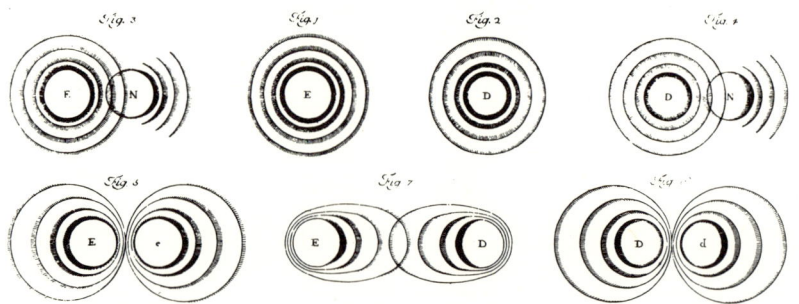

other. This answering electrification does not occur in response to a force acting over macroscopic distances. Only the universal force, gravity, has such power. An electrified body acts by imposing an equal and opposite electrification on the contiguous insulating layer. To fix ideas, consider a horizontal air condenser with the positive plate uppermost. The top layer of air becomes 'charged' (a 'peculiar state' enjoyed only by dielectrics) in a certain manner, imposes the charge on the next layer, and so on, until the lower plate is reached, and rendered negative by the air adjacent to it.

Avogadro, a singlist, pictured the normal air molecule as a little sphere surrounded by a spherical shell of electrical fluid. The surplus fluid of the upper plate distorts the atmospheres of the molecules adjacent to it; a relative defect occurs above, a relative surplus beneath. This surplus in turn decentres (to use Avogadro's term) the atmosphere of the next lowest molecule, whence the distortion propagates to the lowest layer of air. Avogadro clearly stated that 'charge' consists of strain in the air or other insulator that preserves the 'electricity' of conductors, and that all electrical motions arise from very short-range forces propagated through dielectrics.

In support of his views, Avogadro pointed to the behaviour of dissectible condensers and to the complicated phenomena of absorbed or residual charge, which had inspired Beccaria to introduce the cumbersome concept of vindicating electricity.[32] One experiment may stand for them all. A glass plate sliced through longitudinally is armed with metal plates top and bottom and electrified as a parallel plate condenser. Connect the armatures and explode the condenser; remove them; separate the glass slices. The upper glass, the one that had touched the positive armature, shows negative on its top surface and positive on its bottom; the lower glass shows negative on top and positive on bottom. Avogadro inferred that every horizontal layer in the dielectric is polarised; and he used this model to explain what Beccaria and Volta called the 'oscillation of the electricities', the reversal of polarity of the coatings when a Leyden jar explodes. The effect arises from induction in the coatings occasioned by residual charges in the dielectric, much as Avogadro pictured it.

Avogadro's memoir had a good press. It appeared in English and German as well as in French.[33] But it did not receive the attention that its elegant solutions of electrostatic problems deserved, perhaps because most electricians were then fascinated with the Voltaic cell. Perhaps Faraday knew nothing of Avogadro's concept of the role of the dielectric medium. After the publication of Faraday's ideas about electrostatic induction, Avogadro himself called attention to his neglected work, 'for the history of science', and observed that induction could be understood to operate on the ether as well as on ponderable matter.[34]

Electric ethers and galvanic currents, 1800–1820

The electrician of the late eighteenth century, who served sometimes to short-circuit the Leyden jar, had an intimate acquaintance with the discharge: a rushing together through his body of the equal and opposite electricities unable to unite directly through the bottom of the jar. The explosion annihilated the electricities; to reproduce them required labour, such as the cranking of the electrical machine, no less personally felt than the discharge itself.

Physicists tried to apply this understanding to the pile. According to Biot's conception, which came to dominate, the apparently continuous current stemming from the pile in fact consists of a very large number of discharges. The pile has the capacity when open to acquire opposite electricities on its terminals, which show the usual signs of frictional electricity. Close the circuit by taking one terminal in each hand; the old signs disappear; you feel a 'commotion', similar to that caused by a feeble torpedo. Release the terminals; the old signs reappear, the commotion ceases. Apparently the pile has the power of recuperation; and it is easy to imagine that, when closed, it explodes and recuperates so quickly as to appear to act continuously. 'It is becoming more and more probable that [galvanism] is the successively repeated effect of a very feeble electricity possessed of a very great velocity'.[35]

How the pile acquired its polarity was a matter of lively debate. Did it work entirely, as Volta and Biot thought, by the contact of dissimilar metals, or, as William Wollaston and the English argued, by chemical combinations between the metals and the moist conductor separating them?[36] Despite their difference over this major point, both schools accepted Biot's conception of a discontinuous *external* current. Davy, for example, explained that the contact disturbs the electrical equilibrium; closure of the circuit restores it; chemical action revivifies the metals; their contact again generates electricity; and so on.[37] In trying to picture the action of this discontinuous current physicists once again flirted with ethers and mediated forces.

Consider first electrochemical processes such as those instigated by the current in acidulated water. How do the hydrogen particles come to the negative electrode and oxygen to the positive? Davy suggested that the electrodes might polarise the water just as charged conductors segregate the electricities in neighbouring pieces of wire. The molecules touching the electrodes are torn apart; one fragment escapes and the other combines with the oppositely charged portion of the neighbouring polarised molecule, whose liberated portion decomposes the next polarised particle, and so on. In the middle of the fluid, where the molecules remain neutral, the last fragments to be released in the progression from each electrode unite into a normal water particle.[38] The concept that a substance undergoing electrolysis first polarised like a conductor suffering induction and then decomposed as the result of contact forces propagated from particle to particle was not original with Davy, who indeed called it the 'received opinion'.[39] Perhaps its earliest proponent was Theodor von Grotthuss, who reasoned from a fanciful analogy between the polarised pairs in Volta's 'electrical magnet' to the state of the water molecules between the electrodes.[40]

Also outside the pile, in a circuit uninterrupted by an electrolytic cell, phenomena occur that focus attention on the medium. Foremost among them are the heat and light produced in wires of narrow gauge. Many investigators, adapting a research topic in frictional electricity, tried the lengths of wires that the current could melt, and the variety of refractory substances that it could fuse. Elaborate and expensive equipment was built for the purpose, such as the battery with plates of thirty-two square feet put into operation by J. G. Children in 1815.[41] To understand these effects the physicist needed to connect his representations of electricity and of heat and light. Many associations were proposed,[42] the variety arising from disagreement over the number of electrical fluids and over the nature of light and heat. A common opinion built upon dualism and the theory of caloric, which referred heat to a material substance spread through the spaces separating the particles of common matter.[43]

Recall that the current was pictured not as a continuous river, but as a sequence of squirts, which start from the terminals at every destruction of tension. On striking the electrical matter combined in the neutral wire, the squirts create disturbances that propagate in opposite directions towards mutual annihilation in the middle of the circuit. The propagation might be conceived as a wavelike motion in the wire's electrical matter,[44] or, in analogy to prevailing theories of the pile, as disruptions and recombinations of the two electrical fluids. (The possibility that oppositely directed squirts move freely until colliding in the middle of the wire was harder to credit.)[45] By 1820, if not before, the dominant theory of the external current, or rather its very definition, was a series of microscopic displacements of the two electricities: A current, says Ampère, is a 'succession, in all the particles of conducting wires, of decompositions and recombinations of the fluid formed by the union of the two electricities'.[46]

To explain the appearance of heat in and beyond the wire it was assumed that the conflict of the squirts put in motion or freed caloric. For example, if, as many speculated, the electrical fluids consisted of 'bases' plus heat and light, the latter might be detached during the disruptions and recombinations in the current-carrying wire.[47] The passage of an electric current might set up vibrations in this matter as well as disruptions; and these vibrations, communicated to the environing caloric ether, would constitute the propagation of heat.[48] For those who accepted the undulatory theory of light, the glow of high-resistance wires propagated in the same manner as their heat; for others, such as Biot, who retained the emission theory, the vibrations threw off light particles 'in a purely mechanical way'.[49]

A consequential form of this theory was worked out by Ørsted. In 1806, after descending from the heights of *Naturphilosophie* to a level more common to physicists, he published a paper on electricity to which he attached great importance.[50] His purpose was to assimilate the mechanisms of conduction and induction. He observed that insulators, including the air, polarise under an external charge whether the charge stands on or next to them. In the first case the insulator electrifies by conduction: The surface touched takes on electricity of the same type as the external charge; the layer just beneath the surface becomes oppositely electrified by induction; and so on. In the second case, the surface electrifies oppositely to the external charge and the adjacent layer does similarly. Now, according to Ørsted, in an imperfect insulator the situation just described holds for the first instant of induction only; in the next instant, the electricity induced in the second layer separates the normal electricities of the third, combines with its opposite to free, momentarily, its homologue; and so on. Exactly the same process occurs, although in an insensible interval, in conductors, which can be defined as very poor insulators. Ørsted calls the process of conduction and induction 'undulatory', in reference, apparently, to the sequence of unions and divisions of the electrical forces at each point in a body passing a current.

The wire closing the galvanic circuit enjoys a peculiar state. Undulations commence at each of its ends and propagate rapidly towards the centre, where, if they meet no resistance, they annul one another quietly. The pile recuperates instantly to produce the next set of undulations, which performs as the earlier. Ørsted notes that 'the electrical forces are conducted only by themselves', by which he means that the instrument of conduction is the electrical forces supposed to occupy all of space.[51] If the forces in the wire suffer impediment to their reunion after separation, as they do when the magnitude of the current exceeds their powers of recovery, heat results. If the impediment is strong, light may be produced, of a colour dependent upon the violence of the union. Evidently both light and (radiant) heat spread throughout the region around the conducting wire. This propagation also takes place via 'undulations' in the forces occupying space.[52] The electric spark propagates by a very rapid breakdown and recovery of the electric forces in the air;[53] a light ray is nothing but a 'series of immeasurably small electronic sparks'.[54] These views claimed the careful consideration of all physicists, when, in pursuing them, Ørsted discovered the action of a galvanic current on a magnetic needle, and so brought to a close a 'long, sorry time of endless, purposeless, indecisive and fruitless detail' in galvanic researches.[55]

In 1812, Ørsted suggested that to detect an interaction between electricity and magnetism one should employ electricity in a 'bound' or 'latent' state. He reasoned that in frictional electricity the opposed forces are free, as appears from the usual attractions and repulsions; in galvanic electricity, they give heat, light, and chemical action, but stay too tightly together to cause electrical motions; in magnetism they may be so closely bound as to have escaped detection.[56] He proposed to try electricity in the most bound state of galvanism, where large currents and resistances constrain the electric forces. He persevered desultorily with large piles and narrow wires, sustained by the hope that undulations sent forth by a glowing wire might excite a magnet as well as the eye. Success awaited placing the needle parallel to the wire. Ørsted did so early in 1820, and soon learned that the strongest deflections occurred with the least resistance.[57]

Ørsted advertised his results in a Latin pamphlet entitled 'Experiments on the effect of the electric conflict on the magnetic needle'. The 'conflict' in the wire is just the converging undulations starting from the poles of the pile; as expected, it also acted outside the wire, in the 'circumadjacent space'.[58] There, contrary to all analogy, the conflict 'performed circles', driving the needle in one direction when placed below the wire and in the opposite when placed above. Ørsted supposed that negative electricity moves in a right spiral about the wire and acts only upon north poles, while positive electricity moves in a left spiral and acts only on south poles.[59] The intricate dance of the conflict calls up a material image; and, indeed, by giving the 'impetus of the contending powers' as the cause of the needle's motion, Ørsted suggested a process more mechanical than usual in dynamical (*naturphilosophisch*) physics.[60] Many of his readers tried to express his conflict through a material ether.[61]

Electromagnetic fields and ethers in the 1820s

Ørsted's fundamental fact related an apparently heterogeneous pair, a magnetic pole and an electric current. Physicists tried to grasp this hybrid in three different ways: by following Ørsted and supposing a novel, circular force between currents and poles; by assimilating the wire to a magnet; and by assimilating magnets to currents.[62] The last two approaches differed essentially from the first. They aimed at reducing all electromagnetic phenomena to a single elementary *force* (italicized to indicate a hypothetical, unmeasurable attraction or repulsion between infinitesimal *elements*); the first aimed at an exact description of the motions of interacting poles and wires, or of their mutual forces, the measurable integrals of *force*.[63]

The only reductionist theory now remembered is Ampère's. Building on his demonstration that current-carrying wires exert ponderomotive forces upon one another, he argued that the fundamental *force* is a rectilinear push or pull between elements of current: attraction between elements moving parallel, repulsion between ones moving in opposite directions. On this assumption he obtained, by double integrals, the observed ponderomotive forces.[64] Ørsted's fundamental fact, as expressed in the law of Biot and Savart,[65] proved more difficult. To deduce it, Ampère required the bold hypothesis that a magnet owes its power to elementary current loops perpendicular to its axis.[66] Mathematical derivations of the Biot–Savart law from Ampère's hypotheses were accomplished in 1823 by F. Savart and J. F. Demonferrand, 'marking a sort of epoch in the history of electrodynamics'.[67] All understood perfectly the relation between laboratory forces and Ampère's undetectable, hypothetical, electrodynamic *force*.[68]

A second *force* theory, developed mainly by German physicists, reduced the wire to a sequence of magnetic poles placed symmetrically at its surface. They tried to do with two poles in each cross section of wire, then with four, finally with two for every molecule.[69] This last version of the theory of 'transverse magnetism' resembles Ampère's: Both derive elementary magnets from a current at right angles to the magnetic moments. But Ampère's *replaced* magnets by currents, whereas the German theories admitted both electricity and magnetic poles. The replacement enabled Ampère to deal with homogeneous interactions only, to reduce the interactions of currents, of poles and wires, and of magnets to one and the same principle. 'Is it not evident that one must look for the primitive fact in the action of two things of the same nature?'[70] To the more literal Germans, it seemed preferable to take as primitive a *force* of the same character as the force observed in Ørsted's experiments.[71]

A decisive argument against transverse magnetism was developed in England by physicists who also emphasised measurable, macroscopic force. Accepting from Ørsted's account that the chief electromagnetic phenomenon was the tendency of poles to 'perform circles' about wires, Wollaston tried to explain the interactions of Ampère's wires 'upon the supposition of an electromagnetic current passing round the axis of [each]'.[72] Although the public notice of his ideas is unintelligible, he was understood to think that 'a kind of revolution of magnetism [occurred] around the axis of the wire', a 'vertiginous magnetism' that swept poles about in circles.[73] Both Davy, who had inclined toward transverse magnetism, and Faraday, who had tried to do with rectilinear attractions and repulsions, adopted Wollaston's manner of interpreting electromagnetic interactions.[74]

Further experiment convinced Faraday that the primitive fact, the elementary phenomenon, was the rotation of a pole about a wire, or of a wire about a pole. His demonstration of these continuous rotations by ingenious apparatus that, in contrast with the usual arrangements, allowed motions over 360°, made his reputation as a physicist. They also made the theories of transverse magnetism untenable, as Faraday, Ampère, and many others insisted.[75] Ampère had no difficulty in accounting for Faraday's rotations, or in obtaining revolutions of wires and magnets about their own axes, an effect Faraday had sought in vain.[76] Above all, Ampère and Faraday could simulate a magnet by a solenoid, which they could make rotate like a needle, whereas no practicable polygonal deployment of needles could be made to drive a pole in a circle.

The explosion of the German theory left four principal alternatives. One could swallow one's objections to Ampère's 'unnatural assumptions', his arbitrary preference for electricity over magnetism, and his bizarre molecular currents, which have no apparent cause and do not, like other currents, spread through conductors.[77] Or one could patch up the German theory, making the *force* between the transverse magnetic poles and the poles of the needle act at right angles to the line joining them. Or, what comes to much the same thing, one could postulate a transverse *force* between elements of current and elements of boreal or austral fluids in the compass needle.[78] None of these three solutions had much appeal, among other reasons because of 'the difficulty of conceiving the mechanical principles by which such a tangential force . . . can operate', its want of analogy to the rest of physics, and its incompetence to explain all of electrodynamics.[79]

The fourth alternative eschewed these various *forces* in favour of observables. Here the difficulty of phenomena 'sufficient to perplex the mind' recommended the procedure to be followed. The memory needs a crutch, a picture or diagram; 'words alone will not do'.[80] Imagine yourself in the circuit, head towards the zinc end of the pile, face towards the needle: Its north pole will move towards your left hand when the current enters your feet.[81] Or place a picture of the Dioscorides (Figure 6.3) on the wire; the left hand of the right twin points to the direction in which a north pole moves.[82] This is no coincidence. According to the ingenious decypherer of these hieroglyphs,

Castor and Pollux originally represented the contrary electricities. The physical knowledge of the builders of the pyramids may thus be made to serve the perplexed student of electromagnetism.[83]

Such representations put one either directly or vicariously in the space where electromagnetic interactions occur. The very complexity of the phenomena drew attention to the field, to the space considered with respect to test bodies moved about in it, and to the need for maps. T. J. Seebeck pictured concentric circles of iron filings in the 'magnetic atmosphere' of a wire.[84] English physicists, in keeping with their emphasis on macroscopic representations, consumed quantities of filings. Davy used them in his first set of electromagnetic experiments; Faraday sprinkled them about wires and helices, a 'very beautiful indication of [the] course of action'; Roget showed a figure similar to Seebeck's.[85] These 'magnetic curves' had the same significance as similar lines about magnets; namely, they were tangents to the 'positions which an infinitely small compass needle . . . will assume'.[86] Alternatively, physicists drew lines representing the paths of wires around poles, and of poles around wires, filling space with curves indicative of the directions of 'the forces that are drawing' the objects in motion.[87]

As in electrostatics, so in electrodynamics, the possession of clear rules and representations was not enough for physicists. They must know what

Figure 6.3. Schweigger's electromagnetic heiroglyphs. (From Schweigger, 1826, *Journal für Chemie und Physik 46.*)

goes on out there, in the field, where filings clump and the Dioscorides play. What is the 'essence of the Naturkräften'? What are the 'qualitative aspects of experience'?[88] P. Erman stressed the need for an accurate map of electromagnetic motions, which he planned to explain by hydraulics; Seebeck hinted that his filings lined up under a 'magnetische Strömung'; Faraday took Ampère to task for failing to provide a physics in the style of Ørsted.[89] In fact, many physicists of the 1820s, including Ampère, indulged conjectures deriving from Ørsted's conception of electric conflict. Two sets of these conjectures may be distinguished. One, emphasising the spiral force outside the wire, used analogies to fluid flow. The other, interpreting the conflict as a source of disturbance, used analogies to the propagation of waves.

Ørsted's circulating magnetic force, sweeping about poles as a whirlpool sweeps corks, recalled the comfortable physics of the Cartesian vortex. It also saved some phenomena. 'It is still scarcely possible', Schweigger wrote in 1826, 'to describe the chief features [of electrodynamics] more correctly than through the picture of a magnetic atmosphere', by which he meant 'rotatory magnetism', a vortex (*Wirbel*) extending outwards from the wire.[90] Davy and Wollaston were understood to have accepted Ørsted's maelstrom, and Faraday to have demonstrated it. 'Most discoveries made since the publication of Ørsted's works come to support his way of conceiving the general philosophy of nature'. The magnetic fluids that accompany the electric ones in motion rush in opposite vertiginous paths around the axis of the connecting wire.[91]

The other set of analogies deriving from Ørsted's conflict was ether models, mediums for the propagation of action. Ampère saw a deep parallel between electricity in motion and light.

> Everything that has been done in physics since the work of Dr. Young on light and the discovery of M. Ørsted is preparation for a new era . . . Explanations deduced from the effects produced by the motion of imponderable fluids will gradually replace those now accepted . . . I believe that we must look to the motions of fluids distributed in space for the explanation of general effects.[92]

The space-filling medium is formed by the union of the two electricities. Its vibration, or series of separations and combinations, constitute light; interference of disturbances from different wires accounts for electromagnetic motions; the electric spark and the Ørsted effect propagate from the conflict of electric current in the wire.[93] 'M. Ørsted regarded compositions and decompositions of electricity . . . as the unique cause of heat and light, i.e., of the vibrations of the fluid occupying all space . . . This opinion of the great physicist . . . agrees perfectly with all the phenomena'.[94]

Ampère's universal luminiferous and electromagnetic ether pervaded the

Annales de Chimie. Becquerel exploited it for chemistry and heat; Marianini's experiments on strengths of current were referred to the hypothesis that 'the electrical current is propagated by vibrations in the manner of sound'; A. de la Rive used Ampère's ether to elucidate calorific effects of currents; L. Nobili accounted for his rings by 'a law of interference' in the spread of electric currents; Savary traced pecularities in magnetism to attenuation of the 'vibrations transmitted from the wire to the environing media'; E. Becquerel reached a similar conclusion from the appearance of the wire itself.[95] 'Maybe Newton [the Newton of ether] glimpsed the true cause of electrical phenomena'.[96]

Our two sorts of mechanical analogies do not exhaust the range of pertinent proposals made during the 1820s. Were we to look further, we should find oddities like Erman's expansive flowing electrical *Thätigkeiten;* G. F. Pohl's oscillating antagonistic longitudinal electric and transverse magnetic tensions; and C. Hansteen's 'electric-fluid magnets', neutral polar products of the conflict, which surround the wire 'like the circular wave around a rock dropped in water'.[97] But we already have enough to suggest that the phenomena of electromagnetism caused many besides Faraday, including the paladins of mathematical reduction, to conjecture about an ether competent to propagate heat and light, and to mediate electrical interactions. It is striking that Ampère's protégé A. de la Rive, writing shortly after the discovery of magneto-electric induction, offered just such a conjecture as the conclusion to his history of electrodynamics during the 1820s.[98]

Faraday's response

The obvious inverse of the Ørsted effect, the induction of a current in a passive circuit by a current-carrying wire laid next to it, did not work. Nor did an electric current arise in a wire wrapped round a magnet, as one expected on Ampère's theory.[99] Ampère himself desired to know whether the molecular currents to which he attributed the power of magnets could be induced by external currents. He managed to deflect a thin copper disk suspended within a wire coil, an effect he interpreted as a consequence of the induction of one current by another. There he stopped. Although the phenomenon suffered from anomalies, he did not deem them important: There was no room in his *reductionist* theory for an induction of electricity by magnetism different from the straightforward, and spurious, induction of currents that he had detected.[100]

This proposition may be illustrated by Ampère's interpretation of Arago's discovery that a magnetic needle may follow the motion of a metal disk rotated beneath it. Ampère immediately tried whether a solenoid could substitute for the needle. The positive result satisfied his interest in the phenomenon: It fell under his general principle of the equivalence of magnets and current loops. He had only to postulate that the induction of one current by another took time. While the molecular currents grow or die, 'they exert forces that probably produce the singular effects that M. Arago discovered'.[101]

An alternative to Ampère's approach was to allow a magnetic body the power to induce poles in a metal plate rotating in its vicinity. The tangential component of the driving force might then be understood by supposing that the magnetism induced sequentially in each sector of the rotating disk reached a maximum after the sector had passed under the needle.[102] Still, the need for relative motion remained a mystery: Why could a magnet not induce poles in a stationary copper plate if it could do so in a moving one? To this difficulty, Arago added another: The vertical component of force on the needle turned out to be repulsive, not attractive as the magnetisers supposed.[103]

Faraday gave Arago's phenomenon sustained attention. The 'extraordinary character of [the] motion' made it a fine lecture demonstration; the impotence of philosophers to explain it made it a moral lesson as well as a challenge. 'Old obstacles removed – new ones raised – highly progressive'.[104] He puzzled over the facts that copper acted more powerfully than iron and that the drag (or push!) on the needle was much diminished by cutting radial slits in the disk. 'Very curious', he remarked of the first fact, 'and unlike ratios of ordinary magnetic power'. As to the second, it indicated 'a vortex or current'.[105]

One might think that from the base just sketched Faraday sprang to his grand discovery, the principle of magneto-electric induction.[106] A big clue had come to light: Metals acted the more powerfully the better they conducted electricity. Faraday had guessed that currents were implicated; he had only to recognise that the currents generated by the disk's motion would deflect the needle by the Ørsted effect. Then, examining the motion with his usual care, he would discover the principle for which everyone had been searching, the means of producing electricity from magnetism.

Two obstacles closed this line of reasoning. Firstly, the experimental arrangement, the deleterious effect of the radial slits, and a presumption in favour of closed currents would suggest that any current induced in the disk, whether electric or magnetic, would circulate *around* the axis of rotation. Faraday himself spoke of a 'vortex'. Such a current would not exert a tangential force on the needle. Only a radially directed current would do so; but where in the radial direction was there anything resembling a wire closing a Voltaic circuit or Ampère's molecular current? Secondly, a clear idea existed of the expected inverse of the Ørsted effect, and it did not involve motion of ponderable bodies.

Faraday tried three different ways to obtain electricity, or an electric force, from magnetism before he succeeded in 1831. All three were static arrangements interpretable as variations of Ampère's experiments on the induction of currents.[107] The experimental arrangement with which Faraday succeeded, however, the famous transformer ring, is more easily interpreted on Ørsted's than on Ampère's theory. The iron ring, wrapped on either side of a diameter with coils of wire, would be a conduit for circulating or undulating magnetic forces. Just as these forces were created in, and spread from, the conflict of the two electricities in the connecting wire in Ørsted's experiment, so an electric force might arise where oppositely propagating magnetic forces, concentrated in the iron ring, came into magnetic conflict. And so it was. When Faraday brought a pile into the circuit with one coil, say A, a galvanometer connected with the other, B, passed a current.[108] The inverse Ørsted effect had eluded physicists not because it is weak, but because in the usual experiments it was transient.

Faraday referred the transiency to the passage of a 'very short and sudden . . . wave of electricity' in B, 'caused at moments of breaking and completing contacts' in A. 'Hence', he concluded, '[there] is no permanent or peculiar state of [the] wire from B'; it responded passively to the pulse struck in it by the magnetic conflict.[109] Faraday set a high value on the discovery of this magnetic wave and its associated electric tension. On 12 March 1832 he sent a sealed note for deposit at the Royal Society. When opened a century later, this note seemed extraordinarily bold and prescient. 'Certain of [my] results', Faraday had written,

> lead me to believe that magnetic action is progressive, and requires time; i.e., when a magnet acts upon a distant magnet or piece of iron, the influencing cause . . . proceeds gradually from the magnetic bodies and requires time for its transmission which will probably be found to be very sensible. I think also, that I see reason for supposing that electric induction (of tension) is also performed in a similar progressive time.[110]

This progressive motion he compared to waves in the transmission of light and sound.

The informed electrician of 1832 would not have been astounded by these conjectures. The various ether theories of the Ørsted effect assumed the propagation of the conflict in time. Faraday knew these theories: He had studied Ørsted's work carefully and had received letters from Ampère detailing the hypothesis of the electromagnetic and luminiferous ether. Moreover, he knew, and probably favoured, the old ether theory of electrical action. His first authority, the third edition of the *Encyclopaedia Britannica,* ascribed electrical action to stresses in the same medium that propagates light, and illustrated the models of Canton, Beccaria, and Cavallo;[111] a textbook he much admired, Singer's *Elements,* accounted for repulsion between like-charged objects in the manner of Cavallo, placing the seat of ponderomotive forces not in bodies, but in the medium between them.[112]

The claim for priority in Faraday's sealed note does not include a wavelike theory of *electromagnetism,* and quite properly, as many others were there before him. The note might best be construed as a hedge against those who might try to explain his new results on magneto-electric induction by an obvious generalisation of the usual ether theories. It bears strong witness that the ideas about the field and its representation with which Faraday *began* his systematic work on electromagnetism were familiar to the physicists of his time.[113]

Notes

Owing to restrictions in space, it has not been possible to give all pertinent citations to the primary literature examined.

1 'Aether', *Encyclopaedia Britannica,* 3rd ed., 18 vols. (Edinburgh, 1797), 1:216.
2 M. J. Brisson, *Dictionnaire raisonné de physique,* 2nd ed., 6 vols. (Paris, 1800), passage quoted from 3:158; Charles Hutton, *A philosophical and mathematical dictionary,* 2nd ed., 2 vols. (London, 1815), 1:487; J. T. S. Gehler, *Physikalisches Wörterbuch,* ed. H. W. Brandes et al., 20 vols. (Leipzig, 1825–45), 1:271.
3 Cf. M. A. Tonnelat (1959).
4 J. C. Maxwell, *Scientific papers,* ed. W. D. Niven, 2 vols. (Cambridge, 1890), 1:527; W. Thomson, quoted in H. Stein, 'On the notion of field in Newton, Maxwell and beyond', *Historical and philosophical perspectives of science,* ed. R. H. Stuewer (Minneapolis, 1970), 264–86, 299–310, on 306: 'Any space at every point of which there is a finite magnetic force is called . . . a magnetic field'. Cf. ibid., 266; M. B. Hesse (1961), 197–8.
5 J. C. Maxwell, *A treatise on electricity and magnetism,* 3rd ed., 2 vols. (Cambridge, 1891), 2:493 (§866). Cf. Stein, 'The notion of field', 299, 305.
6 L. P. Williams, *The origins of field theory* (New York, 1966), 67.
7 Maxwell, *Treatise,* 1:380 (§260).
8 Cf. H. A. Lorentz, *The theory of electrons,* 2nd ed. (Leipzig, 1915), 22–3, 30–3, 213–15.
9 L. P. Williams, *Origins of field theory,* pp. 31, 67, 78. Cf. L. P. Williams, 'Faraday and the structure of matter', *Contemporary Physics 2* (1960), 92–105, on 95, 100; Berkson (1974), 25, 28, 30, 33.
10 L. P. Williams, 'Michael Faraday and the evolution of the concept of the electric and magnetic field', *Nature 187* (1960), 730–3; Joseph Agassi, *Faraday as a natural philosopher* (Chicago, 1971), 79–80.
11 J. Brookes Spencer, 'Boscovich's theory and its relation to Faraday's researches: an analytic approach', *Archives for History of Exact Sciences 4* (1967), 184–202; T. H. Levere, 'Faraday, matter and natural theology: reflections on an unpublished manu-

script', *British Journal for the History of Science 4* (1968), 95–107; Heimann (1971).
12 Respectively, Berkson (1974), 53; and Agassi, *Faraday,* 28–9, 323.
13 Berkson (1974), 33, 40, 46, 48, 60, 62. 'Faraday's new metaphysical view . . . became clear to him in a rush'. Ibid., 72.
14 Ibid., x.
15 See Heilbron (1979).
16 Cf. C. H. Pfaff, 'Elektricität', in Gehler, *Physikalisches Wörterbuch,* 3:233–389, on 254.
17 Cf. S. P. Thompson, 'The development of the mercurial air pump', *Telegraph Journal 21* (1887), 556–8, 587–90, 610–13, 632–4, 656–65.
18 J. B. Biot, *Traité de physique,* 4 vols. (Paris, 1816), 2:213–14, 254–5, 265, 291–4, 312–13; A. C. Becquerel, *Traité expérimental de l'électricité et du magnétisme,* 7 vols. (Paris, 1834–40), 2:4, 169, 185.
19 M. Faraday, *Experimental researches in electricity,* 3 vols. (London, 1839–55), 1:439.
20 R. J. Haüy, *Traité élémentaire de physique,* 2 vols. (Paris, 1803), 1:361–4. Cf. Becquerel, *Traité expérimental* 2:192–3.
21 Biot, *Traité,* 2:218, 242–3, 281, 316–18, 326. Cf. P. M. Roget, 'Electricity', in *Library of useful knowledge: natural philosophy,* vol. 2 (London, 1832), 12.
22 Canton to Priestley, 5 Apr. 1766, in J. Priestley, *The history and present state of electricity,* 3rd ed., 2 vols. (London, 1775), 1:305–6.
23 T. Cavallo, *A complete treatise on electricity,* 3rd ed., 3 vols. (London, 1786–95), 3:195–7; T. Milner, *Experiments and observation in electricity* (London, 1783), 91–9. Cf. A. Bennet, *New experiments on electricity* (Derby, 1789), x–xi; G. Adams, *An essay on electricity,* 2nd ed. (London, 1785), 56, 76–80.
24 Milner, *Experiments,* 99.
25 Cavallo, *Treatise,* passage quoted from 3:192–5; T. Cavallo, *Elements of natural or experimental philosophy,* 4 vols. (London, 1803), 3:359, 404–5; adapted by J. J. Prechtl as 'Untersuchungen über die Modification des elektrischen Ladungszustandes', *Annalen der Physik (Gilberts Annalen, Poggendorffs Annalen) 35* (1810), 28–104, on 46 ff., and by M. van Marum as 'On the theory of Franklin', *Annals of Philosophy* 16 (1820), 441–53.
26 F. U. T. Aepinus, *Tentamen electricitatis et magnetismi* (St. Petersburg, 1759), 192; T. Young, *Course of lectures on natural philosophy and the mechanical arts,* 2 vols. (London, 1807), 1:664–5; A. Libes, *Histoire philosophique des progrès de la physique,* 4 vols. (Paris, 1810–13), 3:201, 205; Cavallo, *Elements,* 3:353; Roget, 'Electricity', 10, 29; Pfaff, 'Elektricität', 301–4.
27 Cf. G. Miller, 'Observations on the theory of electrical attraction and repulsion', *Transactions of the Royal Irish Academy 7* (1800), 139–50; 'Observations on [van Marum's] memoir', *Annals of Philosophy 1* (1821), 181–6.
28 E.g., G. J. Singer, *Elements of electricity and electrochemistry* (London, 1814), 24, 55–7, 65, 78–9; 'On the supposed repulsion of electricity', *Philosophical Magazine 49* (1817), 208–9; J. A. Deluc, *Traité élémentaire sur le fluide électrogalvanique,* 2 vols. (Paris, 1804), 1:55–9; J. B. van Mons, 'On . . . galvanism and electricity', *Journal of Natural Philosophy, Chemistry and the Arts (Nicholson's Journal) 24* (1809), 179–89; Prechtl, 'Untersuchungen', 46–50.
29 Cf. F. A. C. Gren, *Grundriss der Naturlehre* (Halle, 1820), §§1114, 1125, 1199, 1204–5, 1212, 1400; J. C. Fischer, *Anfangsgründe der Physik* (Jena, 1797), 650–3; E. G. Fischer, *Physique mécanique* (Paris, 1806), 263–9.
30 A. Avogadro, 'Considérations sur l'état . . . d'un corps non-conducteur', *Journal de Physique 63* (1806), 450–62, and *65* (1807), 130–45. Cf. M. Gliozzi, *Fisici piemontesi* (Turin, 1962), 14.

31 G. B. Beccaria, 'De atmosphaera electrica', *Philosophical Transactions 60* (1770), 277–301; G. B. Beccaria, *A treatise upon artificial electricity* (London, 1776), 186, 218, 220, 384–7. Cf. M. Gliozzi, 'Giambatista Beccaria nella storia dell'elettricità', *Archeion 17* (1935), 15–47, on 41–2.

32 Heilbron (1979), 409–12.

33 *Journal of Natural Philosophy, Chemistry and the Arts (Nicholson's Journal) 21* (1808), 278–90; *Journal für Chemie und Physique (Schweiggers Journal)*, 6 (1808), 53–83. Cf. J. J. Prechtl, 'Einige Bemerkungen zu Herrn Avogadros Abhandlung', ibid., 84–115; Prechtl, 'Untersuchungen', 30 ff.

34 A. Avogadro, 'Note sur la nature de la charge électrique', *Archives de l'Electricité 2* (1842), 102–10. Cf. Becquerel, *Traité expérimental*, 2:178–9; the editor's remarks in A. Avogadro, *Opere scelte* (Turin, 1911), 375–7.

35 J. B. Biot, 'Sur le mouvement du fluide galvanique', *Journal de Physique 53* (1801), 224–74; T. M. Brown, 'The electric current in early 19th-century French physics', *Historical Studies in the Physical Sciences 1* (1969), 61–103, on 72–5; M. L. Shagrin, 'Resistance to Ohm's law', *American Journal of Physics 31* (1963), 536–47.

36 J. Bostock, *An account of the history and present state of galvanism* (London, 1818), 100–50; W. Henry, 'On the theories of the excitement of galvanic electricity', *Memoirs of the Manchester Literary and Philosophical Society 2* (1813), 293–312; Singer, *Elements*, 378–85.

37 H. Davy, 'On some chemical agencies of electricity', *Philosophical Transactions 97* (1807), 1–56, on 45–7; H. Davy, *Elements of chemical philosophy*, vol. 1 (London, 1812), 168–70. Cf. J. W. Pfaff, *Der Voltaismus* (Stuttgart, 1803), 69; C. A. Russell, 'The electrochemical theory of Sir Humphry Davy', *Annals of Science 15* (1959), 1–13, 15–25, and *19* (1963), 255–71, on 6–9.

38 Davy, 'Some chemical agencies', 29–30; Davy, *Elements*, 170.

39 According to a notebook dating from *c*. 1810 (Russell, 'Electrochemical theory', *19*, 262). Cf. Pfaff, *Voltaismus*, 35, 40; Bostock, *Account*, 141–2.

40 T. von Grotthuss, 'Mémoire sur la décomposition de l'eau . . .', *Annales de Chimie et de Physique 58* (1806), 54–74. Cf. Faraday, *Researches*, 1:142–3; E. Hoppe, *Geschichte der Elektrizität* (Leipzig, 1884), 278–9.

41 J. G. Children, 'An account of some experiments with a large Voltaic battery', *Philosophical Transactions 105* (1815), 363–74.

42 J. T. Mayer, *Anfangsgründe der Naturlehre*, 3rd ed. (Göttingen, 1812), 476–82; Whittaker (1910), 78–80; Pfaff, 'Elektricität', 350–89.

43 Cf. A. Bennet, 'A new suspension of the magnetic needle . . .', *Philosophical Transactions 82* (1972), 81–98, on 87–8; *Encyclopaedia Britannica*, 3rd ed., 1:216.

44 E.g., Young, *Course*, 1:668–9, 684.

45 J. G. Children, 'An account of some experiments . . .', *Philosophical Transactions 99* (1809), 32–8. Cf. K. Meyer, 'The scientific life and works of H. C. Ørsted', in H. C. Ørsted, *Naturvidenskabelige skrifter*, ed. K. Meyer, 3 vols. (Copenhagen, 1920), 1:xiii–clxvi, on lv; J. F. Demonferrand, *Manuel d'électricité galvanique* (Paris, 1823), 7.

46 A. M. Ampère, in Société française de Physique, *Collection de mémoires relatifs à la physique*, pts. 2–3, *Mémoires sur l'électrodynamique*, 2 vols. (Paris, 1885–7), 1:192 n. Hereafter *Mémoires sur l'électrodynamique*. Cf. A. M. Ampère, in ibid., 1:6–10, 12, 216; A. M. Ampère, letter of 21 Feb. 1821, in A. M. Ampère, *Correspondance du grand Ampère*, ed. L. de Launay, 3 vols. (Paris, 1936–43), 2:566. Similar definitions may be found in Prechtl, 'Untersuchungen', 30–46; Demonferrand, *Manuel*, pp. 5–7; P. Prevost, 'Tentative faite dans le but de concilier deux principes fondementaux de la théorie d'électricité', *Bibliothèque Universelle 21*

(1822), 178–88, on 185; J. D. Colladon, 'Déviation de l'aiguille aimantée par le courant', *Annales de Chimie et de Physique 33* (1826), 62–75; A. de la Rive, 'Recherches sur la cause d'électricité voltaïque', ibid. *39* (1828), 297–324, on 304–5; Becquerel, *Traité expérimental,* 2:3, 214–16.

47 H. Davy, *A syllabus of a course of lectures* (1802), in H. Davy, *Collected works,* ed. J. Davy, 9 vols. (London, 1839–40), 2:327–436, on 390–1. Cf. A. C. Becquerel, *Traité de physique,* 2 vols. (Paris, 1842–4), 1:53–4.

48 E.g., Becquerel, *Traité expérimental,* 2:215–16. Cf. Gren, *Grundriss,* §§1403–4.

49 Biot, quoted in Pfaff, 'Elektricität', 380–1. Cf. Young, *Course,* 1:670–1; T. Young, 'Outlines of experiments and inquiries respecting sound and light', *Philosophical Transactions 90* (1800), 106–50, on 125–6.

50 Ørsted, *Skrifter,* 1:267–73. Cf. Meyer, 'Scientific life', xxv–xxxvi, cxiv; R. C. Stauffer, 'Speculation and experiment in the background to Ørsted's discovery of electromagnetism', *Isis 48* (1957), 33–50.

51 H. C. Ørsted, *Ansicht der chemischen Naturgesetze* (1812), in Ørsted, *Skrifter,* 2:90–100.

52 Ibid., 40–1, 110, 132–5; H. C. Ørsted, 'Theorie over lyset' (1815–16), in Ørsted, *Skrifter,* 2:433–5.

53 As this proposition conflicted with the usual explanation of electric light *in vacuo,* Ørsted hinted that vacuum might insulate. *Skrifter,* 2:126–7.

54 *Skrifter,* 2:435. Cf. T. Thomson, review of Ørsted's *Recherches* (1813), *Annals of Philosophy 13* (1819), 368–77, 456–63, and *14* (1819), 47–50 and esp. 459–63.

55 Thomson, review of Ørsted's *Recherches,* 369; F. W. J. von Schelling, 'Ueber Faradays neueste Entdeckung', in F. W. J. von Schelling, *Sämmtliche Werke,* 14 vols. (Stuttgart and Augsburg, 1856–61), 9:439–52, quoted passage on 443–4. Cf. Bostock, *Account,* 102; W. Whewell, *History of the inductive sciences,* 3rd ed., 2 vols. (New York, 1859), 2:243; Hoppe, *Geschichte,* 200, 222–3.

56 *Ansicht,* in Ørsted, *Skrifter,* 2:147–8.

57 Ørsted, *Skrifter,* 2:223–5, 356–8; J. S. C. Schweigger, 'Ueber Elektromagnetismus', *Journal für Chemie und Physik (Schweiggers Journal) 46* (1826), 1–72, and *48* (1826), 289–352, on 10–14; Meyer, 'Scientific life', xc, xcvii–xcviii; Stauffer, 'Speculation', 48–50; J. P. Gérard, 'Sur quelques problèmes concernant l'oeuvre d'Ørsted en électromagnétisme', *Revue d'Histoire des Sciences 14* (1961), 297–312.

58 H. C. Ørsted, *Experimenta* (1820), in Ørsted, *Skrifter,* 2:214–18; translated in *Annals of Philosophy 16,* 473–6, *Journal de Physique 91,* 72–80, *Annales de Chimie et de Physique 14,* 417–25, *Annalen der Physik (Gilberts Annalen, Poggendorffs Annalen) 66,* 295–304, *Journal für Chemie und Physik (Schweiggers Journal) 29,* 275–81, all in 1820.

59 *Skrifter,* 1:cxiii.

60 Ørsted had eschewed ordinary pictures such as atoms and fluids (*Ansicht,* 149–57), but some of his concepts, such as 'quantities' of forces, might best be understood on a mechanical representation, which he deemed to be compatible with his general theory (ibid., 129–30, 136). He occasionally used one himself. Meyer, 'Scientific life', lviii–lx.

61 E.g., M. Faraday, 'Historical sketch of electromagnetism', *Annals of Philosophy 18* (1821), 195–200, 274–90, and *19* (1822), 107–17, on 107–8; C. H. Pfaff, *Der Elektro-Magnetismus: eine historisch-kritische Darstellung der bisherigen Entdeckungen* (Hamburg, 1824), 206–9.

62 A similar classification occurs in G. T. Fechner, *Elementar-Lehrbuch des Elektromagnetismus* (Leipzig, 1830), 127. Whewell, *History,* 2:247, conflates the first two types.

63 The distinction between suppositious elementary *force* and measurable force is elaborated in Heilbron (1979), 449, 475–7.
64 A. M. Ampère, 'Suite du mémoire sur l'action mutuelle des deux courans', *Annales de Chimie et de Physique 15* (1820), 170–218, on 178.
65 [J. B. Biot and F. Savart], 'Note sur le magnétisme de la pile de Volta', *Annales de Chimie et de Physique 15* (1820), 222–3; J. B. Biot, *Précis* (1823), in *Mémoires sur l'électrodynamique*, 1:80–127.
66 Ampère, in *Mémoires sur l'électrodynamique*, 1:20. Initially, Ampère hesitated between a collection of elementary loops and a current circling the magnet. Dumonferrand, *Manuel*, 117–19; Hoppe, *Geschichte*, 396; Fresnel, in *Mémoires sur l'électrodynamique*, 1:141–7, 219–23; Ampère, letters of 18 Dec. 1820, 25 March 1821, in Ampère, *Correspondance*, 2:563, 568.
67 Ampère to his son, 4 Feb. 1823, in Ampère, *Correspondance*, 2:624. Cf. Ampère to A. de la Rive, 23 Mar., 21 Aug. 1823, and to Faraday, 18 Apr. 1823, in ibid., 628, 630, 635; Demonferrand, *Manuel*, 65–8.
68 Demonferrand, *Manuel*, 43–5. Cf. Fechner, *Lehrbuch*, 20–2, 102; Ampère, in *Mémoires sur l'électrodynamique*, 1:26, 185.
69 C. H. Pfaff, *Elektro-Magnetismus*, 245–51, 268–9; Fechner, *Lehrbuch*, 129–31.
70 Ampère to A. de la Rive, Oct. 1822, in Ampère, *Correspondance*, 2:605. Ampère insisted on this point: e.g., ibid., 570, 646–7, 653; *Mémoires, sur l'électrodynamique*, 1:27, 184, 187, 403–4. Cf. Pfaff, *Electro-Magnetismus*, 254–5, 261–2.
71 Cf. K. L. Caneva, 'From Galvanism to electrodynamics: the transformation of German physics in its social context', *Historical Studies in the Physical Sciences 9* (1978), 63–159, on 78–83, 91; Pfaff, *Electro-Magnetismus*, 254–5, 261–2.
72 Wollaston, quoted in Faraday, 'Sketch', 110. Cf. D. H. Conybeare to Faraday, 4 Apr. 1823, in M. Faraday, *The selected correspondence of Michael Faraday*, ed. L. P. Williams, 2 vols. (Cambridge, 1971), 1:142.
73 Respectively, H. Davy, 'On the magnetic phaenomena produced by electricity', *Philosophical Transactions 111* (1821), 7–19, 425–39, on 14; and M. Faraday, 'On some new electro-magnetical motions, and on the theory of magnetism', *Quarterly Journal of Science 12* (1822), 74–96, on 79; reprinted in Faraday, *Researches*, 2:127–47.
74 Davy, 'Magnetic phaenomena', 14, 17, 427; Faraday, 'Sketch', 198–9; Faraday, *Researches*, 2:161–2; M. Faraday, *Diary*, ed. T. Martin, 7 vols. (London, 1932–6), 1:49; M. Faraday, 'A course of lectures on the philosophy and practice of chemical manipulation' (1827), MS no. 13 at the Royal Institution, 54.
75 Faraday, 'Sketch', 109–10; Faraday to Marcet, 15 Jan. 1822, and to Ampère, 3 Sept. 1822, in Faraday, *Correspondence*, 1:129, 134–5; Ampère to Faraday, 10 July 1822, and to A. de la Rive, Oct. 1822, 23 Mar. 1823, in Ampère, *Correspondance*, 2:590, 606, 629–30; *Mémoires sur l'électrodynamique*, 1:243; Demonferrand, *Manuel*, 190–2; P. Barlow, *An essay on magnetic attractions*, 2nd ed. (London, 1824), 230; J. T. Seebeck, 'Ueber den Magnetismus der Galvanischen Kette', *Abhandlungen der Akademie der Wissenschaften, Berlin* (1820/1), 289–346, on 295–6. Cf. Fechner, *Lehrbuch*, 128; Hoppe, *Geschichte*, 220.
76 A. M. Ampère, 'Exposé sommaire' (1822), in *Mémoires sur l'électrodynamique*, 1:192–204. Cf. Ampère, *Correspondance*, 2:576, 579, 583, 605.
77 The objections, respectively, of Schweigger, 'Ueber Elektromagnetismus', 17–18, and Muncke as reported by Fechner, *Lehrbuch*, 116–18; and of Pfaff, *Elektro-Magnetismus*, 245–6, and Faraday, letters to Marcet, 15 Jan. 1822, in Faraday, *Correspondence*, 1:130, and to G. de la Rive, 12 Sept. 1821, in Williams, *Faraday*, 126. Cf. Davy to Ampère, 26 May 1821, in S. Ross, 'The search for electromagnetic

induction, 1820–1831', *Notes and Records of the Royal Society of London 20* (1965), 184–219, on 197: 'I wish you may be able to furnish some *direct* proof of the existence of electrical currents in the magnet'.

78 Pfaff, *Elektro-Magnetismus*, 261–3; G. F. Pohl, 'Versuch über die Einwirkung des Erd-Magnetismus . . . ,' *Annalen der Physik (Gilberts Annalen, Poggendorffs Annalen) 74* (1823), 389–409, and *75* (1823), 269–322; Barlow, *Essay*, 232–3.
79 Barlow, *Essay*, passage quoted from 255–6; Demonferrand, *Manuel*, 193–4.
80 Respectively, Whewell, *History*, 2:255; and Schweigger, 'Ueber Elektromagnetismus', 58–9, 290 n. Cf. Pfaff, *Elektro-Magnetismus*, 65; Caneva, 'Galvanism', 93–4.
81 This mnemonic is Ampère's: see *Mémoires sur l'électrodynamique*, 1:12; Ampère, 'Suite du mémoire', 203; Hoppe, *Geschichte*, 206–7. It became standard.
82 Schweigger, 'Ueber Elektromagnetismus', 63.
83 J. S. C. Schweigger, 'Ueber die elektrische Erscheinung, welche die Alten mit dem Namen Kastor und Pollux bezeichneten', *Journal für Chimie und Physik (Schweiggers Journal) 37* (1823), 245–342, on 252, 262, 268, 276, 320–1; Schweigger, 'Ueber Elektromagnetismus', 42–3, 66–7.
84 Seebeck, 'Magnetismus', 297–8. Cf. H. Schimank, 'Die Entdeckung der elektromagnetischen Induktion', *Beiträge zur Geschichte der Technik und Industrie 21* (1932), 1–11, on 10 n.
85 Davy, 'Magnetic phaenomena', 11, 14; Faraday, *Diary*, 1:57; Faraday, 'Sketch', 276, 283; Faraday, 'New motions', 88–9.
86 P. M. Roget, 'Magnetism', in *Library of useful knowledge: natural philosophy*, vol. 2 (London, 1832), 19–22; P. M. Roget, 'On the geometrical properties of the magnetic curve', *Journal of the Royal Institution 1* (1830/1), 311.
87 E.g., Faraday, 'New motions', plate 3, figs. 6–8, 11, 14; Faraday, *Diary*, 1:53.
88 Pfaff, *Elektro-Magnetismus*, 201.
89 P. Erman, *Umrisse zu den physikalischen Verhältnisse des von Herrn Oersted entdeckten elektrochemischen Magnetismus* (Berlin, 1821), 53, 58–63; Seebeck, 'Magnetismus', 298; Faraday, 'Sketch', 117.
90 Schweigger, 'Ueber Elektromagnetismus', 3, 54–5, 316–17.
91 A. C. Becquerel, 'Des effets électriques qui se développent pendant diverses actions chimiques', *Annales de Chimie et de Physique 23* (1823), 244–58, quoted passage on 245; P. M. Roget, 'Electromagnetism', in *Library of useful knowledge: natural philosophy*, vol. 2 (London, 1832), 80. Cf. Faraday, 'A course', 55: 'My rotatory apparatus is a striking illustration of the vertiginous nature of the power in the wire'.
92 Ampère to Faraday, 1825, in Ampère, *Correspondance*, 2:674–5. Cf. *Mémoires sur l'électrodynamique*, 1:250.
93 Ampère to A. de la Rive, 2 July 1824, 5 Aug. 1826, in Ampère, *Correspondance*, 2:215–19, 249–50. Cf. Becquerel, *Traité expérimental*, 2:3.
94 Ampère, 'Réponse à . . . M. van Beck' (1821), in *Mémoires sur l'électrodynamique*, 1:216.
95 Becquerel, 'Effets électriques', 248–9; review of S. Marianini, *Saggio di esperienze elettrometriche*, *Annales de Chimie et de Physique 33* (1826), 113–54, on 120; A. de la Rive, 'Recherches', 323; L. Nobili, 'Sur une nouvelle classe de phénomènes électrochimiques', *Annales de Chimie et de Physique 34* (1827), 280–92, 409–38, on 292 (cf. Hoppe, *Geschichte*, 249–50); F. Savary, 'Mémoire sur l'aimantation', *Annales de Chimie et de Physique 34* (1827), 5–57, 220–1, on 55–6, 220–1; E. Becquerel, according to A. Becquerel, *Traité de physique*, 1:71. Cf. Roget, 'Electromagnetism', 57.
96 Becquerel, *Traité de physique*, 1:44. Cf. Heilbron (1979), 223–6, 256–7, 271–2, 293.

The electrical field before Faraday

97 Erman, *Umrisse*, 92; Pohl, 'Beiträge zur näheren Kenntnis des Elektromagnetismus', *Isis* (1820), 390–409, on 393–5; C. Hansteen, 'Zusätze und Berichtigungen zu den Bemerkungen über Polarlichter und Polarnebel', *Journal für Chemie und Physik (Schweiggers Journal) 48* (1826), 360–73, on 366–8.

98 'Esquisse historique des principales découvertes faites dans l'électricité depuis quelques années', *Bibliothèque Universelle 52* (1833), 225–64, 404–47, and *53* (1833), 70–125, 170–227, 315–51, on 225. Cf. Becquerel, *Traité expérimental*, 1:183, and 2:215.

99 Faraday, 'Sketch', 281; Fresnel, in *Mémoires sur l'électrodynamique*, 1:76. Cf. Ross, 'The search', 185–90; Faraday, *Researches*, 1:24.

100 A. de la Rive, in *Mémoires sur l'électrodynamique*, 1:328; Ampère, in ibid., 1:332–4; and Ampère, letter to G. de la Rive, 25 Sept. 1822, in Ampère, *Correspondance*, 2:602. Cf. ibid., 2:744, 760–7, 774. Ampère's disk should have jumped at make and break of the current in the coil; perhaps it also oscillated under the torsion in the wire of suspension.

101 Ampère, in *Mémoires sur l'électrodynamique*, 2:169, 173. Cf. K. R. Gardiner and D. L. Gardiner, 'André-Marie Ampère and his English acquaintances', *British Journal for the History of Science 2* (1965), 235–45; Ross, 'The search', 193–5.

102 C. Babbage and J. F. W. Herschel, 'Account of the repetition of M. Arago's experiments', *Philosophical Transactions 115* (1825), 467–96, on 471–4, 485–90; S. H. Christie, 'On the magnetism developed in copper and other substances during rotation', ibid., 497–509; P. Barlow to Faraday, 4 May 1825, in Faraday, *Correspondence*, 1:148–9; *Quarterly Journal of Science 20* (1825), 385–6.

103 M. Arago, 'Note concernant les phénomènes magnétiques auxquels le mouvement donne naissance', *Annales de Chimie et de Physique 32* (1826), 213–23, on 216–18; Roget, 'Magnetism', 91–6; A. de la Rive, 'Esquisse', 438–41.

104 M. Faraday, 'Friday evenings, 1825–9', MS no. 12 at the Royal Institution, 207 (1825), and 227 (26 Jan. 1827), respectively.

105 Faraday, 'A course', 57–8; and Faraday, 'Friday evenings', 227 (26 Jan. 1827), respectively.

106 Hoppe, *Geschichte*, 397–401.

107 Faraday, *Diary*, 1:178, 279, 310. Cf. T. Martin, *Faraday's discovery of electromagnetic induction* (London, 1949), 45–51.

108 Faraday, *Diary*, 1:367 (29 Aug. 1831); Faraday, *Researches*, 1:7–9; Martin, *Faraday's discovery*, 52–6. The connection of ideas is obscured in Faraday, *Researches*, 1:2–4, which introduces the discovery via an experiment with a *wooden*, not an iron, ring. Cf. Williams, *Faraday*, 177–83.

109 Faraday, *Diary*, 1:369 (30 Aug. 1831); Martin, *Faraday's discovery*, 57. The electrotonic state was therefore a second thought; an analysis of this concept will be given in the paper mentioned in n. 113.

110 The note is quoted in full in Williams, *Faraday*, 181, and in Faraday, *Correspondence*, 1:217. Cf. J. Larmor, 'Faraday on electromagnetic propagation', *Nature 141* (1938), 36.

111 H. Bence-Jones, *The life and letters of Faraday*, 2 vols. (London, 1870), 1:11; [J. Tytler], 'Electricity', *Encyclopaedia Britannica*, 3rd ed., 6:418–545, on plate 177, fig. 75, and pp. 443–5, 452, 455–61.

112 Singer, *Elements*, v, 56–8, 66–8.

113 In a paper in preparation, I show that Faraday's suspicion of material electrical currents, his recurrent positivism, and his usage of the terms *power* and *state* also had roots in contemporary physics. There is no reason to posit that Kant or Boscovich or the Sandemanians gave Faraday his basic concepts of electricity and magnetism.

7

The quantitative ether in the first half of the nineteenth century

JED Z. BUCHWALD

Institute for the History and Philosophy of Science and Technology, University of Toronto, Toronto, Canada M5S 1A1

Throughout the nineteenth century ethers played a major role in many physical theories, particularly the wave theory of light, since, it was thought, light waves travel in some ethereal medium. Between 1830 and 1890 wave optics was primarily concerned with the mathematical laws that govern ether waves. Much effort was expended on attempts to analyse various quantitative physical models for ether in order to determine its laws of motion. The first such model was developed by Augustin Fresnel in the early 1820s; by the mid 1830s it was being carefully investigated. The results of the investigations were frequently useful: Several optical phenomena that had previously escaped explanation were accounted for. However, difficulties also arose with the model, and one in particular led to important changes in it. In this chapter, I shall discuss the origins of this first theory of the optical ether and examine the broad outlines of its subsequent development. First it is essential to recall briefly the major developments in optics during the late seventeenth and the eighteenth centuries.

During the seventeenth century the hypothesis that light was a spherical pulse propagating at a finite rate in a universally present medium was created and developed mathematically, culminating in the *Traité de la lumière* (Leiden, 1690) of Christiaan Huygens. Huygens demonstrated that the common facts of refraction – Snell's law and total internal reflection – could be deduced by assuming that the pulse travelled more slowly in the refracting body than in the bounding medium. Moreover, Huygens was also able to explain the peculiar 'double refraction' of Iceland crystal[1] by using his principle of secondary waves and by assuming that, in the crystal, light was propagated in both spheroidal and spherical pulses.[2] What Huygens's theory lacked was a strong dynamical foundation; his explanation of how the universal medium propagated light was not quantitative – he could not show mathematically that the characteristics of the medium implied the kinds of motion that, his theory supposed, constituted light.[3]

In the 1670s, Isaac Newton created a quite different theory of light that agreed with Huygens's only in supposing that light takes time to travel (not everyone believed that it did until after 1728).[4] Using experiments on prismatic refraction, Newton argued, in the first instance, that light was not a form of motion, whether actual or potential, but had properties much more like those possessed by matter. White light, he argued, consisted of a large number of distinct kinds of light, each of which corresponded to a spectral colour. The passage of white light through a prism separated its component kinds, which could not thereafter be further decomposed. Each kind of light obeyed Snell's law of refraction, but with a unique index of refraction that increased from the red to the blue end of the spectrum.[5]

According to Newton each kind of light consisted of a single type of material 'corpuscle', and he explained refraction by an attractive force exerted between the refracting medium and the corpuscle.[6] Consequently, in Newton's theory the speed of light had to be greater in the refracting than in the bounding medium – a requirement that, of course, conflicted with the requirements of the pulse theory. Newton modified his theory in later years in order to explain partial reflection, the colours of thin films, and the related phenomenon of 'Newton's rings', and he then introduced, in addition to the corpuscles, a universal medium in which pulses were propagated and which acted dynamically upon the optical corpuscles.[7] However, Newton continued to identify light with the corpuscles and not with the pulses.

Comparing the two theories at the end of the seventeenth century, we find that they differed, in the first instance, in the velocity condition: Light moved faster in refracting media according to Newton; it moved more slowly according to Huygens. Neither Newton nor Huygens (who did not consider colours) provided a mathematical, deduced formula for dispersion. Newton, however, could explain the phenomenon, and, perhaps less convincingly, he could also explain the colours of thin films and related reflection phenomena; Newton could also explain Grimaldi's phenomenon, or what we would now call diffraction.[8] Both Newton and Huygens provided formulae for double refraction, but only Huygens's was deduced (from his hypothesis that light spreads in spheroidal, as well as in spherical, pulses in Iceland crystal). However, neither formula was widely accepted throughout most of the eighteenth century. Finally, whereas Newton could explain the polarisation effects associated with double refraction, Huygens could not.[9]

Corpuscular optics in the early nineteenth century

By the middle of the eighteenth century many of Newton's ideas, and other ideas that, though not his, were developed by people who thought of themselves as his disciples, were widely accepted on the Continent as well as in Britain. Corpuscular optics in particular achieved widespread assent, and Huygens's pulse theory was largely disregarded.[10] The powerful impact of Newtonian concepts led, during the Napoleonic era in France, to a programme of physical research that was institutionally organised and carried on by men who thought of themselves as professional scientists.[11] The guiding figure of the programme, Pierre Simon de Laplace, investigated several phenomena within a sharply defined theoretical context. Laplace and his followers assumed that electricity, magnetism, heat and light, as well as matter, consisted of particles between which centrally directed forces acted at a distance. They frequently referred to the particles as 'molecules', which meant simple, uncompounded units. Although they rarely discussed the nature of the molecule, for the most part they treated it as a material 'point' that, although it possessed mass and exerted force, was not extended. The molecules could, however, exist in stable groups, thereby composing a complex structure.[12] Rays of light, according to the Laplacians, consisted of particles that could be acted on by matter through central, short-range forces.

Laplace used this assumption to deduce a formula for atmospheric refraction by calculating the action that the atmosphere exerted on the optical corpuscle.[13] Then, between 1807 and 1811, Laplace and his pupil Etienne Louis Malus were able to show that Huygens's laws of double refraction (which were empirically confirmed by Malus) were consistent with corpuscular optics.[14] They based their demonstration on the principle of least action, which, they believed, applied uniquely to systems governed by forces acting at a distance.[15] In effect, Laplace and Malus assumed an expression for the velocity of the extraordinarily refracted corpuscle; they then demonstrated that, when this expression was substituted into the principle of least action, Huygens's laws resulted. Consequently, by 1811 corpuscular optics had already achieved a high degree of success in accounting for two phenomena that the pulse theory had never treated in detail (specifically, atmospheric refraction) or that had, until recently, been its sole province (double refraction).

During this period corpuscular optics achieved further successes after Malus discovered that polarisation could be produced by reflection, as well as by passing light through a doubly refracting crystal.[16] It was also observed that regular, coloured images were produced when light passed through a thin plate of crystal and then through a double refractor (chromatic polarisation).[17] No one was able to deduce a mathematical explanation of these phenomena from corpuscular principles in the way that Huygens's laws for double refraction could be deduced from them. Malus and another participant in this programme of research, Jean Baptiste Biot, were able, however, to provide convincing qualitative explanations and to predict certain phenomena by using experimentally justified (though not theoretically deduced) laws.[18] The foundation of their explanation was, of course, the optical corpuscle, or, rather, the complex grouping of material points that had replaced the corpuscle by about 1812.[19]

Thomas Young's wave optics

At about the time that the Laplacian programme took shape (*c.* 1800), but shortly before its first major successes in optics, Thomas Young revived Huygens's theory in England,[20] and he added to it the requirement that, like sound, light consisted of a series of regularly spaced troughs and peaks, or waves, which could therefore interfere with one another; the peak to peak, or trough to trough, distance – the wavelength – determined the colour. Aided by his principle of interference, Young was able to explain several diffraction phenomena. Moreover, he used results of eighteenth-century experimental work on sound to claim – without mathematical demonstration – that the waves of light would not diverge more than was necessary to explain diffraction.[21] However, like Huygens, Young did not develop a quantitative account of the properties of the medium in which light propagated, though he frequently speculated upon the nature of the medium (Cantor, 1970). Moreover, again like Huygens, Young initially ignored polarisation since he could not see how to introduce an asymmetry about the axis of a ray. In both France and Britain there was no compelling reason, especially after 1811, to pay close attention to Young's revival and development of the medium theory. Only a quantitative account of diffraction escaped corpuscular optics but not wave optics; however, as far as was then known, this phenomenon was nevertheless consistent with corpuscular hypotheses.[22]

Towards the end of the Napoleonic era (*c.* 1814), then, most French scientists would have had little reason to consider the wave theory a compelling alternative to corpuscular optics. If, however, we look dispassionately at the rival claims of the two theories at this time, we can see that, if quantitative prediction were our sole criterion, there would have been little to choose between them. Both could yield Huygens's construction for double refraction; both also yielded Snell's law for ordinary refraction. Both could be founded on a basic principle (least action for corpuscular optics and least time for wave optics).[23] The wave theory did provide a mathematics – crude though it then was – for diffraction,[24] but the corpuscular theory could at least explain the phenomenon, and not enough was known empirically about diffraction to compel consideration of the wave theory's mathematics in this area. Physically, moreover, the corpuscular theory was unquestionably superior to the wave theory because it could explain polarisation, and it therefore at least held out the hope that it might one day be possible to deduce the laws of polarisation from corpuscular principles. And the physical hypotheses of corpuscular optics – material points with forces acting at a distance between them – were widely accepted in both France and Britain at the time, whereas the concept of a universal medium was rarely used in France, although it was occasionally discussed in Britain. Clearly the wave theory could only begin to command serious attention as a rival to corpuscular optics if, at the least, it could deal with polarisation and could provide a clear physical alternative to the optical corpuscle.

Augustin Fresnel and ether

In 1814 a young French civil engineer wrote to his brother concerning his increasing doubts about the received corpuscular theories of both light and heat. At the time, Augustin Fresnel had only a sketchy knowledge of contemporary optics, but, for reasons that remain unclear, he was disposed to believe that both light and heat were somehow connected with the 'vibrations of a special fluid which extended throughout space'.[25] Within little more than a year, Fresnel, independently of Young, had explained diffraction by means of the principle of interference. He also demonstrated experimentally that the pattern of fringes produced by a rectangular diffractor was inconsistent with what could reasonably be expected on corpuscular grounds, but that the pattern was well explained by the principle of interference. Fresnel soon strengthened his critique of corpuscular optics with an experiment in which interference occurred without the presence of a material body, which would have to be present according to the corpuscular theory in order to deflect the rays of light.[26]

Before 1818, however, Fresnel, like Young, had not provided for wave optics a sophisticated mathematical structure that permitted a completely general calculation of interference patterns.[27] This he accomplished by the end of 1818,[28] and, as is well known, he submitted his paper for the prize competition the Paris Académie des Sciences had announced for a mathematical account of diffraction. Fresnel's memoir won the prize, but it succeeded in convincing none of the judges that the wave theory should be accepted, despite the fact that the Laplacian programme of research had never itself provided a mathematical account of diffraction.

There were, no doubt, many reasons for the feeble impact of Fresnel's mathematically sophisticated treatment of diffraction on most of his contemporaries, but two in particular were probably of especial importance. First, at this time, Fresnel had not published a theory of polarisation, and the old objection to the wave theory in this area – that waves had to be symmetrical about the ray – still stood. Second, Fresnel's memoir was only superficially founded on the dynamics of the wave-propagating medium. In his prizewinning memoir on diffraction, Fresnel used an expression for the force on an ether particle in order to calculate the wave velocity.[29] However, he did not at this time deduce the expression from ether's mechanical properties; he simply assumed it. Indeed, Fresnel did not even describe ether's properties in detail. Consequently, he had not as yet provided a firm physical alternative to the optical corpuscle. Nevertheless, he had been working on polarisation since 1816, and his investigations here inevitably led him to the question of the dynamical properties of the medium.

Fresnel would probably not have penetrated very far into the vexing question of ether dynamics had it not been for the problems posed by polarisation, for his theory of diffraction accounted quite well for the phenomenon of diffraction without utilising ether dynamics. Certainly the existence of a dynamics for ether led, in the 1830s, to a thriving programme of theoretical research, and perhaps furthered the acceptance of the wave theory among those who would otherwise have refused to choose between it and corpuscular optics. Once the dynamics was available, lines of mathematical research arose that led to important progress in explaining phenomena that the kinematic wave theory had not been able to treat. But Fresnel's deep interest in the dynamics of ether was due neither to a clairvoyant perception of events that occurred after his death nor to the demands of his unconvinced contemporaries. Rather, the invention of the new science of ether dynamics – whose purpose was to deduce optical phenomena mathematically from the mechanical properties of the medium – was a solitary affair that derived from questions plaguing no one but Fresnel.

The primary problem that bothered Fresnel was the connection between the polarisation of a ray and the law governing its motion in a doubly refracting crystal.[30] In mid-1816, Fresnel and François Arago, Fresnel's earliest supporter, had shown experimentally that light rays polarised in planes at right angles to one another do not mutually interfere, but that rays polarised in parallel planes do.[31] Using these facts, Fresnel explained chromatic polarisation through the principle of interference (though he did not actually produce a quantitative theory of it). Now, the interference properties of polarised light clearly could not be explained unless a polarised ray possessed some kind of asymmetry about its axis, and that was incompatible with the natural assumption that light waves were longitudinal, on analogy with sound waves.

Fresnel had early (1817) perceived that if an unpolarised ray consisted of two vibratory components, one along the ray (longitudinal), and the other at right angles to it (transverse), then at least the basic facts of polarisation could be explained if the act of polarisation destroyed the longitudinal component.[32] However, he could not understand how such a process of selective destruction could occur mechanically. Nor did he then imagine that all light might be purely transverse, because he could no more understand how a mechanical process could generate only a transverse component than he could understand how it could destroy only a longitudinal one. For several years, therefore, Fresnel tried to construct polarised light out of a combination of longitudinal and transverse components, but he was in this way unable to develop a satisfactory combination. Although he was not explicit about the difficulties of the theory, the major problem was almost certainly the connection between polarisation and the paths of the rays in double refraction.

New ideas frequently contain many elements of the concepts that they were created to replace, and the physical hypothesis Fresnel developed for ether by 1821, which solved the problem of polarisation, was no exception to the rule. He had, of course, completely rejected the optical corpuscle, and perhaps the matters of heat and electricity as well, but he had not rejected material points (molecules) and central forces that act directly at a distance, because these concepts were among the most fundamental in the contemporary store of common notions. Consequently, he argued that ether itself consisted of molecules in the Laplacian sense, between which forces acted.[33] By adopting the additional (and debatable) assumption that two parallel lines of molecules can be readily separated laterally, but strongly resist mutual approach, Fresnel was able to uncouple the transverse and longitudinal vibrations, since the former would now travel much more slowly than the latter. Consequently, all light, polarised or unpolarised, was, Fresnel argued, purely transverse, and the problem of the destruction of the longitudinal component was avoided. Moreover – and here were the tangible fruits of the hypothesis – the velocity of propagation of a transverse wave clearly had to depend upon the intensity of the reaction force that a molecular displacement produced. If, then, the force of reaction depended upon the direction of the displacement, then the velocity of propagation would depend upon it as well. That, of course, was precisely what Fresnel needed to construct an explanation of the polarisation of doubly refracted rays, because the direction of vibration determined the direction of polarisation.

In this momentous 1821 memoir, Fresnel provided his first, essentially qualitative theory of polarisation and propagatory velocity in double refraction. But he did not explicitly deduce the theory from the properties of the molecular ether. Rather, he merely assumed that the force of reaction could depend upon the direction of displacement without demonstrating that the postulated dependence was consistent with the dynamical properties of the medium. During the next six months, as Fresnel progressed towards his most stunning mathematical achievement – the wave surface of the biaxial crystal[34] – he probably did not rely directly on the dynamics of ether. But, in his last memoir, he finally deduced the wave surface from ether dynamics, albeit not without a crucial failure. That deduction was among his most influential analyses, for it inaugurated a new era in physics, the era of quantitative ether dynamics.

Consider a single molecule of Fresnel's ether. If there are no incident waves, then the molecule is in equilibrium, and there is zero force acting. When the molecule is displaced, a restoring force is exerted. Now, if the ether were not homogeneous – that is, if the distribution of molecules differed from one region to another – then, for a displacement of given magnitude and direction, different molecules would experience different forces. Thus calculations that are valid for one molecule would not apply to another; hence the course of a wave could not be predicted. Consequently, Fresnel assumed homogeneity, and, to simplify the calculation further, he also imposed a condition of symmetry on the distribution of the molecules. One further assumption was necessary. To carry out the calculation he had to hold still every molecule but the one that was displaced. That, he admitted, was an incorrect assumption, since it would forbid waves: The displacement could not be propagated from particle to particle if the displacement of one did not move the others. However, the manner in which Fresnel had formulated his mathematics for double refraction required this assumption, and he accordingly used it. As a result he was able to deduce the wave surface in biaxial crystals from ether dynamics; the properties of uniaxial crystals were simply degenerate cases of biaxial properties.

Cauchy's ether dynamics

Fresnel's method inaugurated a programme of research that was continued after his death in 1827 by the mathematician Augustin Louis Cauchy. From the fundamental hypothesis of Fresnel's molecular ether, Cauchy was able to deduce a general equation of motion for any ether particle that, unlike Fresnel's equation of motion, did not presuppose an incorrect calculation of the reaction.[35] Cauchy, that is, allowed every particle to be displaced, and he calculated the net force on any given particle that resulted from its altered

The quantitative ether

distances from the remaining ether particles. The equation of motion he obtained contained summations over the set of remaining ether particles, and these summations involved the differences between the displacements of the given ether particle and the particles included in the summations. This equation, unfettered by special conditions, became the foundation of all subsequent work in the dynamics of the molecular ether.

However, precisely because Cauchy's equation was exactly derived, whereas Fresnel had used a limiting and incorrect assumption, Cauchy could not obtain wave motions as readily as could Fresnel, whose equation yielded wave velocities without requiring integration: Cauchy's equation was not so malleable. That was a difficulty for which Cauchy was particularly well prepared. Intimately familiar, unlike Fresnel, with linear partial differential equations, and rapidly becoming more familiar with them, Cauchy realised that he could transform his basic molecular equation into a differential equation of wave motion by expanding the differences in the displacements into a Taylor's series. Having done so, he then imposed restrictive conditions upon the coefficients in the resulting equations, these conditions being equivalent to arranging the particles of ether in certain symmetrical patterns. As a result he obtained differential equations of motion that yielded Fresnel's wave surface almost, but not quite, exactly.[36] Cauchy's expression for the wave surface differed from Fresnel's by terms that were much too small to detect experimentally. Consequently, his formula was as acceptable as Fresnel's. In effect, Cauchy not only had corrected Fresnel's erroneous deduction, but had also introduced a new mathematical element – the differential equation – into wave optics. Subsequently, the aim of most theoretical research in optics, whether based on an ether dynamics or not, was to discover the differential equation of wave motion obeyed by light passing through matter.

Why, one might ask, was it necessary for Cauchy to embrace the hypothesis of a molecular ether if (as was almost certainly the case) he was primarily interested in discovering and solving differential equations for light? Fresnel was directly concerned with the physical nature of ether, primarily because he could not solve the problem of polarisation without investigating this question; the sequence of his researches betrays a continual, if occasionally uncertain, preoccupation with ether dynamics. Cauchy, on the other hand, was more concerned with the mathematics of the wave theory than he was with ether's physical nature. But without the hypothesis of a molecular ether there would, at the time, simply have been no route at all to the mathematics. For, although the ultimate aim for Cauchy was always a mathematical proposition from which calculable phenomena could be deduced, this aim could only be achieved throughout most of the 1830s by deductions founded on a molecular ether. Moreover, there was then little to object to, because the hypothesis fitted so well into contemporary physical ideas; that is, it utilised the widely accepted concepts of material points and central forces. Cauchy's theory gave more than just the laws of double refraction; it also provided a formula for dispersion – something that had escaped corpuscular optics. By 1835, Cauchy had developed a unified theory of double refraction and dispersion in which both phenomena were explained by assigning the requisite values to the constant coefficients in the differential equations of motion.[37]

Dispersion had previously been a problem for the wave theory. Since light, like any wave motion, must propagate at a speed determined by the 'elasticity' and the 'density' of the medium, and since these were assumed to be fixed quantities for the ether in a given body, dispersion, which required the speed of light to depend upon the wavelength, was difficult to explain. George Biddell Airy was among the first after Fresnel to suggest an explanation, and he recurred to the Laplacian concept of adiabatic compression of a gas, which had the disquieting consequence of granting ether something like a capacity for latent heat.[38] Airy's ideas were, however, of only passing importance, and it was an earlier suggestion of Fresnel's that provided the elements of a quantitative theory.

If, Fresnel remarked in 1822,[39] the range of the molecular force were of the order of the longest visible wavelength, then the velocity of a wave would increase with the wavelength. Fresnel did not prove his claim, nor could he have done so convincingly given the limitations of his ether dynamics. Cauchy's theory of dispersion built upon Fresnel's suggestion by referring dispersion to the values of the constant coefficients in the differential equations of motion; these coefficients, of course, are functions of the molecular force. Thus in 1835, Cauchy took his basic molecular equations and, solving them, obtained an expression for the wave velocity without making approximations (he had made approximations in his theory of double refraction of the late 1820s). The expression contained the constant coefficients of the original differential equation (combined to form other constants), as well as a factor that was a function of the wavelength. In a complicated analysis, Cauchy expanded both the coefficients and the wavelength function into a Taylor's series, and he obtained the following series for the square of the index of refraction, n^2, with a wavelength λ:

$$\frac{1}{n^2} = A_1 + \frac{A_2}{\lambda^2} + \frac{A_3}{\lambda^4} + \frac{A_4}{\lambda^6} + \ldots$$

The constants A_i were sums that contained the masses of ether molecules, the force between them, and their mutual distances in a state of equilibrium.

Now, Cauchy's formula did not directly explain the cause of dispersion, because dispersion depended upon all of the A_i except A_1, and there were several physical conditions that determined their values. Adopting Fresnel's suggestion, however, Cauchy recurred to the problem of the range of the molecular force. He offered a demonstration (unfortunately incorrect) that, if the molecular force were repulsive and of the form $1/r^4$, then dispersion would not occur.[40] That repulsive action therefore governed ether in the void, where dispersion did not occur, whereas, to account for dispersion in matter, the force on an ether particle had somehow to be changed. Cauchy did not here discuss the mechanism that produced the change, but in any case his (incorrect) deduction of the $1/r^4$ repulsion for nondispersive media remained generally unknown for some time, as, in fact, did his dispersion formula itself.

Cauchy's theory of dispersion (with its allied improvement of the theory of double refraction of the late 1820s) had little impact in France when it was published late in 1835.[41] His earlier theory of double refraction had, however, been widely discussed in France, almost certainly because he presented its principles in his lectures at the Collège de France. Cauchy, however, left France after the July Revolution (1830) for political reasons and did not return for eight years. Probably because his dispersion memoir was lithographed in Prague, it was apparently difficult to obtain in France and attracted little attention.

The reception of Cauchy's theory in Britain

In Britain, however, an active programme of research founded on Cauchy's theory was in existence by early 1836. In part, the British programme emerged because Baden Powell, who corresponded with Cauchy, published a series of detailed and favourable articles on the theory in mid-1835 in the widely read *Philosophical Magazine*. The existence of Powell's account explains why Cauchy's theory was well known in Britain, but it does not fully account for the highly favourable reception that the theory there received. One further reason, no doubt, was that the kinematic wave theory was just at this time achieving full recognition in Britain. But that cannot explain the large impact of Cauchy's theory. Indeed, quite possibly Cauchy's ether dynamics was itself a major factor in the acceptance of the wave theory in Britain. Cauchy's theory was influential because for the first time it provided the wave theory with a dynamical foundation to which the tools of modern analysis – themselves introduced in Britain only during the preceding twenty years – could be successfully applied to deduce quantitative propositions. Humphrey Lloyd well expressed the view of the mathematical scientist in 1834, one year before Cauchy's dispersion theory became known in Britain:

> In making this comparison [between the wave and the corpuscular theories of light] it is not enough to rest in vague explanations which may be molded to suit any theory. Whatever be the apparent simplicity of an hypothesis, – whatever its analogy to know laws, – it is only when it admits of mathematical expression, and when its mathematical consequences can be numerically compared with established facts, that its truth can be fully and finally ascertained. Considered in this point of view, the wave-theory of light seems now to have reached a point almost, if not entirely, as advanced as that to which the theory of universal gravitation was pushed by the single-handed efforts of Newton. Varied and comprehensive classes of phenomena have been embraced in its deductions; *and where its progress has been arrested, it has been owing in a great degree to the imperfections of that intricate branch of analysis by which it was to be unfolded.*[42]

The intricate branch of analysis Lloyd mentioned was probably the integration of differential equations, for he was well aware of Cauchy's early work on double refraction and the role that differential equations played in it. Although the analysis remained intricate in Cauchy's 1835 theory, nevertheless Powell's 'Abstract' of its principles gave the mathematical scientist a method for explaining optical phenomena.[43] Within a year several British and Irish mathematicians extended Cauchy's theory to new and previously unanalysed areas of optics. They soon produced an impressive body of results that for about six years dominated optical science. The first results were dictated by the phenomenon that Cauchy had spent half a decade explaining: dispersion. Their initial aim, of course, was to use the theory to calculate dispersive effects for comparison with experiment. This itself posed a number of problems, some theoretical, others experimental, which are worth describing because they are excellent examples of the difficulties frequently faced in nineteenth-century optics in linking theory and experiment.

Foremost among the theoretical problems was the actual deduction of a dispersion formula. Powell had not seen all of Cauchy's memoir (which was issued a year *after* his 'Abstract'), but Cauchy had evidently sent him enough to reproduce preliminary deductions. Unfortunately, the dispersion formula that Cauchy derived required many pages of complicated analysis, and Powell was not a sufficiently competent mathematician to create the analysis by himself. Consequently, he had recourse to an approximation that gave the following formula for the refractive index:

$$\frac{1}{n} = H \frac{\sin(\delta/\lambda)}{(\delta/\lambda)}$$

Here H is a constant determined by the properties of the medium (the force law and the molecular spacing), and δ is directly proportional to the distance between consecutive particles – or, rather (and herein lay one form of the approximation), δ is, in effect, an average molecular distance.

Even this formula was quite difficult to compare with Fraunhofer's measured values of the indices of refraction for the seven (solar) spectral lines in different media. To use the formula as it stands, Powell would have had to find an arc for each wavelength that satisfied two conditions: First, the ratio of the arcs for any two spectral lines in a given medium had to be inversely proportional to the ratio of the wavelengths of the two lines; second, the ratio of the sines of the two arcs had to be directly proportional to the ratio of the respective indices of refraction. To avoid this time-consuming method, Powell recurred to a simpler, though again approximate, method that required the indices of two spectral lines as data points, leaving five for comparison with experiment. The results were quite good for the ten media Fraunhofer had investigated, though, as was usual at the time, there was no error analysis.

Later in 1835, William Rowan Hamilton pointed out to Powell that his formula was based on a questionable approximation (that the molecular force extends only between contiguous molecules), and Hamilton deduced a more exact one (which was, however, still approximate).[44] During 1836, Powell embarked on a series of original experimental investigations aimed at confirming Hamilton's formula, for which Hamilton had developed a particularly simple method of calculation that was not approximate, as Powell's earlier method for his first formula had been. Hamilton's formula, however, required more data points than had Powell's: Three indices and the corresponding periods had to be specified for each medium. Powell and Hamilton calculated the indices for the D line in several different media, using, as data points, Fraunhofer's indices for the B, H, and F lines in these media. Powell then tabulated the differences between theory and experiment for Hamilton's formula, as well as the same differences for his first formula. Hamilton's formula agreed with experiment to the fifth decimal place in the index; Powell's agreed only to the fourth.[45]

Continuing his researches, however, Powell found that theory and experiment did not always agree quite so well. In the fall of 1836 he completed a series of experiments that revealed a number of discrepancies.[46] Although the formula remained well confirmed for media of low dispersive power, problems arose for large dispersions. Powell actually found a fairly constant discrepancy for several media in the E and G lines, theory always predicting somewhat too low an index for E and much too high an index for G, the latter even for several weakly dispersing substances. When the dispersion was large, the deviation from theory was quite marked and, Powell admitted, could not be attributed to observational error.[47] Nevertheless, he felt that neglected terms in Hamilton's formula could possibly explain the discrepancy, though he did not attempt the calculation.[48]

The intransigent indices for the E and G lines are particularly fascinating historically because they almost certainly were due to a phenomenon that, though discovered several times, was ignored until 1870: It is now called *anomalous* dispersion. Anomalous dispersion was later seen as a fundamental phenomenon, though the recognition both of the phenomenon and of its importance was accompanied by a profound change in the theoretical complexion of optics.[49] Powell had observed anomalous dispersion, but he had not discovered it. The theoretical perspective imposed by molecular optics, which implied a specific dispersion formula, impelled Powell to conclude that this phenomenon could be explained merely by manipulating the formula implicit in the molecular theory. In the end, Powell bypassed the difficulty altogether by assigning it to experimental problems associated with the measurement of refraction spectra.[50]

The power of the molecular ether theory: selective absorption

The tenacity with which Powell upheld his dispersion formula in the face of discrepant data is one indication of his continued, and growing, faith in molecular analysis. Others in Britain were also persuaded by the theory as early as 1835, for two scientists, John Tovey and Philip Kelland, also produced dispersion formulae – which were consistent with Hamilton's – between 1835 and 1837.[51] I shall shortly discuss Kelland's work, for it led to an important controversy, but first let us examine Tovey's work, which appeared between 1835 and 1842. Tovey extended the molecular theory to selective absorption, which, like dispersion, had not previously been subject to quantitative analysis. His theory is not complicated, but it is one of the most striking examples of the power of molecular analysis and is therefore worth examining as an epitome of the theory.

In selective absorption, discovered by David Brewster in 1833, the coloured substance absorbs numerous distinct spectral portions from a beam of white light.[52] Tovey began his theory with Cauchy's basic molecular equations of motion.[53] Consider a wave travelling along the x axis. Since it is transverse, it has components η, ζ, respectively, along the y,z axes. Cauchy's equations here have the form:

The quantitative ether

$$d^2\eta/dt^2 = \Sigma b\Delta\eta + \Sigma a(\Delta y\Delta\eta + \Delta z\Delta\zeta)\Delta y$$
$$d^2\zeta/dt^2 = \Sigma b\Delta\zeta + \Sigma a(\Delta y\Delta\eta + \Delta z\Delta\zeta)\Delta z$$

Here a and b are functions of the molecular force. The differences Δx, Δy, Δz over which the summations are carried out are the components of the distances between the ether molecules; $\Delta\eta$, $\Delta\zeta$ are the differences between the values of the components of the displacement for different molecules.

Since ether consists of discrete elements, we must calculate $\Delta\eta$, $\Delta\zeta$ by taking *finite* differences. If η, ζ are represented by the exponentials $ce^{(nt+kx)}$, $de^{(nt+kx)}$, respectively, in which k can be a complex number (if the medium is absorbent), and n is a pure imaginary, then we have, for the difference $\Delta\eta$ between two molecules, one at x_2 and the other at x_1, with $x_2 - x_1 = \Delta x$:

$$\Delta\eta = \eta(x_2) - \eta(x_1) = ce^{(nt+kx_1)}(e^{k(x_2-x_1)} - 1)$$
$$= ce^{(nt+kx_1)}(e^{k\Delta x} - 1)$$

This is an *exact* equation.

Using a similar expression for $\Delta\zeta$, Tovey substituted the two into Cauchy's equation of motion, first defining several constants: s, equal to $\Sigma[b\,a(\Delta y)^2](e^{k\Delta x} - 1)$; s', equal to $\Sigma[b + a(\Delta z)^2](e^{k\Delta x} - 1)$; and s_1, equal to $\Sigma b\Delta y\Delta z(e^{k\Delta x} - 1)$. As a result he found that, setting the imaginary number n equal to $m(-1)^{1/2}$, where m is real:

$$(m^2 + s)(m^2 + s') = s_1 \tag{7.1}$$

Now k can be complex, say $q + r(-1)^{1/2}$, and, if it is, and if q is less than zero, then clearly the wave will be attenuated, since, taking real parts, it will decrease as e^{qx}. Moreover, r is equal to 2π divided by the wavelength. Consequently, according to equation (7.1), the wavelength, the frequency (n), and the coefficient of absorption (q) are linked by a quartic equation in m: Any two of q, r, m determine an equation for the other. The result is, obviously, that the absorption depends in a quite complicated fashion upon the colour of the light and upon the molecular characteristics of the medium.

The existence of a relation between colouration and absorption seems at first to be remarkably puzzling, because it appears to imply that the real and the imaginary parts of a complex number (q and r) must be related to one another. Herein lies the power of molecular theory, for the relationship is a direct result of the discrete character of ether. If the distances between the ether molecules were so small that the discreteness could be neglected, then Tovey might have calculated $\Delta\eta$ by taking the differential, $d\eta$, equal to $k\eta\Delta x$. This would have given him, instead of (7.1), an equation containing only real terms on the left, and the product of the complex number k and a real term on the right. Consequently, k would have to be real; r would accordingly be banished; and therefore no wave motion at all would be implied. In other words, Tovey had concisely demonstrated that, if ether is discrete, then some form of selective absorption *must* occur (though it could still be eliminated by properly adjusting the molecular constants for transparent media and for the void).

Although Tovey did not actually test the formula – it was too complicated and the experimental data were as yet meager – nevertheless the implication was clear and powerful: The molecular theory of ether could in principle even provide a quantitative explanation of a complex phenomenon. Tovey, and Powell, further demonstrated between 1837 and 1842 that, if the molecular constants are properly adjusted, then ether will propagate *only* elliptically polarised waves. Since reflection always produces elliptical polarisation (except, it was – incorrectly – thought, at the 'polarising' angle), they therefore had a physical account of the state of ether at the boundaries between media.[54]

The effect of matter upon ether

As molecular theory was increasingly exploited, controversies arose concerning the causes of particular phenomena. One controversy involved the cause of dispersion. Some – Kelland in particular – attributed dispersion to the finite distances between ether particles, a hypothesis requiring a particular force law. Others attributed dispersion to the direct action of material upon ethereal particles; for example, Matthew O'Brien proposed this kind of theory in 1842. However, Samuel Earnshaw, a Cambridge mathematician, vehemently rejected Kelland's theory; angry public exchanges between Kelland and both Earnshaw and O'Brien resulted.[55]

Although these differences occurred between scientists who, at the time, accepted the fundamental principles of the molecular theory, nevertheless they reflect an important internal difficulty in the theory that was extremely important after 1842. The differences involved the role of matter in optical phenomena. Although most scientists assumed that material particles exerted forces on the ether particles, they differed considerably about the details of the process. Kelland assumed that an extremely short-range force acted between ether and matter, its function being to separate ether from the matter particles and not to affect the motion of the ether particles. The major effect of this force was therefore to cause matter to displace ether, thereby altering the distances between ether particles. According to Kelland, dispersion depended solely upon these distances. Hence in his theory the force between ether and matter only *indirectly* produced dispersion. O'Brien and Earnshaw, on the other hand, used a longer-range force, which they insisted directly caused dispersion by affecting the motion of the ether particles. The issue was a complicated one, but it was generally agreed that the presence of matter had one effect on ether that could not be treated exactly with the current tools of analysis: It rendered ether inhomogeneous.

Every molecular theory of optics in the 1830s and the early 1840s, whether or not it employed matter directly, presupposed that ether remained homogeneous throughout a given material body. That is, as noted previously, no single volume of ether differed in the number of molecules it contained from any other portion. The spatial arrangement of the molecules was, of course, affected by matter, but the density (number of molecules per unit volume) was constant for a given substance. However, it was universally admitted that, in fact, matter *must* render ether inhomogeneous, since in the vicinity of a material particle the ethereal distribution had to be different from the distribution at a distance from the particle, whether the ether–matter force was extremely short range or not. Since inhomogeneity would make ether's equations of motion impossibly difficult to integrate, it was ignored. This type of problem – which is purely internal to the structure of the theory – occurs frequently in the history of physics, and usually the partisans of an ongoing programme of research either claim that the problem does not affect the validity of the researches undertaken, without first solving it, or simply do not write about it. Both tacks were taken in molecular theory.

One kind of problem that cannot be ignored or easily dismissed, however, is an unavoidable conflict between theory and experiment. If the conflict seems to be manageable by including further terms in an approximate formula, or if it constitutes only one or two exceptions within a given class the other members of which are not aberrant, then the discrepancy may be explained by recurring to difficulties of calculation or of experiment. For example, Powell first attempted to explain the difficulty with the indices of the E and G lines by arguing that terms had been left out of his dispersion formula; if the extra terms were included, he claimed, the discrepancy would disappear. He later abandoned this explanation, which would not have worked, and instead blamed the failure on experimental difficulties. But this kind of difficulty is essentially a numerical one: For example, it was not that dispersion was totally incompatible with theory, since only a small proportion of the results did not fit the theory well. There is, however, a second kind of discrepancy between theory and experiment that is much deeper, for the theory may be fundamentally incompatible with an entire class of phenomena. Given the startling successes of ether theory in explaining phenomena as diverse as double refraction, dispersion, selective absorption, and the production of elliptical polarisation by reflection, it is not surprising that theorists strongly believed that the theory could be applied successfully to all optical phenomena. In 1841 they were proved decisively wrong.

The problem of optical rotation

When a beam of linearly polarised light passes through a crystal of quartz in a certain direction (or through certain types of solutions in any direction), it is split into two beams that travel at different speeds but in the same direction. One beam is left circularly polarised; the other is right circularly polarised. A single resultant beam emerges, and it is again linearly polarised, but its plane of polarisation has been rotated. The phenomenon is appropriately called optical rotation. It was widely assumed in the late 1830s and the early 1840s that Cauchy's molecular equations could explain the phenomenon. But in 1841 the irascible Irish mathematician James MacCullagh conclusively and simply demonstrated that optical rotation is *incompatible* with the molecular equations of motion.[56]

A problem of this depth requires a solution of equal depth, and a quantitative one, if the theory as a whole is to remain capable of motivating further research. But in this case the difficulty was so profound that only someone who possessed a penetrating familiarity with modern analysis could even hope to pursue it. Cauchy had that familiarity, and when he realised that the problem was linked to the basic form of his molecular equation, he began to reflect upon another problem, previously dismissed casually, which also affected the equation. This was the problem of the inhomogeneity of ether in the presence of matter. At some time after 1844, Cauchy realised that the latter problem contained the solution to the otherwise fatal conflict between his basic equation and optical rotation. In effect, Cauchy argued, one can imagine that, in crystals and in certain solutions, the inhomogeneity imposed upon ether by matter is not random – as had previously been assumed – but periodic: That is, the alteration in the density of ether repeats at certain small intervals throughout the body. This implies that the coefficients in Cauchy's basic equation are not constants, nor do they vary haphazardly; they are periodic functions of position, and thus the partial differential equations had to be integrated with periodic coefficients. This was a difficult task, and Cauchy never published a complete method of integration. However, he was able to show that periodicity contained the answer to the problem of optical rotation.[57] Uniting an old puzzle (inhomogeneity and the effect of matter upon ether) with a pressing quandary (optical rotation), Cauchy not only saved molecular theory; he gave it a new mathematical foundation. However, the new mathematics was extremely complicated and underdeveloped. During the years after Cauchy's death in 1857, two French scientists – Charles Briot and Emile Sarrau – continued Cauchy's optical work. Their studies resulted finally in Sarrau's concise and rigorous exposition of 1867,[58] which was the last innovative use of molecular principles in optics.

Conclusion

The function of the molecular theory of ether in the development of wave optics was fundamentally mathematical and empirical. It was initially used by Cauchy to create a differential equation for wave motions in anisotropic media. He further developed the mathematics of the theory to deduce a law for dispersion. During the late 1830s and the early 1840s, Cauchy's analysis was further applied to discover the law of selective absorption and to explain elliptical polarisation. During these years the molecular ether was utilised primarily because it yielded a wealth of mathematical and empirical consequences.

Like all hypotheses, however, this one was not complete, in that, in its early form, it failed to specify the link between ether and matter. Divergent opinions on this question arose, but all shared the fundamental hypothesis that both matter and ether consisted of material points between which central forces acted directly at a distance. During the first twelve years of the life of Cauchy's theory (1830–42) most scientists concentrated on deducing laws from the theory under the assumption that, whatever the true effect of matter on ether, one could nevertheless ignore it mathematically.

This situation might have continued for several more years if the theory had been altogether successful in yielding empirical laws. But it was not. It failed to explain optical rotation, and the failure pointed to a deep-seated difficulty. In circumstances like these, scientists often become acutely conscious of lacunae in their work that they had previously ignored, and Cauchy was so affected. The function of matter therefore became a pressing question; by introducing a quantitative hypothesis concerning the ether–matter link, he was able to solve the crisis precipitated by optical rotation.

When, later in the nineteenth century, the concept of ether was generally accepted, it served a number of purposes, many of which were neither mathematical nor empirical. It buttressed philosophical speculation;[59] it salved religious doubt by locating the supernatural in ethereal regions.[60] However, ether always retained a distinctly scientific function, and it frequently yielded important results. We have seen that, in a direct way, the hypothesis of the molecular ether gave birth to the science of the differential equations of wave optics.

Notes

1 In double refraction a ray of light entering the crystal is, in general, split in two. One of the two obeys Snell's law, whereas the other does not. If the two rays, after leaving the crystal, are incident on a second crystal, then there are two positions of the second crystal with respect to the first in which each ray produces only one

ray – rather than two rays – in the second crystal. In one of these two positions, each ray produces a ray that obeys the same law of refraction as the original ray; in the other position, each produces a ray that obeys the contrary law. The two-crystal phenomenon involves the property of light now called polarisation. Huygens was able to deduce a law of refraction for the ray that does not obey Snell's law, but he was unable to explain polarisation. Polarisation eluded him because, since his pulses of light were, like sound, compressions and rarefactions, they were entirely symmetrical about the axis of the ray, and therefore no phenomenon should have depended on the orientation of the refracting body.

2 Huygens assumed that each point on the surface of a wave was itself the source of an expanding front, and that these various secondary fronts determine by their common tangent at a given moment the resultant primary front. In Iceland spar, Huygens assumed, the secondary waves consisted of spheres (for the ray that obeys Snell's law) and spheroids (for the ray that does not).

3 A dynamical theory is one that, instead of simply assuming the possibility of certain motions, deduces them from more basic propositions. For a discussion of kinematic optics, which does not involve dynamics, see A. E. Shapiro, 'Kinematic optics: a study of the wave theory of light in the seventeenth century', *Archive for History of Exact Sciences 11* (1971), 134–226.

4 James Bradley's discovery of stellar aberration and his explanation of it in terms of the finite speed of light (1728) rapidly quelled any lingering opposition.

5 Newton's published articles on optics in the 1670s are conveniently collected in section 2 of *Isaac Newton's papers and letters on natural philosophy*, ed. I. B. Cohen (Cambridge, Mass., 1958).

6 I. Newton, *Principia*, bk. 1, props. 94, 95. The difficulties Newton faced in developing a consistent theory of dispersion are discussed in Z. Bechler, 'Newton's law of forces which are inversely as the mass: a suggested interpretation of his later efforts to normalise a mechanistic model of optical dispersion', *Centaurus 18* (1974), 184–222.

7 I. Newton, *Opticks*, 4th ed. (London, 1730; reprinted, New York, 1952), 347–54.

8 Ibid., bks. 2, 3.

9 Ibid., 354–61. Newton's optical corpuscles could have a shape, in which case there would be an asymmetry about the ray.

10 One of the most striking theoretical successes of corpuscular optics in the eighteenth century was its account of stellar aberration, which Huygens's theory could not readily explain (though no one tried to use it for that purpose until Fresnel, who greatly modified it; Huygens had actually denied the possibility of anything like aberration in his *Traité* by arguing that the motion of the observer does not affect the visual position of the object).

11 R. Fox, 'The rise and fall of Laplacian physics', *Historical Studies in the Physical Sciences 4* (1975), 89–136.

12 The concept of the material point and the idea of grouping points to form complex structures were first developed by R. J. Boscovich, *A theory of natural philosophy*, trans. J. M. Child (Cambridge, Mass., 1966).

13 P. de Laplace, *Mécanique céleste*, trans. Nathaniel Bowditch, 4 vols. (Boston, 1829–39), 4:415.

14 E. Frankel, 'The search for a corpuscular theory of double refraction: Malus, Laplace and the prize competition of 1808', *Centaurus 18* (1974), 222–45.

15 Although usually referred to in the early nineteenth century as the principle of 'least' action, it should be called the principle of extreme action, since, under certain circumstances, the action is a maximum. The principle asserts that the integral of the

velocity of light over distance is an extremum for the path actually traversed. The principle holds for any system that is governed by central forces acting at a distance.
16 E. Malus, 'Sur une propriété de la lumière réfléchie par des corps diaphanes', *Mémoires de la Société d'Arcueil 2* (1809), 254–67.
17 D. F. Arago, 'Mémoire sur une modification remarquable qu'éprouvent les rayons lumineux dans leur passage à travers certains corps diaphanes, et quelques autres nouveaux phénomènes d'optiques', *Mémoires de l'Institut*, pt. 1 (1811), 92–134.
18 E. Frankel, 'Jean-Baptiste Biot', unpublished doctoral dissertation, Princeton University, 1972, 231–86.
19 Material points had to be gathered together in groups to form the corpuscle in order to grant it a shape, which was necessary to explain polarisation phenomena.
20 T. Young, 'Outline of experiments and inquiries respecting sound and light', *Philosophical Transactions 90* (1800), 106–50; T. Young, 'On the theory of light and colours', ibid. *92* (1802), 12–48; T. Young, 'An account of some cases of the production of colours', ibid., 387–97; T. Young, 'Experiments and calculations relative to physical optics', ibid. *94* (1804), 1–16.
21 Newton had urged against all theories of light as a motion that, because the motion would diverge into the shadow, one would be able to see around corners (*Opticks*, 362–3). For Young's reply see the first and second articles cited in n. 20.
22 According to the corpuscular explanation of diffraction, a deflection occurs in the *immediate vicinity* of the diffracting body owing to the short-range force. Moreover, the deflection depends upon the colour, i.e., upon the type of corpuscle.
23 The principle of least time, like the principle of least action (n. 15), is in fact an extremum principle, for under certain circumstances the time can be a maximum. The principle asserts that the integral of the inverse of velocity over distance is an extremum. One can convert from least time to least action, and conversely, by replacing the velocity with its inverse – which, of course, corresponds to the opposite assumptions of corpuscular and wave optics concerning the velocity of light in refraction.
24 Young's calculations went no further than comparing the distances from two optical point sources to a given point; moreover, he could only predict positions of total destructive or total constructive interference: He could not calculate intermediate intensities.
25 Letter from A. Fresnel to L. Fresnel, 5 July 1814, in A. Fresnel, *Oeuvres complètes d'Augustin Fresnel, publiées par MM. Henri de Sénarmont, Emile Verdet et Léonor Fresnel*, 3 vols. (Paris, 1866–70), 2:821. Fresnel's work has yet to receive a comprehensive treatment. See, however, R. Silliman, 'Augustin Fresnel (1788–1827) and the establishment of the wave theory of light', unpublished doctoral dissertation, Princeton University, 1967.
26 A. Fresnel, 'Supplément au deuxième mémoire sur la diffraction de la lumière', in Fresnel, *Oeuvres*, 1:150–5. Fresnel set two plane mirrors facing one another at an acute angle and placed a light source within the angle. He obtained an interference pattern where the rays reflected from one mirror intersected those reflected from the other mirror. This was surprising, on corpuscular principles, because no matter was present to deflect the rays.
27 At this stage, Fresnel, like Young, could calculate only points of total destructive or total constructive interference produced by two point sources.
28 Fresnel used Huygens's principle to effect a general theory of diffraction: Fresnel, 'Mémoire sur la diffraction de la lumière, couronné par l'Académie des Sciences', in Fresnel, *Oeuvres*, 1:248.
29 Silliman, 'Fresnel', 189–90.
30 See n. 1.

31 A. Fresnel and D. F. Arago, 'Mémoire sur l'action que les rayons de lumière exercent les uns sur les autres', in Fresnel, *Oeuvres*, 1:509–22.
32 A. Fresnel, 'Nôte sur le calcul des teintes que la polarisation développe dans les lames cristallisées', in Fresnel, *Oeuvres*, 1:629.
33 Fresnel, 'Nôte sur le calcul', 629.
34 The 'wave surface' is a surface whose radii are each equal to the distance that a ray of light refracted along the radius travels in a unit of time. It is used to find the angle of refraction for a given angle of incidence.
35 A. Cauchy, 'Sur l'équilibre et le mouvement d'un système de points matériels sollicités par des forces d'attraction ou de répulsion mutuelle', in A. Cauchy, *Oeuvres complètes de Cauchy* (2 ser., Paris, 1882–1975), ser. 2, 8:227–52.
36 A. Cauchy, 'Mémoire sur la théorie de la lumière', *Mémoires de l'Académie des Sciences 10* (1830), 293–316.
37 A. Cauchy, *Mémoire sur la dispersion de la lumière* (Prague, 1836); reprinted in Cauchy, *Oeuvres*, ser. 2, 10:195.
38 G. B. Airy, 'On the nature of the light in the two rays produced by the double refraction of quartz', *Transactions of the Cambridge Philosophical Society 4* (1833), 79–123. Airy conceived that, if ether were heated on compression, then the wave velocity would depend on the frequency: The quicker the vibrations, the *less* time there would be for the heat to increase the elasticity of the medium. Hence the slower the vibration – with the implication of more time for the heating to increase the elasticity – the greater the wave velocity. Airy was not proffering the hypothesis as more than a suggestion, for it obviously introduced into optics all of the growing contemporary problems concerning the nature and effects of heat.
39 A. Fresnel, 'Extrait du second mémoire sur la double réfraction', in Fresnel, *Oeuvres*, 2:473.
40 Cauchy's deduction was flawed by a trivial mathematical error (he neglected a differentiation). The correct repulsion to forbid dispersion varies inversely as the sixth power of the distance.
41 See n. 37.
42 H. Lloyd, 'Report on the progress and present state of physical optics', *Annual Report of the British Association for the Advancement of Science 4* (1834), 295.
43 B. Powell, 'An abstract of the essential principles of M. Cauchy's view of the undulatory theory, leading to an explanation of the dispersion of light; with remarks', *Philosophical Magazine 6* (1835), 16–25, 107–13, 189–93, 262–7.
44 B. Powell, 'On the formula for the dispersion of light derived from M. Cauchy's theory', *Philosophical Magazine 8* (1836), 204–11.
45 B. Powell, 'Remarks on the formula for the dispersion of light', *Philosophical Magazine 9* (1836), 116–19.
46 B. Powell, 'Researches towards establishing a theory of the dispersion of light: no. III', *Philosophical Transactions 127* (1837), 19–24.
47 Hamilton carried out an elementary calculation to determine what changes in the wavelength were necessary to explain the discrepancies. The results were much too large for the accuracy of Fraunhofer's experiments.
48 In fact, if Powell had calculated further terms, the discrepancy would have been even greater, not smaller.
49 After the mid-1870s refraction of all kinds was attributed to characteristic vibrations within or of the molecules of bodies. If the frequency of the incident light was less than that of the characteristic vibration, then the index of refraction rose from unity to its maximum as a function of increasing frequency, reaching the maximum just below the characteristic frequency. The index decreased as the frequency of light approached the characteristic frequency from above, reaching a minimum (less than

one) just above it. 'Ordinary' dispersion was explained by characteristic frequencies in the ultraviolet; in anomalous dispersion the body contained, in addition to characteristic frequencies in the ultraviolet, others in the visible spectrum. Finally, absorption occurred at and near the characteristic frequencies.

50 He argued that the G line in the refraction spectrum was probably not single but an assemblage of narrow lines with consequently different wavelengths. The G line produced by an interference spectrum – which was Fraunhofer's technique for measuring wavelengths – was, on the other hand, extremely narrow, argued Powell, with a correspondingly precise wavelength. Consequently, index measurements that used wavelengths found from interference spectra would not be reliable for this line. See B. Powell, 'Researches towards establishing a theory of the dispersion of light: no. IV', *Philosophical Transactions 128* (1838), 67–72.

51 Kelland's and Tovey's memoirs are too numerous to list here. Kelland's work was published in the *Philosophical Magazine* at intervals from 1836 to 1842, and in the *Philosophical Transactions* in 1836. Tovey's work appeared exclusively in the *Philosophical Magazine* from 1836 to 1842.

52 D. Brewster, 'Observations of the absorption of specific rays, in reference to the undulatory theory of light', *Philosophical Magazine 2* (1833), 362–8.

53 J. Tovey, 'Researches in the undulatory theory of light, continued: on the absorption of light', *Philosophical Magazine 15* (1839), 450–5, and *16* (1840), 181–5.

54 See, e.g., B. Powell, 'Remarks on the theory of the dispersion of light as connected with polarization', *Philosophical Transactions 128* (1838), 253–64; B. Powell, 'A supplement', ibid. *132* (1842), 157–9.

55 Again, the exchanges are too numerous to list here; all were printed in the *Philosophical Magazine* between 1839 and 1842.

56 J. MacCullagh, 'Notes on some points in the theory of light', in J. MacCullagh, *Collected works of J. MacCullagh* (Dublin, 1880), 194–217.

57 A. Cauchy, 'Mémoire sur les perturbations produites dans les mouvements d'un système de molécules par l'influence d'un autre système', *Comptes Rendus 30* (1850), 17–24.

58 C. Briot, *Essais sur la théorie mathématique de la lumière* (Paris, 1864); E. Sarrau, 'Sur la propagation et la polarisation de la lumière dans les cristaux', (*Liouville*) *Journal de Mathématiques Pures et Appliquées, 12* (1867), 1–46, and *13* (1868), 59–110.

59 See, e.g., H. Spencer, *First principles*, 4th ed. (New York, 1880).

60 See, e.g., B. Stewart and P. G. Tait, *The unseen universe; or, physical speculations on a future state,* 3rd ed. (London, 1875).

8

Thomson, Maxwell, and the universal ether in Victorian physics

DANIEL M. SIEGEL

Department of the History of Science, University of Wisconsin, Madison, Wisconsin 53706 USA

By the middle of the nineteenth century the wave theory of light had become widely accepted in Britain, and with it had come belief in the existence of the luminiferous ether – an elastic solid that filled space and whose transverse undulations constituted light waves. In the decades after 1850 the luminiferous ether was assigned additional functions: James Clerk Maxwell showed how the optical ether could be fruitfully regarded as the seat of electrical and magnetic effects as well; and William Thomson (Lord Kelvin) argued that atoms of ordinary matter could be viewed as nothing but patterns of vortex motion in a ubiquitous, space-filling medium – 'the Universal Plenum' (Knudsen, 1972:200). These ideas of Maxwell and Thomson, as interpreted and elaborated in the closing decades of the nineteenth century by George Fitzgerald, Oliver Lodge, Joseph Larmor, and others, gave rise to the notion of a truly universal ether. This 'fundamental' and 'primordial medium' was 'assumed to be the ultimate seat of all phenomena', and all phenomena were then seen as 'dynamical', being referred to motions and associated forces in the universal medium. The programme of dynamical explanation based on the ether had great appeal because of the unification it promised, and it enjoyed great success in connection with electromagnetic theory and optics; thus, in spite of its very limited success in the realm of matter theory, the programme compelled allegiance through the end of the nineteenth century.[1]

Thomson, Maxwell, and Faraday's lines of force

That the ideas of Michael Faraday played a crucial role in the development of ether theory is not without irony, for he started out from a position of fairly radical scepticism about subtle fluids and ethers and remained reasonably true to that scepticism. It was Thomson and Maxwell who identified Faraday's lines of force with mechanical conditions in a material medium, thus providing the background for a unified ether theory of electromagnetic and optical phenomena.

Faraday, in brief, rejected the electrical and magnetic fluids and the action-at-a-distance forces associated with them, and instead proposed an approach that he considered less hypothetical. He characterised electrical and magnetic phenomena by the geometrical patterns in space of the vectorial forces that would be exerted on electrical or magnetic test bodies. These spatial patterns were graphically delineated by representative 'lines of force', whose directions represented the directions of the forces, while their spacing indicated the magnitudes of the forces (close spacing indicating strong force, and vice versa). Faraday managed to work out a rudimentary calculus of these lines of force, which enabled him to deal with known electrical and magnetic phenomena effectively, and which proved immensely fruitful in suggesting new experiments that led to the discovery of novel phenomena.[2]

Though Faraday's sceptical stance militated against the identification of these abstract lines of force with some concrete space-filling substratum or ether, he firmly believed in the relatedness of all natural phenomena, and he speculated in 1844 on a possible relationship between his lines of force and light. He proposed that 'radiant phaenomena' – light and radiant heat – be referred to transverse vibrations of lines of force rather than to the luminiferous ether; this 'notion [would] . . . dispense with the aether'. Faraday's speculations concerning a relationship between electromagnetic phenomena and light were given a convincing experimental foundation in the very next year, when he found that a beam of light propagating in a piece of glass situated in a strong magnetic field would experience a rotation of its plane of polarisation. This discovery of an action of magnetism on light profoundly influenced the later work of Thomson and Maxwell, resulting ultimately in the assimilation of Faraday's lines of force to the luminiferous ether. Indeed, Faraday himself later adopted a somewhat more positive attitude toward ether, granting that magnetic force, in particular, 'may be a function of the aether; for it is not unlikely that, if there be an aether, it should have other uses than simply the conveyance of radiations'.[3] This turn of Faraday's thought in the early 1850s was closely tied to what Thomson had done with Faraday's ideas and discoveries in the 1840s (Doran, 1975:162–79; Buchwald, 1977).

Thomson's scientific career spanned the Victorian era, from the 1840s to the opening years of the twentieth century, and throughout he was centrally occupied with questions relating to ether. In the 1840s, initially on an analogical basis, he began to associate Faraday's lines of force with conditions in a medium in space. Paradigmatic for Thomson and subsequently for Maxwell was an analogy that had arisen in the context of Thomson's studies of Joseph

Fourier's theory of the conduction of heat in solids. Thomson found that the relationship between heat sources and heat flow in this case was analogous to the relationship between electricity and electrical attraction in the conventional treatment of electrostatics, as developed by Coulomb, Laplace, and Poisson. Given this analogy, which Thomson developed in strict mathematical fashion in his initial publication of 1842, he was able simply by reinterpreting the symbols in the equations to transfer theorems and calculations from electrostatics to heat flow and vice versa.[4]

The relevance of this analogy to Faraday's ideas did not emerge until 1845, when Thomson spent some time in Paris, working in the laboratory of Victor Regnault. Faraday's work had been received with some bemusement on the Continent, and Thomson was asked if he could explicate Faraday's approach, giving particular attention to the relationship between Faraday's electrical lines of force and the traditional, action-at-a-distance electrostatics. Thomson took on the task and found that some of the mathematical tools he had used in his initial and continuing investigations of the analogy between heat flow and conventional electrostatics were relevant.[5] In particular, Thomson had developed, in part on his own and in part on the basis of his reading of works of George Green and Carl Friedrich Gauss, a mathematical approach to electrostatics that emphasised the spatial distribution and geometrical relationships of electrical forces. This approach made use of the language of partial differential equations and potential theory, including the electrostatic potential function and its derivatives, the associated families of orthogonal curves, and the relevant theorems of Gauss and Green. Using this formalism, Thomson was able to show that action-at-a-distance electrostatics and Faraday's calculus of electrical lines of force could both be cast in the same mathematical form, so that they were equivalent, at least in mathematical and operational terms.[6]

Thomson was now in a position to consider two alternative perspectives on Faraday's lines of force. On the one hand, because a complete equivalence between Faraday's treatment and action-at-a-distance electrostatics had been demonstrated, Faraday's work – including the work on dielectric materials, which Thomson had managed to treat by the methods Poisson had developed for magnetic materials – was no longer a threat to the continued acceptance of action at a distance. On the other hand, Faraday's calculus of electrical lines of force had now been mathematised in such a manner that it could be inserted as one term in Thomson's original analogy, which then became an analogy between Faraday's calculus of electrical lines of force and Fourier's treatment of heat flow. If one took the mathematical analogy to be suggestive of a physical similarity, the implication would be that, just as heat flow depends on an 'intervening medium', so also do Faraday's lines of force depend upon a medium:

> It is, no doubt, possible that . . . [electrical forces] may be discovered to be produced entirely by the action of contiguous particles of some intervening medium, and we have an analogy for this in the case of heat, where certain effects which follow the same laws are undoubtedly propagated from particle to particle. It might also be found that magnetic forces are propagated by means of a second medium, and the force of gravitation by means of a third.[7]

Faraday's discovery of the magnetic action on light in 1845 stimulated further development of Thomson's ideas concerning the dependence of electrical and magnetic forces on a medium or media. In a paper of 1847 entitled 'On a mechanical representation of electric, magnetic, and galvanic forces', Thomson considered states of linear and rotational strain in an elastic solid, showing that their distributions in space would be analogous respectively to distributions of electric force (for the case of a point charge) and magnetic force (for the cases of magnetic dipoles and electric currents).[8] Once again the basic thrust was analogical, but once again there was an undercurrent of physical significance. This was evident in the choice of a rotational mechanical strain to represent magnetic force, in consonance with Faraday's finding that the magnetic action on light was rotary in character. Furthermore, as Thomson wrote to Faraday at the time, though what he had published was merely a 'mathematical analogy', what he looked forward to was a 'physical theory':

> What I have written is merely a sketch of the mathematical analogy. I did not venture even to hint at the possibility of making it the foundation of a physical theory of the propagation of electric and magnetic forces, which, if established at all, would express as a necessary result the connection between electrical and magnetic forces . . . If such a theory could be discovered, it would also, when taken in connection with the undulatory theory of light, in all probability explain the effect of magnetism on polarized light.[9]

Thomson, then, presented his results to Faraday with diffidence; a real theory was a long way off. But it was already apparent that, when that theory came, it would exhibit a connection with the luminiferous ether. Looking back, half a century later, Thomson traced the onset of his 'fits of ether dipsomania', which had left him 'not . . . a moment's peace' over a period of fifty years, to the time of the composition of this paper of 1847.[10]

Thomson soon acquired a close follower in these endeavours, a fellow Scotsman seven years his junior named James Clerk Maxwell. The rudiments of a master–apprentice relationship centering on electromagnetic subjects were established even before Maxwell went up to Cambridge in 1850.[11] In 1854, opening a correspondence that was to continue until his death in 1879, Maxwell wrote to Thomson, asking him to prescribe a course of reading in electricity and magnetism: 'If [one] wished to read Ampère Faraday &c how should they be arranged, and at what stage and in what order might he read your articles?' In 1855, Maxwell reported on his reading, indicating that he had concentrated on the works of Thomson and Faraday and that he had been particularly impressed by Thomson's analogical method: 'Have you patented that notion with all its applications?' asked Maxwell; 'for I intend to borrow it for a season'.[12] Borrow it he did, making it the basis for a paper entitled 'On Faraday's lines of force'. In the introductory section of this paper he discussed 'physical analog[ies]' and indicated his reasons for using them. 'By a physical analogy', he wrote, 'I mean that partial similarity between the laws of one science and those of another which makes each of them illustrate the other'. As a prime example of a physical analogy, he described in some detail the parallels between heat flow and electrostatics that had been pointed out by Thomson. Thomson, it will be recalled, saw such analogies as merely steps along the way to a hoped-for 'physical theory'. Maxwell agreed, giving a clear exposition of this preparatory role of physical analogies, in particular as applied to electricity and magnetism. Maxwell felt that although much was known about the various branches of electricity and magnetism, there were also great gaps in existing knowledge, so that it would be folly to attempt a complete theory at this point.

In this situation, Maxwell felt that there were three possible approaches. First, one might try for a partial physical theory or hypothesis; but this would be dangerous, as it would tend to foster premature commitment: 'If . . . we adopt a physical hypothesis, we see the phenomena only through a medium, and are liable to that blindness to facts and rashness in assumption which a *partial* explanation encourages'. Of course one could avoid any premature conjecture concerning the actual physical basis of the phenomena by seeking refuge in a 'purely mathematical' description, devoid of 'physical conceptions', but such an approach was also to be avoided, since it would inevitably be unfruitful. The third possible approach – indeed, the approach of his choice – was that of physical analogy: Here one could avoid any commitment as to the actual physical nature of the phenomena, for the physical picture entertained was merely an analogy; on the other hand, one also avoided the sterility of the purely mathematical approach. The construction of physical analogies, then, was an appropriate first step in electrical science; the ultimate goal, however, for Maxwell as for Thomson, was a 'mature theory, in which physical facts will be physically explained'.[13]

In discussing physical analogies in general, Maxwell allowed for the possibility that such an analogy might ultimately form the basis for a physical theory. Thus, in the case of the wave theory of light, the original basis was an 'analogy . . . between light and the vibrations of an elastic medium' – a 'resemblance *in form* between the laws of light and those of vibrations' – and ultimately this 'resemblance in mathematical form . . . [gave] rise to a physical theory of light'. In the case of the particular analogies for electromagnetic phenomena that Maxwell was presenting in 'Faraday's lines', however, no basis for a physical theory was evidently being envisioned. Maxwell treated electrical lines of force, magnetic lines of force, and electric currents each by analogy with the flow of an incompressible fluid through a resistive medium, while insisting that this incompressible fluid

> is not even a hypothetical fluid which is introduced to explain actual phenomena. It is merely a collection of imaginary properties which may be employed for establishing certain theorems in pure mathematics in a way more intelligible to many minds and more applicable to physical problems than that in which algebraic symbols alone are used.

Intended as heuristic devices, these analogies in fact proved quite successful, enabling Maxwell to progress with the task begun by Thomson of expressing Faraday's calculus of lines of force in partial differential equations. The resulting equations, with some later modifications, have come to be known as 'Maxwell's equations'.[14]

Within a year of the publication of Maxwell's 'Faraday's lines', Thomson decided that the time had come to go beyond mere heuristic analogy in the mechanical representation of electromagnetic phenomena. In a paper of 1856 he announced that he was now proposing a description of 'reality', of the 'ultimate nature of magnetism'.[15] This new departure must be understood against the broader background of Thomson's developing commitment to a 'dynamical' understanding of physical phenomena. Looking back some years later, Thomson traced the origin of this programmatic commitment to an encounter with James Prescott Joule at the British Association meeting in 1847, when he 'learned from Joule the dynamical theory of heat, and was forced to abandon at once many, and gradually from year to year *all* other, statical preconceptions regarding the ultimate causes of apparently statical phenomena'. Thomson's conversion to the dynamical theory of heat was in fact not quite so abrupt; he continued to defend the caloric theory against Joule's novel views for some years after 1847. In essence, however, Thomson's recollection was correct: He did experience a dramatic conversion to the dynamical theory of heat – by 1851 if not in 1847 – and the general lesson that he abstracted from this conversion experience – that all phenomena are ultimately 'dynamical' – informed his thoughts concerning the ultimate nature of things for the next half a century.[16] For Thomson, a 'dynamical' theory, as opposed to a 'statical' theory, was one in which the forces – and hence effects in general – exerted by a given physical system were referred to internal motions within that system, rather than to primitive attractions or repulsions between its particles. Thus, in Thomson's paradigmatic case of the dynamical theory of heat and gases, gas pressure was the result of internal motions rather than the static repulsive forces between caloric particles within the gas.[17]

Thomson in his dynamical theory of heat followed Humphry Davy and W. J. M. Rankine in assuming that the motions that constitute heat were rotary motions associated with individual molecules – 'molecular vortices', in Rankine's terminology. Surrounding each 'molecular nucle[us]', then, there was vortical motion of the material medium that 'interpermeat[es] the spaces between molecular nuclei'. The nature of this material medium was not precisely specified: It might be 'electricity', a 'continuous fluid', or a molecular fluid.[18] In 1856, Thomson drew a connection between this picture of the motion that constitutes heat and a peculiar feature of the Faraday rotation – namely, that the handedness of the rotation of the plane of polarisation of the light beam depends on the direction of propagation. Thomson argued that this distinguishing feature of the Faraday rotation could be explained *only* on the assumption that the magnetic line of force corresponds to an axis of rotation of some of the material through which the light propagates. He concluded that the actual mechanical condition characterising a region traversed by magnetic lines of force would be one where the axes of the molecular vortices were all aligned in one direction, this being the direction of the line of force. This particular mechanical representation of the magnetic line of force was proposed not as an analogy, but as 'reality'. A few years later, in a talk at the Royal Institution, Thomson asserted that 'a certain alignment of axes of revolution in this [vortical] motion IS *magnetism*. Faraday's magneto-optic experiment makes this not a hypothesis, but a demonstrated conclusion'.[19]

There was more to Thomson's dynamical programme. He hoped that ultimately all magnetic forces would be fully explained in terms of the pressures associated with the centrifugal forces of the vortices, whereas electromagnetic induction would be understood in terms of the rotational inertia of the vortices. The material medium in which the vortical motions existed was identified in some notebook entries of 1858 as 'the Universal Plenum', and Thomson speculated that this 'universal fluid' might be the material substratum of the entire physical universe; ordinary matter would then be explained dynamically, in that its properties would be seen as arising from certain motions, 'vortical or other', in this universal plenum.[20] These speculations on vortical motion in a universal ethereal medium were vague and incomplete, but they were soon rendered quite definite and precise by Maxwell and used as the basis for a unified ether theory of electromagnetic and optical phenomena.

Maxwell's theory of molecular vortices

'Professor Thomson has pointed out that the cause of the magnetic action on light must be a real rotation going on in the magnetic field', Maxwell wrote approvingly; the time had come to go beyond the analogical approach and begin constructing, on the basis of Thomson's picture of molecular vortices oriented along the magnetic field lines, something like the 'mature theory' that Maxwell had envisaged in 1855. Maxwell may have begun thinking seriously along these lines as early as 1857,[21] and the results of his deliberations were published in a series of installments – constituting parts 1 through 4 of a paper entitled 'On physical lines of force' – over a period of eleven months in 1861–2. By the time he had finished, Maxwell had not only accounted for the whole known range of electromagnetic phenomena in terms of a mechanical medium in space – the 'magneto-electric medium' – he had also shown that this medium could be identified with the luminiferous ether.[22]

Part 1 of 'Physical lines' was published in March 1861 and was entitled 'The theory of molecular vortices applied to magnetic phenomena'. In the introductory passages, Maxwell contrasted his new theory with the analogies he had worked out earlier. Thus the theory of molecular vortices was not a 'mechanical illustration . . . to assist the imagination' or an 'analog[y]', but a theory with truth value: 'a theory, which if not *true,* can only be proved to be *erroneous* by experiments which will greatly enlarge our knowledge of this part of physics'. The new theory differed from the earlier analogies also in that it would 'account for' or 'explain' the observed forces. In 'Faraday's lines' the purpose had been to 'lay before the mind of the *geometer* a clear conception of the relation of the lines of force to the space in which they are traced' but 'not to account for the phenomena'. 'Now', however, he 'propose[d] . . . to determine what tensions in, or motions of, a medium are capable of *producing* the mechanical phenomena observed'. A parallel contrast was drawn with the analogies Thomson had presented in 1847. Thomson had 'not attempt[ed] to explain the origin of the observed forces', whereas Maxwell was going to 'consider the magnetic influence as existing in the form of some kind of *stress* in the medium', from which he was going to derive the 'observed forces'.[23]

In part 1 of 'Physical lines', Maxwell conceived the magneto-electric medium as a fluid. In a region of nonzero magnetic field, this medium would be filled with innumerable small vortex tubes or filaments, corresponding in geometrical arrangement to the magnetic field lines; the angular velocities of these 'molecular vortices' would be taken proportional to the field intensity. These rotational motions of the molecular vortices would engender centrifugal forces, causing the vortex filaments to have a tendency to expand equatorially and contract along their lengths; the corresponding magnetic field lines would appear to repel each other and have a tendency to shorten – a behaviour that Faraday had already discussed qualitatively. Evaluating the associated stress tensor in the medium quantitatively, Maxwell was able to account precisely for all classes of magnetic forces. Thus, at least as far as magnetic forces were concerned, Maxwell had successfully shown how to construct an explanatory mechanical theory rather than an illustrative analogy.[24]

In part 2 of 'Physical lines', entitled 'The theory of molecular vortices applied to electric currents', Maxwell began by considering the system of molecular vortices as would a mechanical engineer, and he encountered the following problem:

> I have found great difficulty in conceiving of the existence of vortices in a medium, side by side, revolving in the same direction about parallel axes. The contiguous portions of consecutive vortices must be moving in opposite directions; and it is difficult to understand how the motion of one part of the medium can coexist with, and even produce, an opposite motion of a part in contact with it.

He went on to observe that this kind of problem had been encountered and solved by the designers of mechanical devices:

> In mechanism, when two wheels are intended to revolve in the same direction, a wheel is placed between them so as to be in gear with both, and this wheel is called an 'idle wheel'. The hypothesis about the vortices which I have to suggest is that a layer of particles, acting as idle wheels, is interposed between each vortex and the next, so that each vortex has a tendency to make the neighbouring vortices revolve in the same direction with itself.

The magneto-electric medium was now to be conceived as a cellular medium, each cell consisting of a molecular vortex – a rotating parcel of fluid – surrounded by a cell wall consisting of a monolayer of small spherical particles, which roll without slipping between adjacent vortices, functioning as idle wheels or ball bearings. The cells were taken to be pseudospherical in shape – perhaps dodecahedral – and a string of vortex cells, rather than an extended vortex filament, was now taken to correspond to a magnetic field line. The cells were pictured in cross section as hexagonal (Figure 8.1).[25]

The introduction of the idle-wheel particles had two purposes: It resolved a mechanical problem left over from part 1 of 'Physical lines', and it provided for the further extension of the scope of the theory in part 2. Thus, Maxwell noted that in a region of homogeneous magnetic field, where adjacent vortices had equal angular velocities, the particles making up the cell walls would behave as ordinary idle wheels, rotating but undergoing no spatial translation. In a region of inhomogeneous magnetic field, however, adjacent vortices would have slightly different angular velocities, giving rise to a translational motion of the idle-wheel particles. Characterising the angular velocities of the vortices by a vector field $\boldsymbol{\omega}^*$, and the motions of the idle-wheel particles by

Figure 8.1

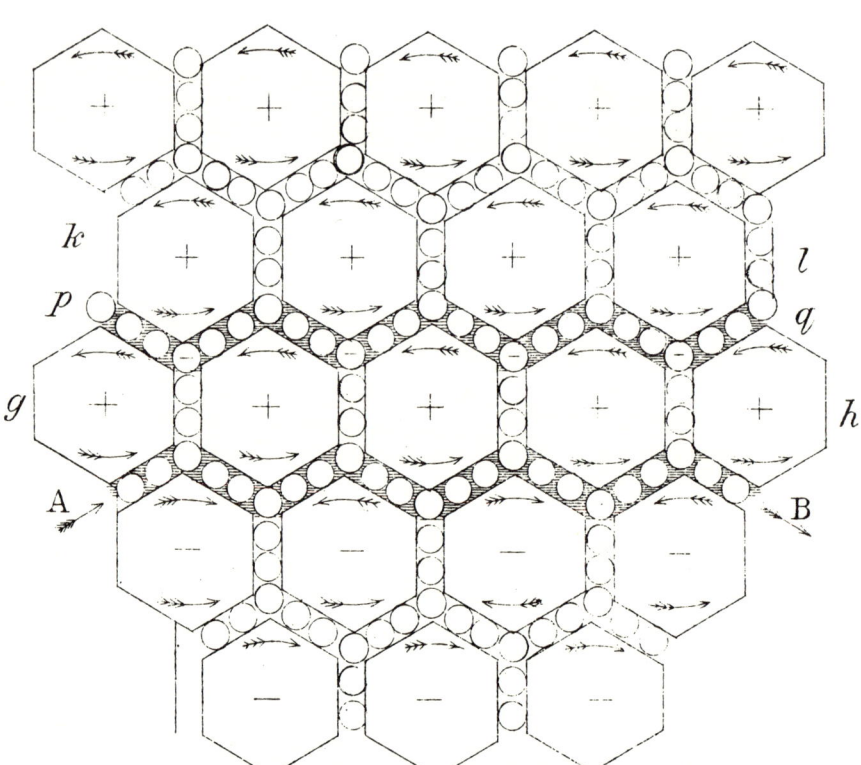

an averaged flux density ι, Maxwell calculated on a purely kinematic basis that[26]

$$\iota = (1/4\pi) \text{ curl } \omega^* \tag{8.1a}$$

As Maxwell duly noted, this equation is similar in form to what he called the 'equation . . . of electric currents' (i.e., Ampère's circuital law in differential form), which relates the electric current density $\mathbf{J}$ to the magnetic field intensity $\mathbf{H}$:[27]

$$\mathbf{J} = (1/4\pi) \text{ curl } \mathbf{H} \tag{8.1b}$$

Having calculated equation (8.1a) on a mechanical basis, and having noted its similarity in form to Ampère's law, (8.1b), Maxwell immediately drew the following conclusion: 'It appears therefore, that according to our hypothesis, an electric current is represented by the transference of the moveable particles interposed between the neighbouring vortices'.[28] The logic of this conclusion can be more explicitly represented as follows:

Given the already established correspondence

$$\omega^* \text{ corresponds to } \mathbf{H} \tag{8.1i}$$

and given the parallel mechanical and electromagnetic relationships

$$\iota = (1/4\pi) \text{ curl } \omega^* \tag{8.1a}$$
$$\mathbf{J} = (1/4\pi) \text{ curl } \mathbf{H} \tag{8.1b}$$

it is concluded that

$$\iota \text{ corresponds to } \mathbf{J} \tag{8.1f}$$

The use of this kind of argument ensured that the coherence of the theory would be maintained as its scope was enlarged: The condition for the inclusion of a new mechanical variable in the theory was that it be properly related to the variables already treated, and this condition was imposed at each step in the extension of the theory.

Maxwell next directed attention to the question of changes in the angular velocities of the vortices, corresponding to changes in the magnetic field. Associated with such changes would be inertial forces, exerted by the vortices on the idle-wheel particles; these forces, tangential to the surfaces of the vortices, were characterised by a vector field $\boldsymbol{\tau}_1$. From energetic considerations, Maxwell calculated that $\boldsymbol{\tau}_1$ would be related to ω^* as follows:[29]

$$-\text{curl } \boldsymbol{\tau}_1 = (\pi \rho_m) d\omega^*/dt \tag{8.2a}$$

where ρ_m is the mass density of the fluid in the cells. This equation is similar in form to Faraday's law of electromagnetic induction in differential form, which relates the relevant part of the electric field, $\mathbf{E}_1$, to the time derivative of the magnetic intensity $\mathbf{H}$:[30]

$$-\operatorname{curl} \mathbf{E}_1 = \mu(d\mathbf{H}/dt) \qquad (8.2b)$$

where μ is the magnetic permeability. Given the established correspondence between ω^* and $\mathbf{H}$ (and an associated correspondence between the fluid density ρ_m and the magnetic permeability μ), Maxwell concluded that the forces $\boldsymbol{\tau}_1$ could be taken to correspond to the electric field $\mathbf{E}_1$:[31]

Given the already established correspondences

$$\begin{aligned} \omega^* \text{ corresponds to } \mathbf{H} \\ \pi\rho_m \text{ corresponds to } \mu \end{aligned} \qquad (8.2i)$$

and given the parallel mechanical and electromagnetic relationships

$$-\operatorname{curl} \boldsymbol{\tau}_1 = (\pi\rho_m)d\omega^*/dt \qquad (8.2a)$$
$$-\operatorname{curl} \mathbf{E}_1 = \mu(d\mathbf{H}/dt) \qquad (8.2b)$$

it is concluded that

$$\boldsymbol{\tau}_1 \text{ corresponds to } \mathbf{E}_1 \qquad (8.2f)$$

Maxwell closed part 2 with what was obviously intended to be a final summary and conclusion to the whole paper. No further installments were contemplated and no extension to electrostatics was promised, though Maxwell was clearly dissatisfied with what he had presented. The problem that he faced and that he did ultimately resolve – further installments of the paper were eventually published – was as follows. The natural route to electrostatics, which Maxwell did ultimately use, can be represented schematically in this way:

Given the already established correspondence

$$\boldsymbol{\iota} \text{ corresponds to } \mathbf{J} \qquad (8.3i)$$

and given the parallel mechanical and electromagnetic relationships

$$\operatorname{div} \boldsymbol{\iota} + d\rho_p/dt = 0 \qquad (8.3a)$$
$$\operatorname{div} \mathbf{J} + d\rho/dt = 0 \qquad (8.3b)$$

it is concluded that

$$\rho_p \text{ corresponds to } \rho \qquad (8.3f)$$

where ρ_p is the excess density of idle-wheel particles and equation (8.3a) follows from the implicit assumption that the idle-wheel particles are neither created nor destroyed; ρ is the electric charge density; and equation (8.3b) is what Maxwell called 'the equation of continuity', expressing the conservation of electric charge.[32]

The problem here was that schema (8.3) was not consistent with schema (8.1): Nonzero values for the terms in equations (8.3a) and (8.3b) are inconsistent with equations (8.1a) and (8.1b), which are applicable only to closed circuits, and do not allow for the accumulation of charge. Maxwell had long been aware that equation (8.1b), Ampère's circuital law in differential form, had restricted applicability, and this was a matter of concern to him.[33] Having incorporated this law into his mechanical theory in the form of equation (8.1a), he now had a theory of restricted applicability, which could not be extended to electrostatics in a straightforward way. This problem halted Maxwell's progress on the theory temporarily, but he eventually found a solution and published two further installments early in 1862.[34]

Just as in part 2, the innovations of part 3 – entitled 'The theory of molecular vortices applied to statical electricity' – were set in train by consideration of a problem in the mechanical functioning of the magneto-electric medium. In electromagnetic induction, when the angular velocities of the vortices are changing in response to torques exerted at their surfaces, how are the changes in motion transmitted from the outer strata of the vortices to the interiors? If the fluid medium were frictionless, neighbouring strata of the vortex would slip aginst each other without interacting, and changes in motion would not be transmitted. If the fluid were viscous, changes in motion would be transmitted but would be accompanied by conversion of some of the rotational energy of the vortex into heat; such frictional losses of magnetic field energy were unacceptable in a realistic theory. Maxwell's solution to this problem was to endow the material in the cells with elastic properties. He now envisaged each 'molecular vortex' as a rotating, approximately spherical parcel or blob of elastic material. The tendency of these rotating blobs to bulge equatorially and flatten at the poles would still give rise to the appropriate magnetic forces, and the account of electromagnetic induction was now much improved, as changes in rotational motion would be transmitted throughout a given vortex by nondissipative elastic shear stresses.[35]

The assignment of elastic properties to the magneto-electric medium also provided for the modification of Ampère's law and thereby made possible the extension of the theory to electrostatics. Consider equation (8.1b), relating the flux of idle-wheel particles ι to the angular velocities of the vortices ω^*. This equation followed from the assumption that each vortex, approximated as spherical in shape, had uniform angular velocity throughout, and thus rotated as a rigid sphere. The molecular vortices were now, however, regarded as rotating elastic spheres, and, as Maxwell put it, the equation would now have to be 'correct[ed] . . . for the effect due to the elasticity of the medium'. This was accomplished by inserting a correction term in the equation:

$$\iota = (1/4\pi) \text{ curl } \omega^* + \iota_{elas\ def} \tag{8.4a}$$

where $\iota_{elas\ def}$ represents the flux of idle-wheel particles attributable to progressive elastic deformation of the elastic vortex spheres. These deformations entailed elastic reaction forces, exerted by the vortex spheres on the idle-wheel particles; characterising these by a vector field τ_2, and denoting the shear modulus of the elastic medium as m, Maxwell calculated the following correction term:

$$\iota_{elas\ def} = -(1/4\pi^2 m)d\tau_2/dt \tag{8.4b}$$

The final form for the equation giving the flux density of idle-wheel particles ι was[36]

$$\iota = (1/4\pi)\ [\text{curl}\ \omega^* - (1/\pi m)d\tau_2/dt] \tag{8.4c}$$

Maxwell apparently regarded this last as primarily a mechanical equation, rooted in the mechanical conceptions of the theory of molecular vortices, and finding its use in the extension of that theory to embrace electrostatics. (A corresponding electromagnetic equation was of course implied, but Maxwell did not stress this point. He did discuss the electromagnetic significance of the correction term itself, but this discussion was somewhat obscure, and its interpretation is problematic. Only later was the electromagnetic significance of this 'displacement current' term clarified [Whittaker, 1951:187–9; Bromberg, 1967, 1968a, 1968b; Hesse, 1973].)

The significance of equation (8.4c) for the theory of molecular vortices lay in the fact that the divergence of ι, the flux of idle-wheel particles, could now take on nonzero values, depending on the spatial pattern of the elastic forces τ_2:

$$\text{div}\ \iota = -\frac{1}{4\pi^2 m}\frac{d}{dt}\ (\text{div}\ \tau_2) \tag{8.4d}$$

Schema (8.3) could now be invoked, with the conclusion that the excess density of idle-wheel particles ρ_p would correspond to the electric charge density ρ. Furthermore, Maxwell was able to calculate, from equations (8.4d) and (8.3b), that ρ_p would be related to the pattern of elastic stresses τ_2 as follows:

$$\rho_p = \frac{1}{4\pi(\pi m)}\ \text{div}\ \tau_2 \tag{8.5a}$$

where m again is the shear modulus of the medium. This equation is similar in form to the electromagnetic equation relating the electric charge density ρ to the electrostatic field $\mathbf{E}_2$:

$$\rho = \frac{1}{4\pi k^2}\ \text{div}\ \mathbf{E}_2 \tag{8.5b}$$

where k is an electromagnetic constant whose signficance will emerge presently. Given the established correspondence between ρ_p and ρ, the conclusion could then be drawn that the elastic stresses τ_2 correspond to the electrostatic field $\mathbf{E}_2$, with an associated correspondence between the shear modulus of the medium m and the electromagnetic constant k:[37]

Given the already established correspondence

$$\rho_p \text{ corresponds to } \rho \tag{8.5i}$$

and given the parallel mechanical and electromagnetic relationships

$$\rho_p = \frac{1}{4\pi(\pi m)} \text{ div } \boldsymbol{\tau}_2 \tag{8.5a}$$

$$\rho = \frac{1}{4\pi k^2} \text{ div } \mathbf{E}_2 \tag{8.5b}$$

it is concluded that

$$\begin{array}{l} \boldsymbol{\tau}_2 \text{ corresponds to } \mathbf{E}_2 \\ \pi m \text{ corresponds to } k^2 \end{array} \tag{8.5f}$$

With this, Maxwell had accomplished his goal of formulating a comprehensive and coherent mechanical field theory: he had identified mechanical correlates for all of the principal electromagnetic variables and constants, and had done this according to a consistent procedure – as in schemata (8.1), (8.2), (8.3), and (8.5) – which ensured the coherence of the theory. Maxwell went on to show that the elastic stresses in the medium that were associated with the electrostatic field would give rise to the appropriate forces on electric charges.[38] Taken in conjunction with his earlier derivation of magnetic forces, this satisfied his goal of accounting for electric and magnetic forces in terms of stresses in the medium. The reward for the fulfilment of these two of Maxwell's basic goals was a theory that encompassed all electromagnetic phenomena and extended to the propagation of light as well.

The key to the treatment of the propagation of light in the theory of molecular vortices was the incorporation of the ratio of electrical units into the theory. This ratio expresses the relative strengths of electric and magnetic forces, as attributable to the presence or passage of a given amount of charge. Having developed a theory that treated both electric and magnetic forces and related them to each other through a chain of corresponding electromagnetic and mechanical equations – schemata (8.1), (8.2), (8.3), and (8.5) – Maxwell was able to connect the ratio of units with the mechanical parameters of that theory. This was accomplished in connection with the constant k in equation (8.5b), and the conclusion was as follows:[39]

$$\sqrt{m/\rho_m} = \textit{ratio of units} = 310{,}740{,}000{,}000 \text{ mm/sec} \tag{8.6a}$$

where m is the shear modulus of the medium, which controls the strength of electric forces; ρ_m is the mass density of the medium, which controls the strength of magnetic forces; the square root arises from the quadratic dependence of force on charge; and the numerical value in millimeters per second was taken from Wilhelm Weber's measurement of the ratio of units, as published in 1857.

Maxwell was now in a position to calculate the velocity, V, of transverse elastic waves propagating in the magneto-electric medium:

$$V = \sqrt{m/\rho_m} = 310{,}740{,}000{,}000 \text{ mm/sec}$$
$$= 193{,}088 \text{ mi/sec} \tag{8.6b}$$

Given the agreement of this result with the measured velocity of light within about 1 percent, Maxwell concluded that these transverse undulations of the magneto-electric medium were to be identified as light waves: 'We can scarcely avoid the inference that *light consists in the transverse undulations of the same medium which is the cause of electric and magnetic phenomena*'.[40]

This identification of the magneto-electric and luminiferous media was not yet an electromagnetic theory of light, for the 'transverse undulations' constituting light waves were not given any definite interpretation in terms of electromagnetic variables; what Maxwell had accomplished was not a reduction of optics to electricity and magnetism but rather a reduction of both to the mechanics of one ether. The unification of optics with electricity and magnetism on this basis bore immediate fruits in Maxwell's explanation of the Faraday rotation and his prediction of a relationship between refractive index and dielectric constant. The success of this unification also served to confirm Maxwell's belief in the reality of his molecular vortices, in three ways: First, the magneto-electric medium, having been identified with the luminiferous ether, now partook of the acknowledged realistic character of the latter; second, certain technical aspects of Maxwell's treatment of the Faraday rotation strengthened the conclusion that 'magnetism is really a phenomenon of rotation'; and finally, the general success of the whole theory, in scope and explanatory power, inspired confidence.[41] It was in this atmosphere of confidence in ether theory that Thomson's speculations concerning 'vortex atoms' were most favourably received.

Thomson's vortex atom and the universal ether

Thomson had broached the plan of understanding the properties associated with ordinary matter as arising in connection with certain patterns of motion, 'vortical or other', in the universal plenum, in the late 1850s, but certain difficulties had presented themselves: 'I see no possibility however of explaining the constancy of the qualities of particular substances on this hypothesis, and I see no opening for a successful investigation on dynamical principles of any of the motion that would result from the supposed circumstances'. One of the chief properties of atoms was their durability, and vortices or 'eddies' in a fluid did not seem to have the requisite permanence. Furthermore, the dynamics of rotational motion in fluids as hitherto developed was not sufficient for investigating this problem. The necessary theorems concerning vortex motion, which showed that certain vortex motions would have the kind of permanence that Thomson was looking for, were published by Hermann von Helmholtz in 1858 and reached Thomson by 1859, but it was not until 1867 that he perceived the relevance of Helmholtz's work to his own concerns. Peter Guthrie Tait, professor of natural philosophy at the University of Edinburgh and Thomson's close friend and collaborator, had taken a great interest in Helmholtz's paper on vortex motion, having translated it for his own use in 1858. Helmholtz had shown that in a perfect fluid having no viscosity or internal friction a vortex filament (i.e., a long thin region of vortical motion, the rotation being around the long axis) would have certain interesting properties. First, the vortex filament could not end within the fluid; it could have termini on the surface of the fluid or it could close back on itself, forming a vortex loop or ring. Second, a vortex filament or loop within such a perfect fluid would persist eternally, with no alteration in the strength of its vorticity. Third, vortex filaments or rings would in general move through the fluid, but vortex filaments would not be able to pass through each other – they would behave as if they repelled each other very strongly at close distances of approach. These laws of vortex motion in a perfect fluid are not subject to direct experimental test, because ordinary fluids have nonzero viscosity. Nevertheless, smoke rings in air exhibit the basic properties. They constitute vortex filaments closing back on themselves; though not eternal they persist for an impressively long time; and they move through the air in such a way that the filaments do not cross, so that two smoke rings approaching each other will be seen to rebound, being set in vibration but not destroyed by the interaction. Tait had devised a demonstration apparatus to display these effects, involving boxes containing smoke generators and fitted with rubber diaphragms, which when struck sharply propelled perfectly circular smoke rings a foot in diameter out of circular holes in the boxes. Thomson visited Tait in January of 1867 and was treated in Tait's lecture room to a 'magnificent display' of the properties of vortex rings. Apparently as a result of seeing this display, Thomson realised that Helmholtz's laws of vortex motion in a perfect fluid provided solutions to the problems of permanence and difficulty of dynamical treatment that had obstructed his earlier speculations about vortex atoms.[42]

Within a few days, Thomson had written to Helmholtz concerning the 'magnificent display' and the conclusions he had drawn:

> Just now, however, *Wirbelbewegungen* [vortex motions] have displaced everything else, since a few days ago Tait showed me in Edinburgh a magnificent way of producing them . . . The absolute permanence of the rotation . . . in a perfect fluid, shows that if there is a perfect fluid all through space, constituting the substance of all matter, a vortex-ring would be as permanent as the solid hard atoms assumed by Lucretius and his followers (and predecessors) to account for the permanent properties of bodies (as gold, lead, etc.) and the differences of their characters. Thus if two vortex-rings were once created in a perfect fluid, passing through one another like links in a chain, they could never come into collision, or break one another, they would form an indestructible atom; every variety of combinations might exist . . . a long chain of vortex rings . . . three rings, each running through each of the other . . .[43]

Expanding on this theme in a paper entitled 'On vortex atoms', which was presented to the Royal Society of Edinburgh and published widely, Thomson wrote that he was 'inevitably' led to 'the idea that Helmholtz's rings are the only true atoms'. Thomson also discussed three further aspects of the theory of vortex atoms. First, primitive elastic or distance forces between molecules, which had necessarily been assumed in the kinetic theory of gases, as developed most recently by August Krönig, Rudolph Clausius, and Maxwell, could now be replaced by the 'kinetic elasticity' of vortex rings, thereby removing anything of an arbitrary or ad hoc nature from the kinetic theory of gases. Second, the complex patterns in the optical spectra produced by heated gases were now to be explained by the vibrations of the vortex-atom rings. Finally, the solid state could also be explained by vortex atoms: An array of 'closely packed vortex-atoms . . . must produce in the aggregate an elasticity agreeing with the elasticity of real solids'.[44]

In the decades after 1867, Thomson and a growing band of followers sang the praises of the vortex atom and discussed various extensions of its explanatory power. Thomson's conjecture that 'closely packed vortex-atoms' would have the properties of an elastic solid was the key to the assimilation of the luminiferous ether to Thomson's universal plenum with its vortex motion, and he reported his continued application to this problem, as well as some progress, in 1887. Gravity would have to be comprehended within the theory if it were ever to achieve completeness, and Thomson in 1871 considered adapting LeSage's hypothesis of ultramundane corpuscles to a theoretical framework based on the vortex atom. Fruitful application of the theory to chemistry was promised by J. J. Thomson's work of 1882, in which he investigated the stability of groupings of vortex rings and found what appeared to be an explanation of the regular variation of valency displayed in Mendeleyev's periodic table. None of these developments of the theory was quantitatively conclusive or physically unassailable, but they were all suggestive of the possibilities of the approach, and were positively received in that vein.[45]

Maxwell was no shallow enthusiast, and he was at his most sober and critical in the articles on various physical subjects that he wrote for the ninth edition (1875) of the *Encyclopaedia Britannica*. It is thus most significant that in the article 'Atom', which was intended to transmit a considered, authoritative view of the subject to a broad scientific and lay public, Maxwell was very enthusiastic concerning the vortex atom, devoting fully one-quarter of the article to it, even while admitting that it was 'an infant theory', in fact more a programme than a theory. As far as Maxwell was concerned, the glory of this theory was its freedom from any arbitrary element; it represented a form of mechanical explanation that would start from the laws of mechanics and the assumption of the existence of an inviscid and incompressible universal plenum, and would proceed to explain all phenomena of the physical universe deductively, without the aid of any auxiliary assumptions. Maxwell is worth quoting at some length:

> But the greatest recommendation of this theory, from a philosophical point of view, is that its success in explaining phenomena does not depend on the ingenuity with which its contrivers 'save appearances', by introducing first one hypothetical force and then another. When the vortex atom is once set in motion, all its properties are absolutely fixed and determined by the laws of motion of the primitive fluid, which are fully expressed in the fundamental equations. The disciple of Lucretius may cut and carve his solid atoms in the hope of getting them to combine into worlds; the follower of Boscovich may imagine new laws of force to meet the requirements of each new phenomenon; but he who dares to plant his feet in the path opened up by Helmholtz and Thomson has no such resources. His primitive fluid has no other properties than inertia, invariable density, and perfect mobility, and the method by which the motion of this fluid is to be traced is pure mathematical analysis. The difficulties of this method are enormous, but the glory of surmounting them would be unique.[46]

Tait was another enthusiastic advocate of the vortex atom; less distinguished than Maxwell, he was also much more prolific in his popular writings and probably reached a much wider audience. A mathematician of some significance, Tait, considering the different ways in which vortex atoms could be linked and knotted, developed a classification of knots that stands as one of the pioneer efforts in modern topology. In his popular writings, Tait presented the idea of the vortex atom as 'by far the most fruitful in consequences of all the suggestions that have hitherto been made as to the ultimate nature of matter'. For Tait as for Thomson, the great goal was an ultimate theory in which there would be no primitive, static forces, but rather only the effects of motion. Tait expressed this by the dictum that all energy is ultimately kinetic, there being no real potential energy in nature. Tait supported this conclusion by a weak argument based on dimensional analysis, and he also invoked the LeSage–Thomson explanation of gravitation, which he apparently took as representative of the nature of all forces.[47]

By the 1880s, the doctrine of the universal ether had become firmly established in Victorian physics, and this doctrine was clearly stated by a number of spokesmen. George Fitzgerald, who was centrally concerned with ether theory through the 1880s and 1890s, saw both matter and ether as manifestations of motion in the universal medium. Two kinds of vortex motion were envisaged: On the one hand, there would be closed vortex rings – localised vortex motions – and these would constitute matter; on the other hand, there would be vortex filaments strung out all through the universal fluid, and these would confer on the fluid as a whole the properties of ether, the bearer of light and various forces. 'This hypothesis', wrote Fitzgerald, echoing Thomson's dynamical programme, 'explains the differences in Nature as differences of motion. If it be true, ether, matter, gold, air, wood, brains are but different motions'. Thus all of nature was to be reduced to motions in the universal plenum.[48]

Oliver Lodge, one of the most vociferous proponents of the universal ether, from the 1880s well into the twentieth century, had a somewhat different articulation of this theory. Lodge placed great emphasis on the requirement that the universal plenum be 'a perfectly homogeneous . . . continuous body incapable of being resolved into simple elements or atoms; it is, in fact, continuous, not molecular'. The treatment of ether at least phenomenologically as a continuous medium whose properties are not to be deduced from considerations of molecular structure had a substantial history in Britain, and Maxwell had argued specifically that the perfect fluid that is the substratum of vortex atoms cannot be molecular, for a molecular fluid would allow for internal diffusion of momentum, and hence viscosity. Now, as Lodge pointed out, a perfectly continuous material medium is something beyond the range of ordinary experience: 'There is no other body of which we can say this [that it is perfectly continuous], and hence the properties of ether must be somewhat different from those of ordinary matter'. Thus, although 'ether is often called a fluid, or a liquid, and again it has been called a solid . . . none of these names are very much good; all of these are molecular groupings, and therefore not like ether'. The universal ether, then, was a material *sui generis,* not classifiable as either solid or fluid, and Lodge had no difficulty in imagining that this universal plenum could have the rigidity required for the transmission of light waves and at the same time the perfect fluidity required for the existence of permanent vortex atoms. Thus, whereas for Fitzgerald both ether and matter were emergent from the universal plenum as a result of its various motions, for Lodge ether was to be strictly identified with the universal plenum itself, and matter was a manifestation of vortex motion in the plenum. Strict definition, then, would allow us to denote only Lodge's as a theory of the *universal ether;* Fitzgerald's would be a theory of the *universal plenum* or *medium*. In any case, Lodge agreed with Fitzgerald that all nature was emergent from this universal medium:

> One continuous substance filling all space: which can vibrate as light; which can be sheared into positive and negative electricity; which in whirls constitutes matter; and which transmits by continuity, and not by impact, every action and reaction of which matter is capable. This is the modern view of the ether and its functions.[49]

In the United States, the scientific community and especially the popular science tradition were quite closely tied to British examples; A. A. Michelson proclaimed the universal ether basically in Lodge's form:

> Suppose that an ether strain corresponds to an electric charge, an ether displacement to the electric current, these ether vortices to the atoms – if we continue these suppositions, we arrive at what may be one of the grandest generalizations of modern science – of which we are tempted to say that it ought to be true even if it is not – namely, that all phenomena of the physical universe are only different manifestations of the various modes of motion of one all-pervading ether.[50]

The retreat

If Maxwell, in his work on the theory of molecular vortices in 1861–2, had been a leader in the movement towards a definite, concrete, and realistically intended mechanical account of the universal ether, he also very soon, beginning in 1864, played a central role in the subsequent retreat from that kind of ether theory. As we shall see, it was a tactical retreat that Maxwell undertook in 1864, but this soon developed into a long-term strategy, a strategy of concentration on *dynamical theories* of ether, which involved much less commitment to specific mechanical pictures.

The theory of molecular vortices had been extremely successful on a heuristic level, and Maxwell remained forever committed to the reality of the molecular vortices themselves, but he had some problems with the idle-wheel particles, whose existence he had been led to postulate in the context of the particular logic of theory construction he had employed in 'Physical lines'. Maxwell had felt, from the outset, that the conception of idle-wheel particles was perhaps too ingenious, too artificial; and he had to admit that this was a 'provisional', 'awkward', and 'temporary' hypothesis. Also, the idle-wheel particles, according to the correspondences established in the theory, were taken to be electrical particles or electrical matter, and the notion of electrical particles or substance was repugnant to the field theories of both Faraday and Maxwell.[51] Maxwell, then, was highly motivated to avoid somehow the hypothesis of idle-wheel particles. Beyond this, as he wrote to a friend at the time, he wanted to 'clear . . . the electromagnetic theory of light from any unwarrantable assumption, so that we may safely determine the velocity of light by measuring the attraction between bodies kept at a given difference of potential, the value of which is known in electromagnetic measure'.[52] Thus, irrespective of the combination of commitment and doubt that characterised his own personal stance towards the concrete mechanical theory of 'Physical lines', Maxwell felt that it would be advantageous to connect the experiments on the ratio of units with measurements of the velocity of light in the most direct way possible, without the intervention of any hypothetical elements. To accomplish this, Maxwell used what he called the approach of 'dynamical theory'; it was presented in rudimentary form in 'A dynamical theory of the electromagnetic field' (1864), and in mature form in the *Treatise on electricity and magnetism* (1873).[53]

In part, Maxwell followed Thomson's definition of a dynamical theory:

> The theory I propose may . . . be called a theory of the *Electromagnetic Field,* because it has to do with the space in the neighbourhood of the electric or magnetic bodies, and it may be called a *Dynamical* Theory, because it assumes that in that space there is matter in motion, by which the observed electromagnetic phenomena are produced.

But there was also another meaning of the term *dynamical,* as Maxwell used it, that is even more relevant in the present context: It referred to abstract or generalised dynamics as developed by Joseph Louis Lagrange and further elaborated by William Rowan Hamilton and also by Thomson and Tait in their *Treatise on natural philosophy* (1867). Lagrangian methods had been used earlier in studies of the luminiferous ether, most notably by James MacCullagh and George Green, whose approach had been to postulate an appropriate potential energy function for ether without specifying the details of its mechanical structure; a wave equation was then derived using the Lagrangian variational principle. In this way, ether could be treated as a mechanical system, without a complete specification of the 'unseen . . . machinery' that gave it the characteristics exhibited in the potential energy function. The phenomenological aspect of this approach was reinforced by parallel trends in thermodynamics, and the energy approach was successfully integrated with the Lagrangian formalism by Thomson and Tait in their *Treatise on natural philosophy*.[54]

In the *Treatise on electricity and magnetism,* Maxwell used the formalism of Thomson and Tait, and expressed his aims as follows:

> What I propose now to do is to examine the consequences of the assumption that [electromagnetic] phenomena . . . are those of a moving system [ether], the motion being communicated from one part of the system to another by forces, the nature and laws of which we do not yet even attempt to define, because we can eliminate these forces from the equations of motion by the method given by Lagrange for any connected system.

Through the Lagrangian formalism, Maxwell was able to maintain his commitment to the mechanical world view and simultaneously avoid any 'unwarrantable assumption' concerning the specific mechanism of ether. He still felt that 'we have good evidence for the opinion that some phenomenon of rotation is going on in the magnetic field' and 'that this rotation is performed by a great number of very small portions of matter, each rotating on its own axis, this axis being parallel to the direction of magnetic force'. But he now preferred not to specify the picture any further, relying instead on the abstract Lagrangian formalism to give a 'dynamical' characterisation of the behaviour of ether. Thus, although he still regarded ether as a material system, he found that the possibility of knowing its detailed structure was receding.[55]

The detailed mechanical picture of ether that Maxwell had constructed still had a role to play, as a 'working model', which furnished a 'demonstration that mechanism may be imagined capable of producing a connexion mechanically equivalent to the actual connexion of the parts of the electromagnetic field'. This kind of model has recently been characterised by Peter Achinstein as an 'imaginary model'; such a model, in general, is not proposed with realistic intent, but rather to show that it is possible to imagine a specific mechanism that will display the appropriate behaviour.[56] Another such model of ether, closely related to Maxwell's, was proposed by Fitzgerald in 1885. It involved, in one version, spinning wheels connected by rubber bands, and in another, paddle wheels coupled by a fluid flowing in connecting canals. Fitzgerald carefully noted:

> I need hardly say that I do not intend it to be supposed that the ether is actually made up of wheels and indiarubber bands, nor even of paddle-wheels, with connecting canals. I think, however, that we may learn several things as to the conditions that the elements of the ether should fulfill if they are to represent Maxwell's equations by motions in ways analogous to those of my model.[57]

Perhaps most prolific in constructing these kinds of models of ether in the closing decades of the nineteenth century was William Thomson. Thomson encountered difficulties in fully developing his programme of explaining all physical phenomena realistically in terms of vortex motions in the universal substratum, and instead had recourse to unrealistic, though illustrative, mechanical models. One of the problems he encountered when attempting to understand the elasticity of ether in terms of vortex motion was that of stability. 'It is exceedingly doubtful,' he wrote in 1887, 'so far as I can judge after much anxious consideration from time to time during these last twenty years, whether the configuration . . . is stable . . . I am thus driven to admit, in conclusion, that the most favourable verdict I can ask . . . is the Scottish verdict of *not proven*'. (Stability problems impinging upon the vortex-atom theory of ordinary matter were also beginning to appear at about this time, and Thomson was beginning to harbour deep doubts concerning it as well.) Two years later, Thomson wrote that 'the difficulties in the way of proving a comprehensive dynamical theory of electricity, magnetism, and light are quite stupendous'.[58] In this situation, he turned to illustrative models to show how the elastic properties of ether could result from motion. In the words of his sympathetic Victorian biographer, 'he proceeded to describe an imaginary model ether or medium . . . To this end he imagined a sort of network, across the meshes of which were set minute gyrostats [gyroscopes] spinning each about its own axis, to give the structure the necessary immobility'. 'I do not admit that [this] is merely playing at theory', wrote Thomson, 'but it is helping our minds to think of possibilities, if by a model, however rough and impracticable, we show that a structure can be produced' having the requisite properties. He proposed various other crude models of ethereal functions, with the intention of providing illumination; however, he merely accentuated the growing gap between the needs of ether theory and the available mechanical models.[59]

Joseph Larmor was the last of the great Victorian ether theorists. A generation younger than Thomson and Maxwell, he synthesised the work of his predecessors and carried existing trends to their logical conclusions. In a series of papers entitled 'A dynamical theory of the electric and luminiferous medium', published between 1893 and 1897, Larmor set out to give an account of 'the primordial medium which is assumed to be the ultimate seat of all phenomena'. Employing an ether similar to MacCullagh's (which had been given a concrete realisation in Thomson's gyrostatic ether) in conjunction with Maxwell's electromagnetic field theory, and incorporating the vortex atom as well, Larmor achieved a monumental synthesis. This synthesis, however, was achieved in the context of a methodological programme that represented an extreme development of traditional ether theory, and that in a sense negated traditional ether theory.[60]

Larmor followed Maxwell in his enthusiasm for the Lagrangian method, which 'allow[s] us to ignore . . . altogether the details of the mechanism; . . . it makes everything depend on a single analytical function representing the distribution of energy in the medium'. Traditionally in the Lagrangian approach to ether theory, the abstract treatment would be followed, sooner or later, by an attempt to give a concrete account of the hidden machinery that gives rise to the effects abstractly treated by the Lagrangian formalism – either in a realistic way or at least by means of an illustrative model. 'The chief representative' of the builders of such models, observed Larmor, 'has been Lord Kelvin'. Larmor, for his own part, however, suggested, 'It may be held that the [abstract treatment] really involves in itself the solution of the whole problem; . . . [the concrete model] is rather of the nature of illustration and explanation, by comparison of the intangible primordial medium with other dynamical systems of which we can directly observe the phenomena'. Thus, as so often happens in the history of ideas, a crucial change in outlook had been brought about by having the tail wag the dog: Whereas the abstract Lagrangian approach had before been regarded as a step towards the ultimate theory, which would display ether as a concrete mechanical system, the abstract theory was now to be regarded as the final and ultimate theory, concrete models being significant only for heuristic or pedagogical purposes.[61]

Larmor soon went even further in relinquishing the fundamental tenets of Victorian ether theory. In 1894 he gave up the vortex atom as the basis of ordinary matter, and instead espoused an electron theory, thus parting company with the tradition of Faraday, Maxwell, and Thomson.[62] By the close of the century, as observed by a sympathetic contemporary, 'chiefly under the influence of Larmor, it came to be generally recognized that the aether is an immaterial medium, *sui generis,* not composed of identifiable elements having definite locations in absolute space' (Whittaker, 1951:1:303).

Conclusion

Viewed in a broader context, Victorian ether theory can be seen as the embodiment of an extreme option within the mechanical world view. Mechanics deals with matter in motion, and mechanical theories could provide for a richness of possibilities by invoking either a variety of matters or a variety of motions. Newton's idea of endowing particles with active forces provided for the utilisation of a variety of matters in the eighteenth century. Thus there were electrical particles of two kinds, bearing specifically electrical forces; magnetic particles of two kinds with their specific forces; heat particles exerting short-range forces; and light particles. In the course of the nineteenth century, especially in the thinking of British physicists, this variety of particles and forces gave way to a variety of motions in one universal substratum. This substratum first appeared in the guise of the luminiferous ether, and, as a result of the work of Faraday, Thomson, and Maxwell, the phenomena of electricity and magnetism were also referred to the motions and strains of this medium. Heat had by midcentury been referred to the motions of particles of ordinary matter, and Thomson in turn viewed these as nothing but patterns of motion in the universal substratum. There resulted a form of the mechanical world view in which all natural phenomena – and perhaps the 'supernatural' as well – were to be explained by the dynamics of the universal ethereal medium.[63]

Acknowledgments

A much expanded version of the section of this chapter entitled 'Maxwell's theory of molecular vortices' will be published elsewhere; an earlier form was presented at the Dec. 1976 meeting of the History of Science Society, and discussions with coparticipants there in a session on the history of field theories, including Donald F. Moyer, Barbara Giusti Doran, and Howard Stein, have been very informative. John Heilbron, Norton Wise, and Geoffrey Cantor have read drafts of this chapter and made extremely helpful suggestions. The research was supported in part by a grant from the National Science Foundation.

Notes

1 J. Larmor, 'A dynamical theory of the electric and luminiferous medium' (1893), in J. Larmor, *Mathematical and physical papers,* 2 vols. (Cambridge, 1929), 1:389–413, on 389.
2 L. P. Williams, *Michael Faraday: a biography* (London, 1965).
3 M. Faraday, 'Thoughts on ray-vibrations' (1844), in M. Faraday, *Experimental researches in electricity* (1839–55; reprinted, 3 vols. in 2, New York, 1965), 3:447–52, first two quoted passages on 447; Williams, *Faraday,* 380–1; M. Faraday, 'On the magnetization of light and the illumination of magnetic lines of force' (1845), in Faraday, *Experimental researches,* 3:1–26; Williams, *Faraday,* 383–92; Spencer (1970); M. Faraday, 'On lines of magnetic force' (1852), in Faraday, *Experimental researches,* 3:328–70, final quoted passage on 331.
4 W. Thomson, 'On the uniform motion of heat in homogeneous solid bodies and its connection with the mathematical theory of electricity' (1842), in W. Thomson, *Re-*

print of papers on electrostatics and magnetism, 2nd ed. (London, 1884), 1–14. See also Buchwald (1977).
5 J. Z. Buchwald, 'Sir William Thomson (Baron Kelvin of Largs)', in *Dictionary of scientific biography*, vol. 13 (1976), 374–88. See also S. P. Thompson, *The life of William Thomson, Baron Kelvin of Largs*, 2 vols. (London, 1910), 113–27, esp. 127. Buchwald (1977), 105, 122–3, and Thompson, *Kelvin*, 19–20, disagree in emphasis concerning Thomson's prior knowledge of and commitment to Faraday's approach.
6 W. Thomson, 'On the elementary laws of statical electricity' (1845), in Thomson, *Electrostatics and magnetism*, 15–37; Thomson 'On the uniform motion of heat', n. on 1–2; Buchwald (1977), esp. 106–7, 119–34.
7 Thomson, 'Elementary laws', 37. The nuances of Thomson's paragraph 50 have not been exhausted in my account here.
8 W. Thomson, 'On a mechanical representation of electric, magnetic, and galvanic forces' (1847), in W. Thomson, *Mathematical and physical papers*, 6 vols. (Cambridge, 1882–1911), 1:76–80.
9 Thompson, *Kelvin*, 203–4.
10 Ibid., 1062–5.
11 L. Campbell and W. Garnett, *The life of James Clerk Maxwell*, with a new preface and appendix with letters by R. H. Kargon, 2nd ed. (1882; reprinted, New York, 1969), 144–6; Thompson, *Kelvin*, 222–3.
12 J. Larmor (ed.), 'The origins of Clerk Maxwell's electrical ideas, as described in familiar letters to W. Thomson', *Proceedings of the Cambridge Philosophical Society 32* (1936), 695–750, esp. 697–705.
13 J. Clerk Maxwell, 'On Faraday's lines of force' (1855–56), in J. Clerk Maxwell, *The scientific papers of James Clerk Maxwell*, ed. W. D. Niven (1890; reprinted, 2 vols. in 1, New York, 1965), 1:155–229, on 155–9 (my italics). See also Siegel (1975); Hesse (1973); Kargon (1969); d'Agostino (1968); J. Turner (1955); Olson (1975), 299–302.
14 Maxwell, 'Faraday's lines', 156, 160 (Maxwell's italics). See also Olson (1975), 287–321; J. Clerk Maxwell, 'Are there real analogies in nature?' (presented to the Apostles at Cambridge in 1856), in Campbell and Garnett, *Maxwell*, 235–44.
15 W. Thomson, 'Dynamical illustrations of the magnetic and the helicoidal rotatory effects of transparent bodies on polarized light', *Proceedings of the Royal Society 8* (1856), 150–8; reprinted in *Philosophical Magazine 13* (1857), 198–204, on 199. The paper is discussed in Knudsen (1976), esp. 244–7 and 273–81.
16 Thomson, *Electrostatics and magnetism*, 423 ('Note added Jan. 1872') (my italics); W. Thomson, 'On an absolute thermometric scale founded on Carnot's theory of the motive power of heat, and calculated from Regnault's observations' (1848), in Thomson, *Mathematical and physical papers*, 1:100–6, on 102–3; W. Thomson, 'An account of Carnot's theory of the motive power of heat; with numerical results deduced from Regnault's experiments on steam' (1849), in ibid., 1:113–55, on 116–17; M. J. Klein, 'Gibbs on Clausius', *Historical Studies in the Physical Sciences 1* (1969), 127–49; Thompson, *Kelvin*, 263–83; W. Thomson, 'On the dynamical theory of heat, with numerical results deduced from Mr. Joule's equivalent of a thermal unit, and M. Regnault's observations on steam' (1851), in Thomson, *Mathematical and physical papers*, 1:174–232.
17 See esp. W. Thomson, 'On a universal tendency in nature to the dissipation of mechanical energy' (1852), in Thomson, *Mathematical and physical papers*, 1:511–14, on 511, for a direct opposition of 'dynamical' to 'statical'. See also, e.g., Thomson, 'Dynamical theory of heat', 174–5; Thomson, *Electrostatics and magnetism*, 423 ('Note added Jan. 1872'); and [W. Thomson], 'Dynamics', in [*Nichol's*] *Cyclopaedia of the physical sciences* (London, 1860), 212–15.

18 Thomson, 'Dynamical theory of heat', 174–5; W. J. M. Rankine, 'On the centrifugal theory of elasticity, as applied to gases and vapours', in *Miscellaneous scientific papers*, ed. W. J. Millar (London, 1881), 16–48, esp. 16–18; Thomson, 'Dynamical illustrations', quoted passages on 199–200. Joule at first espoused the hypothesis that heat was rotatory motion but then gave it up – see Brush (1976), 1:161.
19 Thomson, 'Dynamical illustrations', 198–200; W. Thomson, 'Atmospheric electricity (Royal Institution Friday Evening Lecture, May 18, 1860)', in Thomson, *Electrostatics and magnetism*, 208–26, quoted passage on 224 (Thomson's capitalisation and italics).
20 Thomson, 'Dynamical illustrations', 199–200, quoted passage on 200; Knudsen (1972), quoted passages on 47; Doran (1975), 179–90.
21 Campbell and Garnett, *Maxwell*, 199 n.
22 J. Clerk Maxwell, 'On physical lines of force' (1861–2), in Maxwell, *Papers*, 1:451–513, quoted passages on 505, 489.
23 Ibid., 452–3 (my italics, except for *'stress'*). In modern terminology – cf. P. Achinstein, *Concepts of science: a philosophical analysis* (Baltimore, 1968), 203–25 – the mechanical representation of 'Physical lines' might be denoted a *theoretical model*, in contrast to the *analogue models* of 'Faraday's lines'; Maxwell used the term *model* only in later work (see 'The retreat' in this chapter).
24 Maxwell, 'Physical lines', 454–66; M. Faraday, 'On the physical character of the lines of magnetic force' (1852), in Faraday, *Experimental researches*, 3:407–37, on 419, 435–6.
25 Maxwell, 'Physical lines', 467–9, quoted passages on 468; ibid., figure facing p. 488. Pseudosphericity of the cells is evidently assumed in the calculations in ibid., 469–71, and is explicitly used at 492 ff.
26 Ibid., 468–71, resulting in equations (33) and (34) on 471; my ι represents Maxwell's p, q, r, and ω^* his α, β, γ.
27 Ibid., 471, equations (33) and (34). This differential form of Ampère's law had appeared before in Maxwell, 'Faraday's lines', 194. Cf. Maxwell, 'Physical lines', 462, 496. My ι and $\mathbf{J}$ both correspond to Maxwell's p, q, r, reflecting explicitly his shifting interpretation of this vector; similarly, both ω^* and $\mathbf{H}$ correspond to his α, β, γ.
28 Maxwell, 'Physical lines', 471.
29 Ibid., 472–5, resulting in equation (54) on 475; my τ_1 represents Maxwell's P, Q, R, and ρ_m his ρ on 456, to within a multiplicative constant. Cf. ibid., 457.
30 Ibid., 475, equation (54). My τ_1 and $\mathbf{E}_1$ both correspond to Maxwell's P, Q, R, reflecting his shifting interpretation of this vector.
31 Ibid., 475–6: 'The forces exerted on the layers of particles between the vortices' represent, 'in the language of our hypothesis, . . . electromotive forces'.
32 Ibid., 485–8, quoted passage on 486. Schema (8.3) is adumbrated in ibid., 477: 'The [idle-wheel] particles . . . in our hypothesis represent electricity'. For equation (8.3b), see Maxwell, 'Faraday's lines', 191–2, where the symbol ρ was used for charge density; and Maxwell, 'Physical lines', 496, where e was used.
33 See Maxwell, 'Faraday's lines', 195; Siegel (1975).
34 Larmor, 'The origins', 728–9; C. W. F. Everitt, *James Clerk Maxwell: Physicist and natural philosopher* (New York, 1975), 98–9.
35 Maxwell, 'Physical lines', 489. Cf. ibid., 486.
36 Ibid., 490–6, quoted passage and final equation, equation (112), on 496. Cf. equation (108), ibid., 495.
37 Equation (8.4d) from equations (113) and (114), ibid., 496–7. Equation (8.5a) from equation (115), ibid., 497; cf. (108) on 495; my ρ_p corresponds to Maxwell's e. Equation (8.5b) also from (115), ibid., 497, interpreted electromagnetically, my k

corresponds to Maxwell's E. Schema (8.5) esp. from (118), ibid., 497; cf. (108) on 495.
38 Ibid., 497-8.
39 Ibid., 495 (equation [108]), 499 (equation [133] and stipulation '$\mu = 1$').
40 Ibid., 499 (equation [136]), 500 (Maxwell's italics).
41 Ibid., 500-13, quoted passages on 500, 505. Cf. Bromberg (1968a), esp. 219, 229-30.
42 Thomson, 'Dynamical illustrations', 200; Knudsen (1972), 47-8; H. Helmholtz, 'Ueber Integrale der hydrodynamischen Gleichungen, welche den Wirbelbewegungen entsprechen', *Journal für die Reine und Angewandte Mathematik* 55 (1858), 25-55; Thompson, *Kelvin*, 402; C. G. Knott, *Life and scientific work of Peter Guthrie Tait* (Cambridge, 1911), 127, 176-204; W. Thomson, 'On vortex atoms' (1867), in Thomson, *Mathematical and physical papers*, 4:1-12, quoted passage on 2; Thompson, *Kelvin*, 510-15; Silliman (1963).
43 Thompson, *Kelvin*, 513-15.
44 Thomson, 'Vortex atoms', 1-4, quoted passages on 1-2. Cf. a report of Thomson's talk in 'the *Scotsman* of February 19, 1867', in Thompson, *Kelvin*, 517-19, quoted passage on 519. 'Vortex atoms' was published in the *Philosophical Magazine* as well as the *Proceedings of the Royal Society of Edinburgh*.
45 W. Thomson, 'On the propagation of laminar motion through a turbulently moving inviscid fluid', *Philosophical Magazine* 24 (1887), 342-53; W. Thomson, 'On the ultramundane corpuscles of LeSage . . .', ibid., 45 (1873), 321-45, discussed in Thompson, *Kelvin*, 1029; J. J. Thomson, *A treatise on the motion of vortex rings* (London, 1883).
46 J. Clerk Maxwell, 'Atom', in Maxwell, *Papers*, 2:445-84, discussion of vortex atom on 466-77, quoted passages on 471-2.
47 Knott, *Tait*, 105-9; P. G. Tait, *Lectures on some recent advances in physical science*, 2nd ed. (London, 1876), 290-300, 362-3, quoted passage on 290-1.
48 G. F. Fitzgerald, 'On a model illustrating some properties of the ether' (1885), in G. F. Fitzgerald, *The scientific writings of the late George Francis Fitzgerald*, ed. J. Larmor (Dublin, 1902), 142-62, on 154-6; G. F. Fitzgerald, 'Electromagnetic radiation' (1890), in ibid., 266-76, quoted passage on 276; G. F. Fitzgerald, 'Helmholtz memorial lecture' (1896), in ibid., 340-77, on 345-54.
49 O. Lodge, 'The ether and its functions', *Nature* 27 (1883), 304-6, 328-30, quoted passages on 305, 330; Maxwell, 'Atom', 466-7; I. Todhunter and K. Pearson, *A history of the theory of elasticity and of the strength of materials*, 2 vols. (Cambridge, 1886-93), 1:496-505; E. M. Parkinson, 'George Gabriel Stokes', in *Dictionary of scientific biography*, vol. 12 (1976), 74-9; Wilson (1971).
50 A. A. Michelson, *Light waves and their uses* (Chicago, 1903), 161-2, cited in Silliman (1963), 473.
51 Maxwell, 'Physical lines', 486; J. Clerk Maxwell, *A treatise on electricity and magnetism*, 2 vols. (1873; 3rd ed., 1891; reprinted, New York, 1954), 1:380.
52 Campbell and Garnett, *Maxwell*, 340.
53 J. Clerk Maxwell, 'A dynamical theory of the electromagnetic field' (1864), in Maxwell, *Papers*, 1:526-97; Maxwell, *Treatise*, 2:199-262.
54 Quoted passages in Maxwell, 'Dynamical theory', 527 (Maxwell's italics); and J. Clerk Maxwell, 'On the dynamical evidence of the molecular constitution of bodies' (1875), in Maxwell, *Papers*, 2:418-38, on 419. A definitive treatment of this whole matter may be found in Moyer (1977). See also Klein (1973), 69-70.
55 Maxwell, *Treatise*, 2:198, 470. Cf. Maxwell, 'Dynamical theory', 533 (¶ [16]).
56 Achinstein, *Concepts of science*.
57 Maxwell, *Treatise*, 2:470; Achinstein, *Concepts of science*, 203-26, esp. 218-21;

Fitzgerald, 'A model', 142–51, quoted passage on 151. Cf. Fitzgerald, 'Electromagnetic radiation', 270–1.

58 Thomson, 'Propagation of laminar motion', quoted passage on 352 (Thomson's italics); Thompson, *Kelvin,* 1043, 1046–7, quoted passage on 1043.

59 Thompson, *Kelvin,* 1044–6, quoted passages on 1044–5. See also W. Thomson, 'On a gyrostatic adynamic constitution for "ether" ' (1889–90), in Thomson, *Mathematical and physical papers,* 3:466–72; W. Thomson, 'Ether, electricity, and ponderable matter' (1889), in ibid., 484–511, esp. 500–11; W. Thomson, 'Steps toward a kinetic theory of matter' (1889), in W. Thomson, *Popular lectures and addresses,* 3 vols. (London and New York, 1889–91), 1:225–59, esp. 242–50; Whittaker (1951), 145; Schaffner (1972), 68–75, 194–203.

60 A. E. Woodruff, article "Joseph Larmor" in Dictionary of scientific biography; J. Larmor, 'A dynamical theory of the electric and luminiferous medium (*abstract of memoir following: and general discussion*)' (1893), in Larmor, *Papers,* 1:389–413, quoted passages on 389. Cf. J. Larmor, 'A dynamical theory of the electric and luminiferous medium: part I' (1894), in ibid., 414–535, on 414–15. For a somewhat different view of Larmor in particular and British ether theory in general, see Doran (1975).

61 Larmor, 'Dynamical theory (*abstract*)', 389–90. Cf. Larmor, 'Dynamical theory: part I', 417.

62 Larmor, 'Dynamical theory: part I', 514 ff.

63 Cf., e.g., Doran (1975); Schofield (1970); Hesse (1961), 126–225; Buchwald (1977), 134–6; B. Stewart and P. G. Tait, *The unseen universe; or, physical speculations on a future state,* 7th ed., (London, 1886); Heimann (1972); Wilson (1971), esp. 34–41.

9

German concepts of force, energy, and the electromagnetic ether: 1845–1880

M. NORTON WISE

Department of History, University of California at Los Angeles, Los Angeles, California 90024 USA

The goal of unity in nature

The notion that there should exist a single pervasive ether uniting all natural phenomena was not new in the middle of the nineteenth century. Descartes, Leibniz, and Kant long before had provided respectability for the idea of primitive matter, *Urstoff,* or the *Weltäther* as a possible ground for systematic natural philosophy. Their views may be taken to stand as well for the ultimate goals of some practising scientists of the eighteenth and early nineteenth centuries, as indicated for the British in Chapter 1 of this book. Nevertheless, precise description of empirically distinguishable phenomena, such as electricity, magnetism, and heat, had generated more specific referents than simply 'ether'. Many ethereal media – electric, magnetic, caloric, luminiferous, and gravitational – populated different regions of specialised investigation. Not until the late 1840s did the general philosophical and the specialised claims find a common foundation acceptable across the broad spectrum of physical scientists. And only then did 'ether' turn into a generally recognisable object of research.

No doubt a major factor in the acceptance of a single *Weltäther* was the explanatory power of the wave theory of light, which seemed to require a medium throughout space. By itself, however, the wave theory was too limited in its range of applications to compel general assent to, or even general concern with, a role for ether throughout nature. That situation changed rapidly around midcentury when the wave theory of light was coupled to the theory of conservation of force and the mechanical theory of heat. This newly interrelated set of ideas formed a qualitatively different context for physical theory.

Conservation of force arose from and further motivated the search for unity in nature, for the ultimate identity and interconvertibility of all natural powers. And this search had by 1845 been legitimised by manifold empirical interconversions.[1] Yet the only adequate foundation available for a physical theory of conservation was mechanics: conservation of *vis viva* (or, as the concept acquired independence from force, conservation of kinetic plus potential 'energy') and the mechanical theory of heat. The mechanical theory of heat proposed that heat in a body consists in internal *vis viva* and potential energy, and usually that heat radiated between bodies consists in some form of transmitted motion. The latter view presupposed an ether throughout space – already required by the wave theory of light – that could transmit the motions of heat as well as the closely related motions of light. As an ideal mechanical medium it would conserve *vis viva*, again a basic tenet of the wave theory. By postulating an ether, physicists could incorporate heat and light into mechanics under conservation of *vis viva*. Reversing the argument, if ether did not exist, then the mechanical foundation of light, heat, and conservation was destroyed, and with it the satisfying vision of a unified physical science. Many investigators considered the existence of a unifying medium throughout space a necessity for the rational comprehension of science as it stood at midcentury.

General assent to the existence of a single ether, however, did not imply agreement on its nature. Those concerned primarily with one set of phenomena – for example, heat, electricity, physical optics, or philosophy – constructed ether primarily to satisfy the demands of their own area, often without paying close attention to other areas but hoping eventually to subsume them. A comprehensive history of unifying ether theories after midcentury would have to begin by considering this wide variety of bases for constructing ether and then show how most of them lost significance; for by about 1880, electromagnetism, including the electromagnetic theory of light and radiant heat, had emerged as dominant. I shall not attempt that complex analysis here. I shall consider only the electromagnetic bases themselves and only as they developed in the German-language community, in which I include authors whose work was disseminated in German-language journals or who received their training in German universities. Restricting discussion of the search for unity in this way will allow a focus, first, on the reception and reformulation in Germany of French mathematical action-at-a-distance theories of electricity and magnetism and, second, on the development of mathematical field theories as alternatives to action at a distance. Both topics will serve to elucidate concepts of force that were unique to German natural philosophy. These general concepts will form the third, and overriding, theme of my discussion. A few introductory remarks on fields and on German physics at midcentury are necessary.

Fields defined

A distinction between action at a distance and field action, as it came to be generally understood in the nineteenth century, is basic to much that follows. The distinction, in the first instance, separates action directly over finite distances from action only between contiguous elements, that is, immediate (*unmittelbar*) from mediate (*mittelbar*) action between separated objects (Heilbron, this volume). In a field view, local action depends directly on local conditions. It is related to changes in conditions at a distance indirectly, through the mediating action of the field existing in the intervening space. The field, moreover, has an existence in space independent of its sources. It carries in itself the power to effect action, that is, quantity of force or energy, and propagates that power in time from point to point. A space in which force is defined at every point merely as a resultant of sources acting from all distances, such as in Laplace's gravitational theory or Poisson's electrostatics, does not qualify as a field. More stringently, an action between two objects that is merely modified (rather than mediated) by an intervening substance, where every point of the substance interacts directly with the objects, is not a field action. This stipulation is necessary in order that descriptions of polarised media in terms of forces acting directly at a distance should be distinguished from polarisation in a field by contiguous action. Examples of the former are Poisson's 1822 theory of induced magnetism,[2] William Thomson's 1845 theory of dielectric inductive capacity (Wise, 1977; Siegel, this volume), and Helmholtz's 1870 electromagnetic ether theory of light (described in a later section of this chapter). All of these authors presented their theories in opposition to contiguous action or field theories.

Fields in the nineteenth century were basically of two kinds: force fields and ether fields, depending on whether force itself was taken to be a power distributed in space (as in modern electromagnetic theory) or whether the power was carried in the state of a medium, ether. Faraday's mature theory of lines of force provides a classic example of a pure force field. Such theories, however, were less acceptable around midcentury than they had been earlier or would be later. This was, if anything, more true in Germany than elsewhere, partly for reasons that form an integral part of my story and that lead to a few general considerations on German concepts of force and ether.

Natural philosophy in German physics

German physicists[3] at midcentury continued to express intellectual concerns traditional in German Idealist philosophy while vehemently rejecting its excesses in speculative *Naturphilosophie*. The significance of this love–hate relationship with tradition is difficult to specify with precision. In an effort to sharpen its relevance for concepts of force and ether, I shall employ as a foil a man who expressed many of the values of physical science but who was not a physicist and whose work remained unacceptable to practising physicists. J. R. Mayer, a medical doctor, self-educated in physics, became well known after 1850 for his pioneering analysis of the mechanical equivalent of heat and of conservation of force.[4] His first paper on conservation was published in 1842 in Liebig's *Annalen der Chemie und Pharmacie,* following rejection by Poggendorf's physics-oriented *Annalen der Physik und Chemie*. Along with several subsequent papers, it remained largely unknown until Helmholtz began to announce widely the priority of Mayer's paper over his own classic conservation paper of 1847. In treating Mayer as a foil I shall consider first concepts that he and the practising physicists continued to employ, then aspects of their common rejection of *Naturphilosophie,* and finally the physicists' rejection of Mayer's theoretical style.

In company with many German scientists of the nineteenth century, Mayer concerned himself not merely with the physical coherence of material nature but also with the relation between matter and mind, or better, between *Natur* and *Geist,* where *Geist* implies both mind and soul. From Spinoza and Leibniz in the seventeenth century, through Kant and Hegel in the nineteenth, most German natural philosophers would have agreed with Mayer's rationalist judgement: 'What subjectively is correctly thought, is also objectively true'.[5] Matter occupied the world of nature as ideas occupied the world of mind. And where logic governed the rational relations of ideas, forces governed the interactions of matter. In the wide variety of constructions available by the mid-nineteenth century for these ideas, no single aspect is more common than the notion of force as an entity occupying a middle position between inert matter and *Geist,* between nonliving nature and the region of purposiveness and beauty, progress and freedom. Mayer posited 'three categories of existence: 1. matter, 2. force, and 3. soul, or the *geistig* principle'. Although he was among those physically oriented physiologists who separated sharply the conserved forces of nonliving nature from *Geist,* force was for him still a stepping-stone between matter and *Geist*. 'Having once attained the insight that there are not merely material objects, that there are also forces, forces in the narrow sense of the new science, just as indestructible as the matter of the chemists, then it is only a further step to the assumption and recognition of *geistig* existence'.[6]

Not all physical scientists, or even biological scientists, were willing to admit the existence of a separate spiritual realm; some, like Helmholtz, sidestepped the issue; others, like the materialist Büchner, explicitly attempted to reduce the realm of mind to that of matter and force.[7] Even such agnostic and reductionist theorists, however, carried in Germany the weight of the tradition that Mayer more directly represented. Forces, they generally agreed, played the role in nature that relations of ideas played in mind. Forces expressed the rationality of nature. They expressed causality (Mayer, Riemann, Helmholtz) or law (Weber, Fechner, Helmholtz) or, generally, relation.

As relations, forces were often seen as something beyond the things related; they were taken not as seated in isolated matter but as arising only in the interrelations of matter. The significance of this begins to emerge when one considers that it refers force more nearly to chemical affinity – *Verwandschaft* = relationship – than to mechanical push or pull. Mayer provides an illustration for falling objects grounded in the principle of sufficient reason. 'A cause, which effects the raising of a weight, is a force; its effect, *the raised weight,* is therefore likewise *a force.* More generally expressed this is: *spatial separation [Differenz] of ponderable objects is a force'.*[8] The concept of force as a relation (and relationship) will provide the overriding theme in my discussion of German electromagnetic ether theories.

In his understanding of conservation of force, Mayer expressed another traditional idea common among his contemporaries: *'There is in truth only a single force',*[9] a force conserved through all the transformations of nature. In Germany before midcentury, however, this idea had its home in the speculative systems of *Naturphilosophie,* which Mayer, as well as most practising scientists, was at pains to combat. *Naturphilosophie* is often associated in its negative connotations with the natural philosophies of Schelling and Hegel, for whom nature did not exist independent of *Geist* and who therefore animated all nature with *Geist.*[10] But seen as the attempt to construct a priori a system of nature that would reflect the operations of mind, *Naturphilosophie* should include as well the less subjective philosophies of Kant and Herbart, who preserved an external – though unknowable – *Ding an sich* as a constraining ground for conceptualisations. As their means for construction of a system of nature the *Naturphilosophen* sought to employ an ultimate logic of mental activity, often a dialectical logic that would allow the mind to construct its concepts through irony, that is, through a mutual conflict of primitive opposites, leading to a synthesis at a higher level. The fundamental opposites in the realm of nature were powers or forces, typically attractive and repulsive, and there was given, as both product and ground of the original conflict of attraction and repulsion, a primordial ether filling space as a continuum. It served as the ground of all other forms of matter and force. In Kant's words:

> The elementary system of the moving forces of matter depends upon the existence of a *substance* which is the basìs (the primordi-

ally originating moving force) of all moving forces of matter, and of which it can be said as a postulate (not as an hypothesis): There exists a *universally distributed all-penetrating* matter within the space it *occupies* or fills through repulsion, which *agitates* itself uniformly in all its parts and endlessly *persists* in this motion.[11]

The secondary forms of matter and moving force were typically conceived either as modifications of the primitive ether motions or as higher powers of the original conflict of forces. Thus they arose either dynamically or dialectically, with no clear separation between the two modes. Also, since matter and moving force were the grounds for each other, they too were not clearly distinct. Here then is a striking difference from action-at-a-distance theories. Primitive matter, like force, is distributed continuously – as matter in action-at-a-distance theories was not. Concomitantly, force, like matter, occupies space – as action-at-a-distance forces did not. One sees immediately the conceptual precedent for both continuous ether fields, filling space with energy of motion, and pure force fields, filling space with energy of attraction and repulsion.

In all systems of *Naturphilosophie,* finally, the dynamical processes continued uninterruptedly, as reflections of the rational operations of *Geist*. Kant, for example, claimed that 'the primordial forces of motion, as originally agitating, cannot bring themselves to rest, for a state of rest itself presupposes a counteraction of the agitating forces in actuality, not merely in potentiality, so that the hindrance of these motions in universal rest is self-contradictory'.[12] Ideas of this sort, widespread in early nineteenth-century Germany, form the basis in Idealist traditions for attempts such as Mayer's to enunciate conservation more precisely.

While continuing to seek the unity of nature through the unity and interconvertibility of forces, and while continuing to believe in a close parallel between rational relations of ideas and physical relations, Mayer and many of his contemporaries nevertheless rejected the notion that one could construct the true system of nature by merely developing the structure of thought. One could not know enough about *Geist*. They rejected in particular the dialectical process, with its conflict of opposites leading to a higher synthesis. And they required that all legitimate science be closely tied to empirical observations. Of particular relevance here, through an emphasis on the empirical materiality of objects, physicists separated matter from force and, like Mayer, required their independent conservation.

Though Mayer thus represents, in the traditions he rejected as well as in those he continued, a variety of common goals, he did not reject enough to suit the dominant mood of practising physicists. In its objectionable aspects his work remained explicitly metaphysical. He employed empirical results primarily as confirming instances and attempted to establish conservation of force a priori, relying on the principle of sufficient reason, and without a rigorous treatment of *vis viva*. Helmholtz expressed the common attitude when he called this a 'metaphysically formulated pseudoproof'.[13]

Mayer fell outside the mainstream of physics in other ways. As noted earlier, most physical scientists sought to unify nature by constructing realistic physical models, usually on a mechanical basis; but even though Mayer had himself discovered the mechanical equivalent of heat, and even though he accepted a mechanical wave theory of light, he did not accept the theory that heat was mechanical. He argued that, although force in the form of heat could be interconverted with mechanical forces – with *vis viva* or a raised weight – the fact of interconversion did not justify taking any particular form of force as the fundamental one or the basis of transformations. The obvious advantage of a completely mechanical theory was that it provided just such an explanatory foundation for conversion and conservation of forces. In refusing, furthermore, to reduce heat (and electricity) to matter, motion, and the forces acting between parts of matter, Mayer maintained the *naturphilosophisch* notion of force as a sort of substance, now independent of matter but still having the same status as matter. To describe both light and heat transmitted through space, for example, he required not only a material ether to carry light waves but an independently transmitted and apparently immaterial force of heat.[14] That the rejection of *Naturphilosophie* meant precisely the rejection of such 'metaphysical' force substances in favour of the mechanics of matter helps to explain why space-occupying force fields did not receive serious consideration in Germany much before 1880. A mechanical ether provided the only legitimate basis for unity. In the last decades of the century, however, as electromagnetic fields and energy relations, described on a positivist basis, came increasingly to be considered the proper foundation for physical theory, Mayer's views were resurrected as precursors of the new trends, particularly by so-called Energeticists who made energy the basis of all reality.[15]

The following discussion of electromagnetic ether theories covers the period from 1845 to 1880, from enunciation of conservation of energy to the period just preceding full incorporation of the energy idea into pure force fields. I attempt to show how general philosophical concerns both conditioned and were conditioned by several specific electromagnetic ether theories. The result is less a coherent history of such theories in the period than a small collection of conceptual histories taken as representative of the full story.

The great questions of electromagnetism as they arose in the German context were two: How were electromagnetic forces, which seemed to require a velocity-dependent relation between portions of electrical matter, to be comprehended; and how were electromagnetism and light, apparently quite closely related, to be unified? Could ether actually perform its role as the unifying medium? Three primary examples serve to develop these questions and the major sorts of answers proposed for them. The questions arose through Wilhelm Weber's velocity- and acceleration-dependent law of electrical forces; and Weber attempted to resolve them through a comprehensive ether theory based on an action-at-a-distance interpretation of the force. Bernhard Riemann outlined, preliminarily, an equally comprehensive reinterpretation utilising a variety of ether field pictures and emphasising propagation of force in time. Hermann von Helmholtz, finally, attempted to explain apparent propagation of force within an action-at-a-distance framework, but his ether theory served instead only to mediate between the demise of action-at-a-distance theories and the rise of mature field theories. All three of these examples point up the significance in German physics of the concept of force as relation.

Weber's ether and Fechner's metaphysics

In the historiography of nineteenth-century physics we have become used to the notion of a Continental action-at-a-distance tradition formalised by early nineteenth-century Frenchmen, typically Laplace and Poisson, and adopted somewhat later by Germans such as Gauss, Wilhelm Weber, and Helmholtz (Woodruff, 1962). As the culmination of this tradition we are likely to think of Weber's law for forces acting at a distance between particles of electrical fluids. This is the law announced by Weber in 1846 that made the force between two electrical particles, e and e', depend not only on the inverse square of the distance r between them but also on their relative velocity v and relative acceleration a:

$$F = \frac{ee'}{r^2}(1 - \frac{1}{c^2}v^2 + \frac{2r}{c^2}a) \tag{9.1}$$

(Here c is a constant that Weber and Kohlrausch showed in 1857 to be approximately $\sqrt{2}$ times the velocity of light.)[16] The action-at-a-distance tradition indeed explains much about Weber's law, but in travelling from France to Germany the concept of action at a distance was transformed. Whereas Laplace and his associates described force as though it emanated from one particle of matter and acted on another particle at a distance – a description that implicitly tied force to a particle as its source – Weber insisted that force existed only as a pairwise relation between particles. The pair of particles, therefore, formed his fundamental unit of analysis. I shall begin to develop the implications of this idea through a description of Weber's programme for unity in physics, proceeding then to metaphysical foundations of the programme as presented by Gustav Theodor Fechner.[17]

Weber's electrical ether

From the first presentation of his force law in 1846, Weber conceived it as the core of an incomplete theory of an electrical ether that might eventually unify many or all natural phenomena, in the sense that it would reduce many forces to a single law of force. By 1848 he had shown that the force law could be derived from a potential, a function describing a system of particles that always acquires the same value when the particles acquire their initial position and velocities.[18] Existence of such a function guaranteed conservation of *vis viva* in the system and, therefore, conservation of all those natural powers that could be subsumed under the law of force. Thus unity and conservation were both early elements of Weber's programme. But for Weber and other empirically minded physicists unity under a specific law of force was the primary goal, conservation somewhat secondary. Only gradually did the conserved and generalised quantity of force, measured by *vis viva* and work, acquire conceptual independence as kinetic and potential 'energy'.

This view of Weber's project is consistent with the French tradition that he partially adopted. Laplace, among others, suggested that chemical affinities might possibly be explained as modifications of the inverse square force of gravity.[19] (As will appear later in this section, Weber's force law may be seen as such a modification.) Apparently following Boscovich in the attempt to make Newton agree with Leibniz, Laplace also reduced Newton's finite atoms of matter to material points so that they could never collide and so that only attractive and repulsive forces could act between them. That guaranteed conservation of *vis viva*, implying that the universe could never run down and that equality of cause and effect, or the principle of sufficient reason, would always be observed.[20] Ampère carried the argument considerably further when he attempted to construct chemical elements from geometrical arrangements of material points interacting through attractive and repulsive forces. After adopting the wave theory, Ampère also incorporated light and radiant heat as vibrations in a self-repulsive ether.[21] This ether formed an atmosphere around every point atom by attraction and extended through all space. It therefore served to transmit, as waves of heat and light, the *vis viva* of vibrating atoms, which he thought might constitute the internal heat of bodies. The same neutral ether could decompose to form positive and negative electrical fluids that, when flowing in opposite directions, would constitute an electrical current. Such double currents when flowing around an atom or molecule would form a little magnet or electrodynamic molecule, thereby accounting for magnetic materials.[22] In this physical picture, Ampère assumed that all of the ultimate forces were forces of material points that would conserve *vis viva*, but he did not attempt to formulate a theory based on interchanges of conserved quantities. His best-known law of force – the ponderomotive force between any two abstracted elements, or short sections, of current – he presented as a purely positive description of empirical observations.

Weber and his friend Fechner followed this French tradition when they reduced Ampère's abstract law of force between two currents to a physical action between point atoms of the two electrical fluids supposed to constitute both currents. The net force resulted from four interactions, all governed by the same law of force,[23] namely, the velocity- and acceleration-dependent law of force displayed in equation 9.1. The first term described Coulomb's inverse square electrostatic force between two separated electrical particles, attractive between unlike particles and repulsive between like particles. The remaining two terms modified the static law for electrodynamics, or electricity in motion, including both electromagnetic moving forces between constant velocity currents (Ampère's law) and electromotive forces induced between accelerating currents (Faraday's law). All major phenomena of electricity and magnetism were thus included under Weber's fundamental law of attraction and repulsion.

Again following Ampère, Weber believed in 1846 that the two electrical fluids in their normal unseparated state formed a neutral ether surrounding ponderable molecules and extending through all space. He hoped to be able to explain the wave theory of light on the basis of oscillations in this ether governed by the electrodynamic force law.[24] His project gained considerable credibility through association with Michael Faraday's discovery of diamagnetism in 1845 and Faraday's related discovery that diamagnetic bodies, when placed in a strong magnetic field, would rotate the plane of polarisation of light transmitted through them. Diamagnetism is the induction in normally nonmagnetic material, when it is subjected to a strong magnetic force, of a magnetic polarity opposite to that induced in normally magnetic substances, or paramagnetics. Weber explained the phenomenon successfully as induction of Ampèrian double currents in the neutral electrical atmosphere surrounding ponderable molecules.[25] These induced diamagnetic currents were similar, but opposite in direction, to the permanent currents around paramagnetic molecules. They offered a natural ground for explanation of magnetic rotation of light, particularly when the luminiferous ether was identified with the neutral electrical ether.

Following such early explanatory successes, Weber began to develop in the

1850s an increasingly comprehensive picture of the interaction of ponderable molecules with ether, a picture that became by 1880 a nearly complete electrical theory of matter, including even gravitation. It was in the course of these extensions that he departed markedly from French traditions, when he isolated and reified into a physical structure a notion that had been central even to his original force doctrine: the pairwise relation. This reified relation was the atompair, to which all actions were to be reduced. But the pair itself in Weber's conception could not be reduced; it was more than a sum of two atoms. A single 'physical point, or atom', Weber later insisted, could possess only mass and motion. More complex properties of matter, even extension, had to be considered as arising, not from properties attributed to individual atoms, but from the independent properties of pairs, existing only in their relation.[26] In 1846, Weber had thought that the acceleration-dependent term in his law of force might indicate a mediating action of ether between two atoms or even the existence of irreducible three-particle relations, because the relative acceleration of two atoms would depend directly, through Newton's second law of motion, on the action of any third particle on the first two.[27] In chemical reactions, Berzelius had named such forces 'catalytic forces', and their appearance in the electrical law of force heightened Weber's awareness of the possibility that his law might contain the reduction, long sought by Faraday and others, of chemical to electrical action. If three-particle and higher-order relations were fundamental, however, nature would be infinitely complex. Weber soon found relief from this complexity in his potential function, from which the force law could be derived in the conventional way as a gradient. The potential contained purely pairwise relations: relative position r and relative velocity v,

$$V = \frac{ee'}{r}(\frac{v^2}{c^2} - 1)$$

Believing already that the whole of a two-particle system was greater than the sum of its parts, Weber considered it extremely important that according to this potential function the 'totality of many bodies' would not give rise to new forces and new properties that could not be reduced to the properties of single particles and pairs.[28]

The simplicity alone of Weber's expression for potential probably argued for its precedence over the force law, but it is indicative of the changes taking place in physical theory around midcentury that he expressed that precedence as he assimilated his unifying force law to the general unity of energy conservation. Under the energy doctrine, potential was not only a simplifying expression for the force between two atoms; it was potential energy, the work expended against those forces in assembling the pair with given relative position and velocity. Total energy, thought of as contained in the pair, was this potential energy plus the *vis viva* (kinetic energy) of the pair. As energy was separated from moving force within the general concept of force or power, Weber came to recognise potential energy as the extra physical something in the abstract relation of a pair. In 1871 he quoted August Beer approvingly: 'In many respects one can speak with more justification of the *physical existence of work, expressed through the potential,* than of the *physical existence of a force,* of which one can only say that it *seeks to change physical relations of bodies*'.[29] A brief summary of the evolution of Weber's programme for ether will help clarify both this changing perspective and his grand plan for the unity of nature based on the atompair.

In 1852, Weber developed an explanatory model for resistance to electrical conduction in which he first described atompairs.[30] The model raised concrete problems in the relation between electrical and ponderable particles that ultimately suggested their unity. If free ether consisted of neutral pairs of positive and negative particles, Weber reasoned, then the particles of each pair should orbit about each other under the action of their attractive force. When subjected to an applied electromotive force, as inside a wire connected to a battery, such orbiting pairs would undergo successive breakup and recombination into new pairs, resulting in opposite motions of positive and negative particles along the wire. Resistance to this double current would derive from the force required to divide electrical pairs. (He did not yet discuss how the expended force disappeared.) Since all pairs were identical, the number of divisions per unit time would distinguish resistances in different materials. Conductors, nonconductors, and free ether, supposedly, would decrease in resistance with decreasing ether density. Though heuristically interesting, this model of double currents and resistance was clearly inadequate. It provided no ground for explaining different densities of ether in different materials. More problematically, it required that the double currents about ponderable molecules, those responsible for magnetism, have completely independent positive and negative components, for otherwise the component currents would resist each other and stop. Weber could only suggest that the positive and negative molecular currents moved in circles of different radii. He offered no explanation of the difference nor even an account of why electrical particles would orbit about ponderable molecules in the first place.

He did not stop long to ponder these obvious difficulties. By 1855 he had outlined the more profound problems that would occupy the remainder of his career.[31] Electrical resistance, he thought, must derive somehow from the connection of electrical particles to ponderable particles: What was the connection? Electrolysis showed a close relation between specific chemical elements and specific quantities of electrical charge: What then was the relation between chemical affinity and charge? One could add the velocity- and acceleration-dependent terms of the electrical force law to the gravitational force law without altering observable results (because $1/c^2$ was small): How close was the analogy thereby revealed between gravitational and electrical forces? By 1862, Weber had added to his unifying list the problem of converting work to heat, specifically, of converting the work done to produce electrical motions in a current into the heat of ponderable molecules, thereby conserving energy. He had also seen that this relation, as well as many other relations between ponderable molecules and electricity, might be discovered in a modified version of molecular currents.[32]

Rather than being surrounded by two currents in smaller and larger circles, molecules might have one kind of electricity, negative say, adhering to their central mass, while only positive electricity circulated about this negative nucleus and also constituted ether. That would eliminate the problem of resistance to molecular currents while simultaneously explaining resistance to conduction currents. To move positive particles from one molecule to another would require work from an applied electromotive force. Motion in conduction currents, furthermore, could be transferred to motions in the molecular currents, appearing as the heat of molecules or the heat of a ponderable body. Any disturbances of the molecular currents would cause oscillations in them at the frequency of the molecular orbits, and the oscillations would produce waves in the surrounding ether, or heat radiation. Similar wave disturbances would constitute light.

This was the model for which Weber, during the sixties and seventies, sought to build a mathematical foundation. Responding directly to the claims of energy conservation, he now made energy and energy exchange the basis of his analysis. By reexamining the states of motion of electrical pairs under the action of his fundamental force law, he found that not only two unlike particles, but also two like particles, could form a stable atompair.[33] In fact, any system of two like particles would possess two different states between which no transitions could occur except under the action of an external force. In a bound or attractive state (*Molekularbewegung*) the two particles remained always closer together than a small limiting value r_0, whereas in an alternate unbound or repulsive state (*Fernbewegung*) they moved between r_0 and infinity. Similar states existed no matter how large the masses of the particles. If, therefore, negative electrical particles were somehow united with ponderable atoms, pairs of the resulting atoms could form stable massive configurations (*Molekularbewegung*) under electrical forces alone. Similarly, positive pairs of negligible mass in the state of *Fernbewegung* could constitute ether.

With that rigorous result, Weber had broken open his set of problems, for now he could envision constructing all ponderable matter out of identical negative atoms, thereby uniting chemical atomic theory with electrical theory. The chemical identity of elements would depend on the number of negative atoms united in a massive nucleus and on their organization; magnetic, electrical, and thermal properties would depend on the states of binding between the nucleus and the light, positive atoms surrounding it.[34] Thermal radiation and light, as before, would interact with matter as waves in the positive ether. Weber conceived this ether first as a gas and later as a stable mass, but in either case it consisted of only positive pairs in the repulsive unbound state.[35] All phenomena of nature seemed to be subsumable under the new mathematical results; all, that is, but one – gravitation.

In 1875, Weber wrote the law of gravitational force in a form corresponding to his chemical atomic theory, where the mass m and charge e of ponderable atoms were proportional, $m = \alpha e$, so that

$$f = \frac{mm'}{r^2} = \alpha^2 \frac{ee'}{r^2}$$

This proportionality between the electrostatic and gravitational forces acting between all ponderable bodies suggested that the two might be integrally related, especially since the usual velocity- and acceleration-dependent terms could be added without affecting appreciably the long-range interactions of gravity. To reduce gravity to electrical forces, however, a force was required that at large distances would act attractively between *neutral* molecules, that at smaller distances would become repulsive (as in a gas), and that at yet smaller distances would again become attractive (as in chemical bonding). At the suggestion of a close associate at Leipzig, Karl Friedrich Zöllner, Weber attempted to encompass all of these factors within his force law, by employing two hypotheses: (1) that each of the identical ponderable atoms combining to form chemical elements was a neutral system consisting of a negative, but highly massive, central particle with a positive satellite of much smaller mass; and (2) that the attractive force between equal but opposite electrical particles was *greater than* the repulsive force between like particles, by a factor $(1 + \alpha)$.[36] On these assumptions the net force between two neutral ponderable atoms (four interactions between two negative–positive pairs) would appear as a net attractive force between the negative component e of each pair and the positive component e' of the other:

$$F = 2\alpha \frac{ee'}{r^2} (1 - \frac{1}{c^2}v^2 + \frac{2r}{c^2}a)$$

This could be the 'gravitational' force at large distances r if α were the ratio between the large mass of a negative particle and the small mass of a positive particle.

Weber had here transformed his original neutral electrical ether of 1846 into a system consisting still of only positive and negative atoms but encompassing now all of the properties of both ether and matter. He was tantalisingly close to a single force law uniting all nature – with one qualification. The final scheme is coherent only with the understanding that force is a relation of two atoms, and its intensity depends on the character of both, for otherwise there is no basis for hypothesis (2).

Fechner's philosophical basis for action at a distance

The preceding discussion of Wilhelm Weber's evolving ether theory has displayed the coherence of form that marked his theorising throughout at least forty years of research. This coherence cannot be fully appreciated, however, without recognition of the philosophical goals he intended the theory to fulfil. Weber wrote with the traditions of German Idealism in mind. He apparently considered it a prerequisite of any valid theory that it make provision for a close connection between *Geist* and *Natur* but that it not degenerate into Materialism by reducing *Geist* to the mechanics of matter. The pairwise relation, which contributed properties to the whole beyond those of its parts, fulfilled this criterion.

> If [two] material essences, which are spatially and temporally separated, interact, then the ground of this interaction lies in the essence of both *as a whole*. The interdependent parts of this *whole* exist in different spatial and temporal points. If there are *material* essences which as *wholes* cannot be reduced to a point in space and time, that applies even more to *spiritual* [*geistig*] essences.[37]

Weber did not publish many of his thoughts on this connection between *Geist* and *Natur,* but throughout his life he associated himself most closely with people who made it their central concern. In the cases of his brother Ernst Heinrich Weber and of Fechner, both celebrated professors at Leipzig, it led to research in psychophysics. For Zöllner, also at Leipzig as professor of astronomy, it led eventually to experiments in spiritualism, experiments in which the Weber brothers and Fechner took part.[38] Weber's philosophical concerns seem to be reflected most closely in the writings of Fechner, with whom Weber first developed his ideas of an electrical double current and neutral ether, and whose 1855 defence of atomism, *Ueber die physikalische und philosophische Atomenlehre,* relied heavily on interaction with Weber.[39]

In the *Atomenlehre,* Fechner sought to found metaphysical atomism on contemporary experimental and mathematical physics, on the 'presentable connections of objects [*Weltdinge*], directly and compactly summarised in their ultimate points and knots [*letzte Spitzen und Knoten*]'.[40] The ultimate points and knots were, to Fechner, physical atoms and forces. Arrayed as enemies against this atomism of the physicists he saw the dynamicists and dialecticians described earlier: Kant, Schelling, Hegel, and Herbart. To them matter was a constructed concept. Even space was phenomenal, an appearance deriving from the process of perception, and time too they ascribed to perception. In that way, *Natur* had been filled with *Geist*, in contrast to the crude, materialistic, and spiritless conceptions of atomism.

The closest approach of the dialecticians to a physicist's atomism appeared in Herbart's 'monadology',[41] a metaphysics somewhat like Leibniz's, based on a hierarchy of discrete, simple essences, called reals or monads. A high-level monad would be the soul of an organism, whereas a low-level one would appear in the physical world as an atom. All of our knowledge, for Herbart, both of external (physical) and of internal (psychological) phenomena derived from the relations between monads. Modelling his view of these relations after a notion of force as chemical affinity (*Verwandschaft*, implying kinship or sympathy in a relationship), Herbart asserted that the relations arose as a process of conflict between monads, each striving to preserve itself. The dialectic of nature, therefore, occurred not between forces but between monads, and one could have knowledge of this conflict through the conflict between one's own soul and external monads, which gave rise to discrete presentations (*Vorstellungen*) in the mind. The relations between presentations mirrored the relations of external monads and produced one's perceptions of the objective world. Space and time appeared as general aspects of the process of relation, whereas the properties of bulk matter arose from more specific relations.

Into this, Fechner proposed to inject some commonsense reality, the reality of empirical appearances reduced to their essence. Most real of all, he argued, were material objects in space and time, which could be tasted, smelled, touched, and heard, not merely once but repeatedly, and not merely by one person but by many.[42] Matter, then, was an empirical reality and should lie at the foundation of any philosophy of nature. But additionally, physical research in its modern state required that matter be conceived as atomic. The successes of chemical atomic theory, particularly, established this view for ponderable matter; and the wave theory of light, to the degree that it explained on the basis of a particulate ether such phenomena as polarisation and differential diffraction of different wavelengths, established the atomic constitution of ether. These claims could hardly have been considered the consensus of physical scientists, but Fechner had in mind specifically as representatives of the 'new mathematical physics' Weber and the French analysts: Laplace, Poisson, Fresnel, and Cauchy. Their work in particle mechanics was only beginning to receive a serious challenge from mathematically developed continuum mechanics, as Jed Buchwald (this volume) has shown for physical optics.

Modern physics established for Fechner the practical necessity of atomism, both for normal matter and for ether. But he had still to construct for the atomistic world view a satisfactory philosophical foundation in order to present it as a coherent idea, and ideal, in the context of German traditions. Most important, he wished to establish, contrary to the dialecticians and dynamicists, that atomism was neither materialistic nor spiritless. The materialism of traditional atomistic mechanics, Fechner reasoned, resided in two notions: that atoms consisted fundamentally of extended gross matter and that forces inhered in this matter, or that forces were properties of matter.[43] On that view all phenomena of nature would derive from independently acting atoms of matter. There would be no unifying tissue tying all of nature together as an organic whole, such that the whole was more than the parts; and there would be no room for the interrelation of mind and matter.

To elevate atomism towards the realm of *Geist,* Fechner began by supposing that material atoms stripped of their relations to other atoms were actually only physical points, or real physical monads, existing in real space and time. These atoms possessed no properties of their own other than mobility. All motions of atoms and all properties of assemblies of atoms, or matter, derived from relations between atoms.[44] To the degree, then, that the concept of matter referred to properties of matter, matter was constructed from immaterial relations of point atoms. That concept allowed an immediate translation from the realm of matter to the realm of mind. Relations between atoms in inorganic nature were laws, physical laws of force; these same laws, however, represented the rule of *Geist:* '*Geist* steps up and asks, what have I to do with you? And the atoms say: we spread our individualities under your unity; the law [of force] is the commander of our band, but you are the king in whose service he leads'.[45]

In the new ordering of concepts – space, time, atoms, laws, *Geist,* and ultimately God – one moved from what one could know objectively to what one could know subjectively. At the interface between the external world and the perceiving subject, objective and subjective were only two references to the same thing.[46] That was the basis of Fechner's and Ernst Weber's well-known pioneering work in empirical psychology, labelled psychophysics by Fechner. Their investigations rested on the assumption that a determinate increase in objective stimulus should produce a corresponding intensity of subjective response (e.g., pain). The entire line of argument involved a simple inversion, similar to Herbart's, of the usual dialectic of conflicting attractive and repulsive forces, which had produced continuous matter throughout space. Fechner supposed instead that the 'dialectic' of discrete atoms gave rise to forces, that forces were the objective relations between atoms, and that this relation either constituted or measured subjective perception.[47]

Fechner's world view provided not so much a complete philosophy of nature as a programme for quantitative research, the programme that he and Ernst Weber pursued primarily in psychophysics and that Wilhelm Weber pursued in physics. Beginning with an empirically and theoretically derived law of pairwise atomic interactions, the force law or the law of potential energy, one would attempt to construct the empirical properties of matter, of ether, and of their interactions. Fechner believed that a single kind of atom, or monad, would serve to explain all of these phenomena; qualitative differences would depend only on different groupings of the atoms. Already in 1855 one could conceive of reducing magnetism, heat, and light to motions and relations of electrical matter; but were both attractive and repulsive forces, and both positive and negative electricities, necessary? What of gravitation; and what of mass itself as a quality of atoms? Fechner took hope in the increasing range of Weber's pairwise force law and attempted to uncover its ultimate implications. If the pairwise law alone proved insufficient, however, he was prepared, like Weber, to consider higher-order relations of three or more particles and higher-order relations of relations.[48]

I have described the Weber–Fechner action-at-a-distance theory of matter and ether at some length in order to show, first, the broad range of applications of the theory envisioned by Weber and, second, its metaphysical significance as presented by Fechner. Both considerations should indicate, third, that when critics of action at a distance charged that such action was incomprehensible, because one could not imagine how the action could reach from one body to another, they were missing much of what action at a distance meant to its German adherents. One body did not simply act on another – they interacted in a relationship – and the force existed as the law of relation. Nevertheless, Weber's law of force severely strained the plausibility of transforming an abstract relation into a real entity. It made of force a relation dependent not only on the distance between two atoms but also on the rate of change of that distance (velocity) and on the rate of change of the rate (acceleration). It is one thing to imagine a relation between two atoms at an instant; but it is quite another to imagine that relation depending implicitly on just prior and just following instants, as Weber's law required. Weber and Fechner found little difficulty in this, apparently because they so closely identified force with a logical relation in which space and time had similar status. A number of other physicists, however, sought a more physical explanation for time dependence. It indicated to them that force did not act instantaneously between two bodies but took time to propagate from one to the other.

Already in 1846, upon reading Weber's first electrodynamics paper, Carl Friedrich Gauss, Weber's close associate at Göttingen, remarked that he had earlier made a similar investigation himself but had lacked what he regarded as the keystone: 'namely, *the derivation* of the additional forces [added to electrostatic force] from *non-instantaneous* actions, actions propagating in time (in a manner similar to that of light)'.[49] This association between electricity and light became all the more pressing with realisation that the constant c in Weber's law was close to the velocity of light. Alternatives to Weber's theory of the electrical ether, therefore, often proposed that electrical action propagated at the speed of light and, in fact, constituted light.

Propagation of force and early field theories

For force to propagate from one place to another, on analogy with the wave theory of light, it must be describable not simply as an abstract relation but as something existing objectively in space and distributed throughout space. That was the judgement of those who saw in Weber's implicitly time-dependent law of action at a distance the denial altogether of such action. Their alternatives were what soon came to be called field theories: both ether fields, in which force actually consisted in a form of motion in ether, and pure force fields, in which force itself propagated independently through space. Although physically quite different alternatives, ether fields and force fields arose in the same context and emphasised the same descriptive foundation, namely direct descriptions, dynamic relations, and partial differential equations.

A partial differential equation in space and time may be seen as describing the behaviour in time of any and all infinitesimal elements of space, elements taken to be characterised by one or more properties. Because the equation describes the change of these properties across an element it also describes their change between elements, and it traces that change in time. It therefore expresses naturally any continuous process of propagation from element to element through a field of properties, which might be properties of ether or properties of force. Partial differential equations, then, served as a broad path for describing directly the processes of nature while ostensibly avoiding speculation. When taken also as the most fundamental description, they served as the ground for continuum theories of ether and force. And they sometimes served as a mathematical rationale for maintaining the emphasis of *Naturphilosophie* on nature as continuous dynamic process.

Field theory was not established independently as an ongoing research tradition in Germany much before 1880; it existed more as an undercurrent that surfaced along with the notion of energy as an independent, conserved entity and in response to an action-at-a-distance interpretation by Helmholtz in 1870 of Maxwell's field theory. For that reason I present here only a cursory summary of several early field theories in order to establish their character. I consider Helmholtz's theory in a following section as a critical articulation that helped to motivate mature field theory.

Ether fields

The ether field for electromagnetism arose first, and with enduring characteristics, directly under the gaze of Weber himself in the work of his young student, assistant, and friend at Göttingen, Bernhard Riemann.[50] Riemann was also a student of Gauss and a friend of one of the most luminous exponents of differential equations applied to physics, Lejeune Dirichlet. Riemann shared Fechner's taste for metaphysics and for unifying all natural phenomena on a common physical basis, though from a different perspective. Basing his position on the epistemology and psychology of Herbart – as opposed to Herbart's *Naturphilosophie* of monads, which had been important for Fechner – Riemann argued that only relations or states could serve as causes of action, or forces, for only something subject to change of degree could itself be the cause of such change: 'What an agent strives to effect must be determined through the concept of the agent'.[51] Being (of things) was not subject to change of degree; therefore things could not be causes, only relations or states could be.

Some perspective on Riemann's view can be obtained by observing that Herbart had considered perception to consist of reception by the mind of discrete presentations, followed by 'sinking' of the presentations below the threshold of consciousness. This sinking derived from mutual suppression of opposed presentations through the forces they exerted on one another. Applying to perception the notion of force that he employed in his discussion of monads, Herbart regarded the forces between presentations as affinities, grounded not in the presentations but in their relations. He supposed also that the relations established themselves in time, approaching only asymptotically a static balance of forces. Perception was thus a dynamic process, and Herbart attempted to describe it through differential equations in time. Riemann seized on this aspect of Herbart's psychology and made continuity and differential equations the starting point for both psychology and physics. 'We observe', noted Riemann in 1853, 'a continuous activity of our soul'. This activity consisted in presentations continuously disappearing from consciousness and yet remaining as part of the substance of the soul. By analogy, he reasoned, gravitation might consist in a continuous flow of imponderable space-filling *Stoff* (substance or ether) into ponderable atoms. In fact, 'both hypotheses may be replaced by the one, that in all ponderable atoms substance perpetually enters the spiritual world from the world of body'.[52] In this way action at a distance would not exist; gravitation might be thought of as dependent on the pressure of ether immediately surrounding ponderable atoms, where pressure in turn depended on velocity of the ether. The same ether could serve to propagate the oscillations that we perceive as light and heat.

Riemann attempted initially to analyse the processes of gravitation and light in terms of resistance of a homogeneous ether to change of volume (gravitation) and to change of shape (light), the latter reducing to resistance to change of length in any physical line. He hoped also to be able to include electricity and electromagnetism in the schema, with electrostatic inverse square force depending on change of volume and electrodynamic force on change of length. All actions, therefore, were to be actions only between 'neighbouring' or 'immediately surrounding' elements. Although Riemann did not carry much further this first glimmer of a field theory, he preserved its essential characteristics in several later attempts. He always sought to unify nature on the basis of a geometrically conceived system of continuous dynamic processes in ether, founding his description on differential equations that described the processes of relation that one could perceive directly, that is, forces and interconversions of forces. His project was probably the first attempt at a mathematically founded unified field theory, much in the spirit of Einstein's later attempts, and he assigned it more weight even than his now-famous efforts in pure mathematics.[53]

Riemann's more sophisticated field descriptions in the late 1850s rested on his general assumption that the cause of both motion and change of motion (inertia and accelerating force) of a body at any point should be sought in 'the form of motion of a substance spread continuously through the entire infinite space . . . This substance can therefore be conceived as a physical space whose points move in the geometrical one'.[54] Neglecting now any explicit correlation with the sinking of presentations in perception, Riemann proposed certain motions of a homogeneous ether that would reproduce the partial differential equations of gravitation and of light propagation. The first of these equations was a continuity equation, with a nonrotational velocity system **u** standing for force:

Divergence **u** = $-4\pi\rho$

It stated that the net flux of ether into any volume element should be given by the ponderable mass ρ per unit volume of the element. The second equation was a wave equation for transverse oscillations in velocity w (not oscillations in displacement) propagating at the speed of light c:

$$\frac{\partial^2 w}{\partial t^2} = c^2 \left(\frac{\partial^2 w}{\partial x^2} + \frac{\partial^2 w}{\partial y^2} + \frac{\partial^2 w}{\partial z^2} \right)$$

Combining the two motions for gravity and light produced a well-behaved velocity function, which confirmed the possibility of uniting the two processes.

Riemann had difficulty incorporating electrostatic and electromagnetic effects into the system, but in a series of lectures delivered in 1861 he indicated that variations of density in ether might be the key. If V were the usual electrostatic potential and **u** an electrodynamic potential (vector potential), the latter could be chosen to satisfy the equation

$$\frac{\partial V}{\partial t} = \text{Divergence } \mathbf{u}$$

In that case, 'V may be regarded as the density, **u** as the flux of this ether'.[55] A stable gradient in density would then correspond to electrostatic force, whereas a time rate of change in density, and associated fluxes, would correspond to electrodynamic effects. Actually, only the rotational part of the flux system in ether was required for electrodynamics, so that the nonrotational part was presumably still left over for gravitation.

Riemann never published his speculations on the unified ether field, no doubt because he did not succeed in fully integrating the various effects of gravity, light, electricity, and magnetism. He did, however, present a paper to the Göttingen Society of Science in 1858 (published only in 1867, after his death), which helps to complete this schema. Here he proposed an electromagnetic theory of light. It returns us to Weber's electrodynamic force law and to the problem of explaining velocity- and acceleration-dependence in a relation at a distance.

'I have found', Riemann announced,

> that the electrodynamic effects of galvanic currents may be explained if one assumes that the action of one electrical mass on the rest does not occur instantaneously but propagates to them with a constant velocity (equal to the velocity of light within the limits of observational error). The differential equation for the propagation of electrical force, according to this assumption, will be the same as that for the propagation of light and radiant heat.[56]

His assumption meant that one electrical mass experienced the action of another always as an electrostatic force (inverse square force or inverse first power potential), but it experienced at time t an action produced at an earlier time t', where the time lag was the propagation time from the position of the acting particle to that of the affected particle. The potential for such a 'retarded' force obeyed a wave equation rather than the usual continuity equation of electrostatics. And Riemann provided a proof that the retarded potential, when referred only to the time of action t, would correspond to the potential of Weber's force law. There are certain inadequacies in the proof, however, and perhaps for that reason Riemann withdrew the paper from publication.[57]

In summarising the general characteristics of Riemann's programme for unifying physics, we note that he sought to replace Weber's discrete relations at a distance with continuous action between neighbouring elements, which circumvented the problem of time-dependent forces. More important, he replaced the abstract relation, force, with states and processes in ether, providing a physical basis for the relation. Electrostatic potential, for example, he ascribed to density in ether. This made variations of density the reality of force, a shift of considerable importance for the energy concept, for it spread force throughout space as potential energy. Most profoundly, by reducing causality (forces or dynamics) in the observable world to the problem of describing motions in ether (kinematics), Riemann reduced physics to descriptive mathematics, or to the geometry of motions in a 'physical space'. That pushed the problem of causes one step farther back into metaphysics, and into the dynamical principles of ether. Riemann recognised this explicitly in regard to his idea of gravitation and light as two forms of motion in ether:

> The further development of this hypothesis divides into two parts insofar as one seeks:
> 1. The laws of ether motion [*Stoffbewegungen*] which must be assumed for explanation of phenomena,
> 2. The causes from which these motions can be explained.
>
> The first task is a mathematical one, the second a metaphysical.[58]

He employed as a foundation for the metaphysical dynamics of ether the principle of continuity of motion and a maximum–minimum condition on the velocity potential integrated over all space (the latter exhibited only mathematically). These ideas should be compared with the very similar ideas of William Thomson, Maxwell, and other midcentury British physicists. They formed the core of what the British called 'dynamical theory' (Moyer, 1978; Wise, 1980).

In this regard, Riemann's emphasis on differential equations as the basis for positive description, coupled with energy as the basis for force, is particularly significant. Both ideas made the study of fluid motion, or hydrodynamics, a pressing research topic, even when a theorist did not set out to reduce all forces to ethereal states and motions. Helmholtz, for example, established in 1858 several important theorems in hydrodynamics while rejecting such a reduction.[59] Nevertheless, his proof that a vortex in a frictionless fluid would last forever became the basis of William Thomson's ether theory of vortex atoms (Siegel, this volume). G. R. Kirchhoff developed hydrodynamics considerably further, apparently with the intention of ultimately ridding physics of forces. Adopting basic theorems of Thomson's and Tait's *Principles of natural philosophy*, he made maximum and minimum conditions on fluid motion the foundation of his hydrodynamics.[60] This suggested a quite general approach to the dynamics of an ether that would explain observable forces. At the same time, however, Kirchhoff spread the new positivism of differential equations, denying the need for specific hypotheses on the reality of nature.[61] He did not pursue directly a generalised ether theory.

Perhaps the most sustained programme of hydrodynamic reduction was carried on by a Norwegian mathematical physicist working in near isolation at the University of Oslo. Carl Anton Bjerknes attended Dirichlet's lectures at Göttingen in 1855–6 and studied there also with Riemann.[62] Impressed with Dirichlet's proof that a solid sphere in a uniform stream of frictionless, imponderable fluid would not be dragged along by the flow, Bjerknes set out to discover whether oscillations and pulsations of such spheres, producing oscillations and pulsations in the fluid, might be affected by forces of the type usually ascribed to action at a distance. In a series of mathematical papers stretching from 1863 to the end of the century, but largely unknown outside Scandinavia, he gradually extended the scope of his theory until by 1880 he could reproduce hydrodynamically the basic forces of magnetism (including induced diamagnetism and paramagnetism), electrostatics, and electromagnetism.[63] Isolated magnetic poles and electric charges appeared as pulsations in volume, polar magnets as linear oscillations, and electric currents as rotational oscillations about the axis of the current. One troublesome aspect of this description was that it produced forces opposite to those observed in nature, giving always repulsions where attractions should have appeared. More important, however, Bjerknes's description did not keep up with the demands of electromagnetic theory as it developed after about 1870, when Maxwell's electromagnetic theory of light became widely known on the Continent.

Although hydrodynamics offered a ready avenue for reducing forces to propagated motions in an ether field and, thereby, for unifying all forces, one could equally well begin, not with the nature of ether, but with the differential equations one wished to represent, and then construct whatever kinds of ether motions the equations seemed to require. That was the approach of Maxwell in Britain and of his nearest competitor on the Continent, Ludwig Lorenz. A professor of engineering at Copenhagen and, like Bjerknes, relatively unknown, Lorenz developed during the 1860s sophisticated mathematical descriptions of processes in physical optics, particularly of reflection and refraction at boundaries. He conceived his optical equations originally on the basis of an elastic ether, but he soon became convinced that elasticity theory was incompatible mathematically with the boundary requirements of optics. Abandoning the elastic ether, and ostensibly all ethers, he developed phenomenologically a set of wave equations for the propagation of light that behaved correctly at boundaries. But Lorenz was not content with a mathematical theory of optics; like many others he sought the physical unity of nature.

'As is well known', Lorenz began a paper in 1867, 'science in our century has succeeded in demonstrating so many connections between the different forces, between electricity and magnetism, between heat, light, molecular, and chemical forces, that one is led with a certain necessity to regard them all as *manifestations of one and the same force*'. Despite this necessity, unity still seemed distant: 'Generally, the two electricities are still viewed as electrical *fluids*, light as vibrations of the *ether*, and heat as motions of the *molecules of bodies*'.[64] Lorenz presented his own first step towards remedy with a demonstration that the motions of light waves were actually motions of electric currents and that both were propagated by contiguous action.

Beginning his analysis with general equations directly relating electric currents and electromotive forces (equations of conduction, previously derived by Kirchhoff from Weber's law), Lorenz argued that no change in observable consequences would follow if these equations, representing instantaneous action at a distance, were replaced by a similar set incorporating the assumption that electrical action traveled at the velocity of light. That reproduced Kirchhoff's equations in terms of retarded potentials, like Riemann's. From the latter equations he then derived a set of partial differential equations of the same form as his phenomenological wave equations for light, but more comprehensive. Reversing the derivation, he obtained the conduction equations by integration from the wave equations, assuming boundary conditions that he had previously found necessary for light at an interface. This proof of mathematical identity Lorenz took to be sufficient evidence for identifying light physically with electric currents and also for assuming the differential wave equations for electric currents to be the 'original and generally valid ones'. With the characteristic emphasis of field theory, he argued on the basis of this priority of the differential description for the further priority of continuously propagated action over action at a distance: 'Every action of electricity and of electrical currents in reality depends only on the electrical condition of the *immediately surrounding* elements'.[65]

The form of the interconvertible equations of light and of currents, Lorenz continued, showed that light could be regarded as *'rotational* oscillations [of current] in the interior of bodies around axes whose direction is the same as that which we regard as the direction of vibration according to the elasticity theory',[66] that is, around the axis of linear polarisation, transverse to the direction of propagation. Whereas a steady current would be a steady rotation continued along its own axis, oscillations in the current would be propagated by electromagnetic induction perpendicular to the current axis, constituting light. One sees immediately the family resemblance of this hypothesis to Maxwell's picture, apparently unknown to Lorenz, of magnetic vortices in the ether. That raises the question of what medium Lorenz imagined to be rotating. He preempted the question unambiguously: 'This conception gives scarcely any basis for maintaining the hypothesis of an ether, since one may very probably assume that in so-called empty space there is contained so much material substance that it offers sufficient substratum for the motion'.[67] But of course this is just another ether hypothesis, one in which ether and normal matter differ only in aggregation or density and not in substance.

Force fields

Ether fields provided a means for spreading force in space and thereby for escaping the problems of Weber's velocity- and acceleration-dependent law of force. But they also transformed push–pull notions of force at a point into energy states in ether, reinforcing a more general recognition of energy as an independent, conserved, entity. Coupled with emphasis on direct, positive description, this suggested an alternative field conception: Treat energy as the reality and ignore the unobservable ether. As discussed previously, such views only gradually overcame the stigma of *Naturphilosophie,* but characteristics of the emergent transition appear in a very short paper presented by Carl Neumann in 1868 to the Göttingen Society of Science.[68]

Neumann developed in a new way Riemann's earlier theory of propagated forces and retarded potentials. 'I take the liberty', Neumann asserted, 'of regarding potential [potential energy] as primary, as the characteristic motive power *(Bewegungs-Antrieb)*, while conceiving forces as secondary, as the form in which that power manifests itself'. He then demonstrated that if potential propagated with the velocity of light, then the effective action of an electrostatic 'emissive potential' of one electrical mass m on another m', given by $mm'/r,$ would be changed when the two were in relative motion into a 'receptive potential', corresponding exactly to Weber's law of force. All of the known laws of electromagnetic action (known only for *closed* currents) likewise followed. As an indication of the generality of his approach, Neumann showed that the results were independent of whether currents were considered from a 'unitary' perspective – Weber's view, with the positive electrical fluid in motion and the negative tied to the ponderable mass of the conductor – or from a 'dualistic' perspective, with both fluids in motion. This generality would not pertain, he noted, for open circuits, a consideration soon to be emphasised by Helmholtz.

Neumann's intent seems to have been not so much to replace Weber's atomistic electrical ether with a pure force field as to defend Weber and his law against a charge by Helmholtz that velocity-dependent potentials were unacceptable from the perspective of energy conservation. He had recently taken a position at Leipzig, where he came into close association with Weber's circle of friends, and in several later publications he fortified Weber's defences. Nevertheless, the thrust of Neumann's defence ran quite counter both to Weber's premises and to his style of analysis. By taking potential (energy) as fundamental and making it propagate through space he gave to it the independent status of a force field, somewhat like J. R. Mayer's force of heat. With good reason the later Energeticists, who took energy to be the only ultimate reality, looked back to Neumann as well as to Mayer as pioneers of their viewpoint.[69] In typical fashion also for all field theorists, Neumann usually based his descriptions on differential equations rather than force laws, justifying that approach as the most positive and nonhypothetical, and looked to maximum and minimum principles for the foundation of dynamics. Here that meant that he took Hamilton's principle in mechanics to have 'unlimited validity' and used it both to derive Weber's force law from the 'receptive potential' and to show that this potential would obey, approximately, conservation of *vis viva*. Even though Neumann employed his field description only to describe the force between particles of Weber's electrical ether, he made that ether largely superfluous for the propagation of light; and even though he may have had in mind a material basis, or ether field, for his propagating potential, he made no reference to it and thereby undermined the necessity of such a material foundation for a force field.

Helmholtz: energy and electrodynamics

Field theories of electromagnetism, as stressed previously, emerged to replace action at a distance for the general community of German physicists when energy was recognised as an entity different from and more fundamental than moving force. Field theories also emerged only when it seemed apparent that no action-at-a-distance theory could avoid satisfactorily the velocity- and acceleration-dependence of Weber's law. The problems of both conceptual shifts can be observed in two famous papers by Helmholtz, his 1847 formulation of conservation of force and his 1870 alternative to Maxwell's electromagnetic field theory.

In 1847, Helmholtz set out quite consciously to build a bridge between German metaphysics (Kant's) and French physics, much as Weber and Fechner were doing at the same time.[70] Helmholtz, however, sought to exclude from nature all considerations of *Geist,* in the sense of soul and purpose, which signalled to him the excesses of *Naturphilosophie*. He wished to establish the physics of point atoms and attractive and repulsive forces as necessary, on both rational and empirical grounds, and in so doing to unite all causes in nature under the determinism of conservation of *vis viva*. Helmholtz's primary conceptual tools in this effort, I shall argue, were the concept of force as relation and a distinction between quantity and intensity.

To Helmholtz, as to Weber and Fechner, forces existed as relations: 'Motive force, as the cause of change, can be predicated only in cases involving at least two bodies spatially related to one another and is then to be defined as the effort of the two bodies to change their relative positions' (cf. the discussion of Weber in the section 'Weber's electrical ether'). To Helmholtz again these relations were rational relations, but in a restricted sense. Following Kant in the view that our knowledge of nature is scientific only to the degree that we understand its laws as necessary, he argued that forces must be invariable: 'As ultimate causes, forces which do not vary in time should be found'.[71] Even implicit time dependence, he supposed, would vitiate the comprehensibility of nature. If force did not always return to the same value for the same spatial relation, *vis viva* would not necessarily be conserved, but might be produced continuously from nothing. That would violate not only the principle of sufficient reason but also, from the empirical side, the impossibility of a perpetual motion machine. The same sort of argument applied to extended atoms, for which continuously acting rotational forces could be imagined, so that only attractive and repulsive spatial forces were allowable.

To formulate his argument verbally and mathematically, Helmholtz had continually to move back and forth between concepts of force as moving force and as *vis viva* and to reconcile the two ideas. This he accomplished with a distinction, drawn in the Kantian tradition, between quantity and intensity as two different categories of understanding. We know concepts as quantities or intensities depending on whether they are considered to add up extensively – that is, side by side (*nebeneinander*), like space – or intensively – on top of one another (*aneinander*), like density. In his construction of continuous matter from forces, Kant applied this distinction in order to demonstrate how a given quantity of matter could completely fill different spaces with different intensities, in contrast to the full-or-empty conception of Newton and Descartes.[72] Helmholtz, conceiving forces already as rational relations, took the logical distinction over into the realm of forces acting between discrete, point atoms, making Newtonian moving force the measure of intensity of force and, under various circumstances, either *vis viva*, potential, or work the measure of the conserved quantity of force.

The intriguing aspect of Helmholtz's description is that he ascribed two aspects to the single notion of a tensional force between atoms at a given spatial separation. The relation possessed at any instant an intensity of tension producing changes in spatial relation, but it also possessed a quantity of tension connecting the past history of the relation to its future, to which the present action either added or subtracted.[73] This quantity was potential. Helmholtz described it as the quantity of tensional force, or work, available in a spatial relation for future consumption, thinking of the quantity consumed between two separations R and r as 'the sum total of the intensities of all the forces which act at all distances between R and r'.[74] Tensional force consumed became *vis viva*, and conversely, *vis viva* lost became a quantity of tensional force available again for consumption.

Helmholtz was able to maintain mathematically his dual conception of the single entity force only through a nonrigorous understanding of an integral. Development of the independent energy concept removed this problem while separating intensity and quantity into force and energy. The point here, however, is that already in Helmholtz's conception potential was a real physical quantity that became increasingly real through myriad theoretical and experimental applications of the conservation law. And as it became an entity the rational notion of force as an abstract relation of atoms became less acceptable. Weber's notion of velocity- and acceleration-dependent action at a distance was unacceptable to Helmholtz on rational grounds and Helmholtz's own notion of action at a distance proved unacceptable on physical grounds.

When, in 1870, Helmholtz turned his attention to electromagnetism he faced two successful challenges to his conception of force, each of which offered a conceivable physical basis for the unity of nature (Woodruff, 1968; Hirosige, 1969). On the one hand, implacably, sat Weber's law for forces acting at a distance, which Helmholtz had finally to admit could conserve *vis viva* formally in spite of its implicit time dependence. He set as his first task uncovering its deeper flaws. On the other hand loomed the propagation theories, particularly Maxwell's elegant mathematical system, but also the several less extensive field theories already discussed. Helmholtz could not accept ether fields like Maxwell's and Riemann's, which replaced force altogether with motions in ethereal matter, apparently because he considered that matter could act only through force;[75] but neither could he accept pure force fields, because a propagating pure force, taking time to act, would be again a time-dependent force like Weber's. Yet field theories of both kinds did succeed in explaining light propagation electromagnetically, as by now seemed necessary. The retarded potential theories reproduced correct empirical laws of electrodynamics from the assumption that inverse square forces travelled at the velocity of light, and Maxwell had derived the correct velocity of light from purely electrodynamic laws. Helmholtz's second and major task, therefore, was to provide an action-at-a-distance alternative to Weber's law that would rely on non–time-dependent forces and still provide an electromagnetic theory of light propagation.

Helmholtz began his search for flaws in Weber's conception by comparing Weber's potential function with a well-known phenomenological potential of Franz Neumann (Carl Neumann's father), describing interactions between closed circuits, and with Maxwell's theory. For this purpose he put Carl Neumann's propagating potential on an equal footing with Weber's mathematically equivalent, direct one. The differences among Weber's, Franz Neumann's, and Maxwell's expressions had not previously been easy to define, but Helmholtz showed in a lucid analysis that all three could be subsumed under a single potential describing the interaction between any two current elements of lengths ds and $d\sigma$, carrying currents i and j, and separated by a distance r:[76]

$$-\frac{ij}{2c^2 r}[(1+k)\cos(d\sigma, ds) + (1-k)\cos(r, ds)\cos(r, d\sigma)]\,ds\,d\sigma$$

Here k is a variable parameter. When $k = -1$, Weber's potential results, assuming that a current consists of electrical particles in motion. Franz Neumann's potential corresponds to $k = +1$ when the expression is integrated over closed circuits; and $k = 0$ leads to results similar to Maxwell's. When applied only to closed circuits, k vanishes from the general expression and all three cases reduce to Neumann's. Only an investigation of open circuits could distinguish among them.

As mentioned previously, Kirchhoff had derived from Weber's law general equations for currents in extended conductors. Helmholtz now found in the same way equations of electricity in motion corresponding to his generalised potential function. From the new equations he showed that Weber's potential, and only that one among the three choices, would in certain cases of open circuits lead to unstable motions of electricity. For example, if two like electrical particles approached each other at a very high speed, their relative velocity would become infinite in a finite distance. Weber and his supporters soon suggested ways of circumventing the difficulty, and a bitter dispute followed, but Helmholtz found no reason to doubt his original judgement that 'the inadequacy of the Weberian law here brought to light is founded deep in its nature'.[77]

Helmholtz still attributed the inadequacy of Weber's law to velocity dependence in the potential between any two electrical atoms. He attempted to establish his own potential solely on the basis of spatial separation between any two points where currents existed (phenomenologically rather than as moving atoms of electricity), this being 'the only spatial magnitude which is completely determined by two points'.[78] Current velocity, nevertheless, remained hidden in the strength of the current at any point. Helmholtz simply ignored it, even though he himself considered currents to be electricity in motion.

Such mathematical manoeuvres could not impress physical theorists like Weber; indeed, Helmholtz later stated that his law was 'no elementary expression of the ultimate acting forces'.[79] It did allow him, however, to employ an instantaneous action-at-a-distance force between currents while avoiding either explicit velocity-dependence or propagation of force.

The latter issue brings us to Helmholtz's second task, explaining propagation of light electromagnetically. For that he appropriated Maxwell's theory to his own ends. Maxwell had shown in 1863 that an elastic displacement in a medium possessing the electric and magnetic properties of free space would propagate at the velocity of light, through a process of mutual induction between electric and magnetic polarisations (Siegel, this volume). The polarisations were attributed by Maxwell to motions in ether; they constituted electric and magnetic forces, and no actions except between contiguous elements of ether occurred. Few on the Continent pretended to grasp how this contiguous action was supposed to function, nor what electricity might be, and Maxwell offered little aid. But his equations provided a thorough description of electromagnetic effects and of electromagnetic waves propagating at the speed of light. The critical element in those equations was the assumption that electric displacement, or polarisation, acted during the displacement exactly like an electric conduction current.

Considering these features from his own perspective, Helmholtz assumed ether to be an electrically and magnetically polarisable medium in which all forces – electrostatic, electromagnetic, and magnetic – acted directly at a distance between its parts (presumably atoms, but not explicitly). The polarisation at any point, an elastic response, was proportional to the sum of the polarising forces acting on that point from all distances. Helmholtz's basic equations, therefore, were two elasticity equations for electric and magnetic polarisations, both of the form $\mathbf{F} = a\mathbf{P}$, where $\mathbf{F}$ is force, $\mathbf{P}$ polarisation, and a an elastic constant. The forces, however, were complex and interconnected between the two equations. Electric forces resulted from free electricity, electricity of polarisation, changing conduction currents, changing polarisation currents, and changing magnetic polarisations, the last three being sources of electromagnetic induction. Magnetic forces, similarly, resulted from magnetic polarisation and from electric conduction currents and polarisation currents.

Employing his own electromagnetic potential, but including both conduction and polarisation currents, Helmholtz derived from the elasticity equations partial differential equations for purely electric and for purely magnetic polarisations, having eliminated cross terms. These new equations were wave equations with solutions describing both transverse and longitudinal waves. The choice $k = 0$ in the potential made the longitudinal waves vanish (infinite velocity), leaving only transverse waves of electric and magnetic polarisation with perpendicular oscillations. Assuming the electric polarisability of ether to be very large, Helmholtz showed that the transverse waves would spread – *propagate* is not quite right – with the velocity of light. In the limit, therefore, and with the choice $k = 0$, his theory reduced mathematically to Maxwell's theory of light.[80]

In this derivation the original assumption of instantaneous action at a distance disappeared, to be replaced by equations apparently describing successive action between contiguous elements of a medium. The result, however, was only mathematical and not physical; it did not imply that polarisations actually propagated, but only that they spread in time. In each original elasticity equation – through differentiations, transformations, and substitutions – Helmholtz had replaced the summed forces of currents and polarisations, acting *from all distances* on a local region, by differentials of their effects (polarisations) *in the local region*. Spreading of polarisation, or the wave equation, arose from the replacement process as a coincidental result of the relation between spatial and temporal derivatives in the laws of electromagnetic action. No approximations were necessary, not even neglect of long-range actions over short-range ones.

Helmholtz was fully aware of the coincidental nature of his agreement with Maxwell and turned the coincidence to support his own view: 'The remarkable analogy between the motion of electricity in a dielectric and that of the luminiferous ether does not depend on the particular form of Maxwell's hypothesis, but follows in essentially the same way if we maintain the older view of electrical action at a distance'.[81] But though he used his own formulation of spatial action at a distance against Maxwell's ether field, Helmholtz also turned Maxwell's field against Weber's time-dependent action at a distance. 'Maxwell's hypothesis appears to me to be very important', he said in 1872, 'because it furnishes proof that nothing is implied in electrodynamical phenomena that forces us to reduce them to an anomalous [*ganz abweichend*] kind of natural forces, to forces that depend, not merely on the positions of the corresponding masses but also on their motions'.[82] With strong physical theories on either hand, both alien, Helmholtz stood on the neutral territory of uninterpreted mathematics and argued that neither alternative possessed the force of necessity, the necessity he had attempted to establish in 1847 for direct spatial action between point atoms.

Conclusion

What was the historical significance of Helmholtz's ether theory? This question has sometimes been answered with the observation that Helmholtz translated Maxwell's theory into terms comprehensible to Continental physicists; and that is certainly correct so far as it goes. By providing a double-ended theory that displayed both poles from which previous electromagnetic theories of light propagation had been sought – atoms and forces *versus* differential field equations – Helmholtz made clear what propagation would have to mean from the perspective of forces acting at a distance. But in supplying this illumination, I suggest, the theory also illuminated the inadequacies of any action-at-a-distance theory, especially for energy considerations.

By 1870 propagation meant propagation of energy. Energy had become the symbol of unity in nature and the quantity that had to be followed through any series of conversions and transmissions: for example, from a chemical battery, to an electric current, to polarisation of the ether, to induced currents, to heat. Any acceptable electromagnetic theory of light, therefore, had to supply a direct explanation of the process of propagation of the energy of light. Ironically, Helmholtz, the most notable author of energy concepts on the Continent, could not meet this demand in his own theory, primarily because it did not distribute energy independently through space; it was not a field theory.

Wave theorists of light ever since Fresnel had considered light to be *vis viva* in ether. In the usual action-at-a-distance theories, which treated ether as an elastic solid, one supposed that direct actions over several molecular distances were negligible in comparison with actions between adjacent molecules. Even here, therefore, *vis viva* was effectively propagated by successive action. Time for transmission derived from inertia of the molecules, each taking time to respond to the force exerted by its neighbours, and the process was conceived mechanically under Newton's laws of motion. In Helmholtz's phenomenological ether theory, however, actions occurred at all distances, there were no masses to carry *vis viva,* all energies were potential energies, and no mechanics was supplied. Propagation of light could almost be conceived as a spreading of potential energy of polarisation, but one had to regard the energy of polarisation in any local element of ether as a temporary receptacle for energy of abstract spatial relation between this element and all other elements at all distances. The local polarisation, upon relaxation, was again transmitted to all distances. Energy did not propagate between contiguous elements of ether as seemed necessary for phenomena of light, particularly when the wave equation was considered fundamental.

Problems of this sort are not merely retrospective evaluations of Helmholtz's theory. They were recognised and discussed immediately and formed the background for the reception and development of Maxwell's field theory in Germany. Joseph Stefan, for example, published in 1874 a detailed analysis of the energy relations in Helmholtz's theory, intending to illuminate the question

> whether one does not in general have to conceive magnetic and electric phenomena as conditions of a medium, perhaps the luminiferous ether, and whether particularly one does not have to assume that magnetic and electric forces are only apparent actions at a distance, being in fact immediate actions of the medium, dependent on its momentary states and therefore also propagated, just as these states, with finite velocity.[83]

Although Stefan preferred the latter theory he recognised that it was not yet a necessity but only more coherent. He limited his critique to a development of the energy relations that would be required for translating between the two views. He obtained in this way a clear perception of a problem soon widely recognised as basic to the development of Maxwell's, or any other, field theory of electromagnetism: What was the relation between energy in the ether field and the source of that field? His own preference was to treat energy in ether as the independent reality and to make source strength dependent on the state of the ether.

Although Stefan's specific views were apparently not influential, the problems that he recognised in Helmholtz's interpretation of Maxwell's theory were also appreciated by the acknowledged giants of Continental electromagnetic theory, H. A. Lorentz and Heinrich Hertz. Lorentz originally followed Helmholtz's own path and tried to generate a more detailed mechanical description of the role of electricity in matter that would allow an explanation of phenomena of physical optics in terms of the interaction of an electromagnetic wave of polarisation with electricity bound to molecules (Hirosige, 1969). Hertz, however, represented the new generation of physicists who believed that differential equations presented the most perspicuous view of reality and that theories should be based directly on them. After bemoaning the incomprehensibility of the relation in Maxwell's theory between sources and propagating effects he produced in 1890 his purely mathematical theory of an electromagnetic field.[84] He eliminated the distinction present in both Maxwell's and Helmholtz's theories between polarising forces and polarisation – thereby eliminating the distinction between forces and ether – and made of the field an entity unto itself. By employing this independent field and relating it to sources, Lorentz created his electron theory in the 1890s (Schaffner, 1969). Lorentz believed that some sort of ether had to be imagined as the basis of the field, but others, such as Wien and Abraham, actively sought a pure force field that would reduce even matter and mechanics to electromagnetic processes (McCormmach, 1970). This electromagnetic view of nature matched the goal of the Energeticists – Wilhelm Ostwald and Georg Helm are notable – who sought to reduce all of nature to energy alone. Although their programme foundered on the second law of thermodynamics, their rejection of atomism and belief in the ultimate continuity of nature was widely shared, even by critics such as Max Planck, to the degree that atomists like Ludwig Boltzmann despaired of having any impact in Germany at all.

Developments in the electromagnetic theory of light only partly caused this shift to continuity in nature, but they well exemplify general trends. The atomistic ethers of Weber and Helmholtz had given way to the electromagnetic ether field, which came increasingly to mean a field of electromagnetic energy. Energy and continuous flux seem best to symbolise late nineteenth-century views of nature in Germany.

Notes

1 T. S. Kuhn, 'Energy conservation as an example of simultaneous discovery', in *Critical problems in the history of science*, ed. M. Clagett (Madison, Wis., 1959), 321–56; P. M. Heimann, 'Conversion of forces and the conservation of energy', *Centaurus 18* (1974), 147–61.

2 S. D. Poisson, 'Mémoire sur la théorie du magnétisme', *Mémoire de l'Académie 5* (1820–2), 247–338, 488–533.

3 *Physicists* will refer here to practising scientists holding institutional positions in physics, mathematics, astronomy, or related fields when they devoted significant effort to *Physik*, as recognised in the abstracting journal *Fortschritte der Physik*.

4 R. S. Turner, 'Julius Robert Mayer', in *Dictionary of scientific biography*, vol. 9 (1974), 235–40.

5 'Ueber notwendige Konsequenzen und Inkonsequenzen der Wärmemechanik', a lecture delivered in 1869 to the Versammlung deutscher Naturforscher und Aerzte, in

Die Mechanik der Wärme in gesammelten Schriften von Robert Mayer, ed. J. J. Weyrauch, 3rd rev. ed. (Stuttgart, 1893), 357.
6 Ibid., 356.
7 H. von Helmholtz, 'The aim and progress of physical science', opening address in 1869 to the Versammlung deutscher Naturforscher und Aerzte, in H. von Helmholtz, *Selected writings of Hermann von Helmholtz*, ed. and trans. Russell Kahl (Middletown, Conn., 1971), 223–45; F. Gregory, *Scientific materialism in nineteenth-century Germany* (Boston, 1977), 100–21.
8 'Bemerkungen über die Kräfte der unbelebten Natur', *Annalen der Chemie und Pharmcie 42* (1842); reprinted in Weyrauch, *Mechanik der Wärme*, 24.
9 *Die organische Bewegung in ihrem Zusammenhange mit dem Stoffwechsel: ein Beitrag zur Naturkunde* (Heilbronn, 1845); reprinted in Weyrauch, *Mechanik der Wärme*, 48.
10 B. Gower, 'Speculation in physics: the history and practice of *Naturphilosophie*', *Studies in History and Philosophy of Science 3* (1973), 301–56.
11 I. Kant, *Kant's gesammelte Schriften*, ed. Preussischen Akademie der Wissenschaften, vols. 21–2, *Opus postumum* (Berlin and Leipzig, 1936), 21:593, lines 7–15, quoted in W. K. Werkmeister, 'The Critique of Pure Reason and Physics', *Kant Studien 68* (1977), 41.
12 *Opus postumum*, 22:583, lines 23–7, quoted in Werkmeister, 'The Critique and Physics', 45.
13 In *Zusatz* 5 (1881) to 'Ueber die Erhaltung der Kraft: eine physikalische Abhandlung' (1847), in H. von Helmholtz, *Wissenschaftliche Abhandlungen von Hermann von Helmholtz* (Leipzig, 1882), 1:73; reprinted in Helmholtz, *Selected writings*, 53.
14 *Beiträge zur Dynamik des Himmels: in populärer Darstellung* (Heilbronn, 1848); reprinted in Weyrauch, *Mechanik der Wärme*, 162, 176.
15 G. Helm, *Die Energetik: Nach ihrer geschichtlichen Entwicklung* (Leipzig, 1898), 16–27.
16 R. Kohlrausch and W. Weber, 'Elektrodynamische Maassbestimmungen insbesondere Zurückführung der Stromintensitäts – Messungen auf mechanisches Maass', *Abhandlungen der Königlichen Sächsischen Gesellschaft der Wissenschaften zu Leipzig 3* (1857); reprinted in *Wilhelm Weber's Werke*, 6 vols. (Berlin, 1892–4), 3:652. See also the 'Vorwort' to Kohlrausch and Weber's paper, *Berichte über die Verhandlungen der Königlichen Sächsischen Gesellschaft der Wissenschaften zu Leipzig 17* (1855); reprinted in *Werke*, 3:594ff.
17 My description may be taken as complementary to Kenneth Caneva's presentation of Weber's and Fechner's methodology as hypothetico-deductive. K. Caneva, 'From galvanism to electrodynamics: the transformation of German physics and its social context', *Historical Studies in the Physical Sciences 9* (1978), 63–159. See also K. H. Wiederkehr, *Wilhelm Eduard Weber: Erforscher der Wellenbewegung und der Elektrizität, 1804–1891* (Stuttgart, 1967); A. P. Molella, 'Philosophy and nineteenth-century German electrodynamics: the problem of atomic action at a distance', unpublished doctoral dissertation, Cornell University, 1972.
18 'Ueber ein allgemeines Grundgesetz der elektrischen Wirkung', *Abhandlungen bei Begründung der Königlichen Sächsischen Gesellschaft der Wissenschaften zu* (Leipzig 1, 1846), reprinted in Weber, *Werke*, 3:157; 'Auszug', *Annalen der Physik und Chemie 73* (1848), reprinted in ibid., 245.
19 'Reflections on the law of universal gravitation', in P. de Laplace, *The system of the world*, trans. H. H. Harte (Dublin, 1830), bk. 4, chap. 15.
20 'Of the motion of a system of bodies', in Laplace, *System*, bk. 3, chap. 5; *Essai philosophique sur les probabilités*, 6th ed. (Paris, 1840), 2–4.
21 'Lettre de M. Ampère à M. le comte Berthollet, sur la détermination des proportions

dans lesquelles les corps se combinent d'après le nombre et la disposition respective des molécules dont leurs particules intégrantes sont composées', *Annales de chimie 90* (1814), 45–86; 'Note de M. Ampère sur la chaleur et sur la lumiere considérées comme résultant de mouvemens vibratoires', *Annales de chimie 58* (1835), 432–44.

22 Summarised by L. Pearce Williams, *Michael Faraday, a biography* (New York, 1971), 142–51.
23 G. T. Fechner, 'Ueber die Verknupfung der Faraday'schen Inductions-Erscheinungen mit den Ampère'schen elektro-dynamischen Erscheinungen', *Annalen der Physik und Chemie 64* (1845), 337–45.
24 'Allgemeines Grundgesetz', 3:213 ff.
25 'Ueber die Erregung und Wirkung des Diamagnetismus nach den Gesetzen inducirter Ströme', *Annalen der Physik und Chemie 73* (1848); reprinted in Weber, *Werke*, 3:255–68.
26 'Ueber das Aequivalent lebendiger Kraft', *Annalen der Physik und Chemie 152* (1874); reprinted in Weber, *Werke*, 4:302. See also fragment of letter from Weber to Fechner in G. T. Fechner, *Ueber die physikalische und philosophische Atomenlehre* (Leipzig, 1855), 73.
27 'Allgemeines Grundgesetz', 3:212 ff.
28 'Aequivalent lebendiger Kraft', 4:303.
29 'Electrodynamische Maassbestimmungen insbesondere über das Princip der Erhaltung der Energie', *Abhandlungen der Königlichen Sächsischen Gesellschaft der Wissenschaften zu Leipzig 10* (1871); reprinted in Weber, *Werke*, 4:255 n.
30 'Elektrodynamische Maassbestimmungen insbesondere Widerstandsmessungen', *Abhandlungen der Königlichen Sächsischen Gesellschaft der Wissenschaften zu Leipzig 1* (1852); reprinted in Weber, *Werke*, 3: esp. 400–5.
31 'Vorwort', 3:595 ff.; Kohlrausch and Weber, 'Zurückführung der Stromintensitäts-Messungen auf mechanisches Maass', 3:652–67.
32 'Zur Galvanometrie', *Abhandlungen der Königlichen Gesellschaft der Wissenschaften zu Göttingen 10* (1862); reprinted in Weber, *Werke*, 4:91–6.
33 'Princip der Erhaltung der Energie', 4:268–78.
34 Ibid., 278 ff.; 'Ueber die Bewegungen der Elektricität in Körpern von molekularer Konstitution', *Annalen der Physik und Chemie 156* (1875); reprinted in Weber, *Werke*, 4:334–57.
35 'Elektrodynamische Maassbestimmungen insbesondere über die Energie der Wechselwirkung', *Abhandlungen der Königlichen Sächsischen Gesellschaft der Wissenschaften zu Leipzig 11* (1878), reprinted in Weber, *Werke*, 4:389–95; 'Elektrodynamische Maassbestimmungen insbesondere über den Zusammenhang des elektrischen Grundgesetzes mit dem Gravitationsgesetze', first published in ibid., 4:524 ff.
36 See Zöllner's editorial contributions to *Abhandlungen zur atomistischen Theorie der Elektrodynamik von Wilhelm Weber*, vol. 1, bk. 1, of J. C. F. Zöllner, *Principien einer elektrodynamischen Theorie der Materie* (Leipzig, 1876); Weber, 'Zusammenhang des elektrischen Grundgesetzes mit dem Gravitationsgesetze', 4:481–5.
37 'Aphorismen', in Weber, *Werke*, 4:630 ff.
38 J. C. F. Zöllner, *Transcendental physics*, trans. C. C. Massey (London, 1880).
39 *Atomenlehre*, 23, 53, 73, 187, 206–9.
40 Ibid., ix.
41 See J. F. Herbart, *Johann Friedrich Herbart's sämmtliche Werke* (Leipzig, 1850–2), ed. G. Hartenstein, vol. 5, *Lehrbuch zur Psychologie* (1816), and vols. 3–4, *Allgemeine Metaphysik, nebst den Anfängen der philosophischen Naturlehre* (Köngsberg, 1828–9), esp. pt. 5, 'Umrisse der Naturphilosophie'.
42 *Atomenlehre*, 90–1, 95.

43 Ibid., 63-73.
44 Ibid., basic viewpoint developed 106-18.
45 Ibid., 65.
46 Ibid., 95.
47 Ibid., 113. Cf. G. T. Fechner, *Elemente der Psychophysik* (Leipzig, 1860), esp. chaps. 1, 5.
48 *Atomenlehre*, 181-210.
49 C. F. Gauss, *Werke* (Göttingen, 1877), 5:629.
50 Riemann's associations and interests at Göttingen are described by R. Dedekind in 'Lebenslauf', *Bernhard Riemann's gesammelte mathematische Werke*, ed. H. Weber, 2nd ed. (Leipzig, 1892), 539-58.
51 'Fragmente philosophischen Inhalts', in H. Weber, *Riemann's Werke*, 524.
52 Ibid., 528 ff.
53 Ibid., 503.
54 Ibid., 533.
55 *Schwere, Elektricität, und Magnetismus: nach den Vorlesungen von Bernhard Riemann*, ed. K. Hattendorff (Hannover, 1876), 330.
56 'Ein Beitrag zur Elektrodynamik', *Annalen der Physik und Chemie 131* (1867); reprinted in H. Weber, *Riemann's Werke*, 288.
57 See note by H. Weber in *Riemann's Werke*, 293.
58 Ibid., 533 ff.
59 'Ueber Integrale der hydrodynamischen Gleichungen, welche den Wirbelbewegungen entsprechen', *Journal für die reine und angewandte Mathematik 55* (1858); reprinted in Helmholtz, *Wissenschaftliche Abhandlungen*, 1:101-34.
60 *Vorlesungen über mathematische Physik: Mechanik* (Leipzig, 1876), lectures 15-26.
61 Ibid., 'Vorrede'.
62 Bjerknes's career is described by V. Bjerknes in C. A. Bjerknes, *Hydrodynamische Fernkräfte: fünf Abhandlungen über die Bewegung kugelförmiger Körper in einer inkompressiblen Flüssigkeit (1863-1880)*, trans. A. Korn, ed. A. Korn and V. Bjerknes (Leipzig, 1915), 212-23. See also Dedekind, 'Lebenslauf', *Riemann's Werke*, 551.
63 'Hydrodynamische Analogien zu den elektrostatischen und magnetischen Kräften', *Naturen* (1880), reprinted in Bjerknes, *Hydrodynamische Fernkräfte*, 176-211; G. Forbes, 'Hydrodynamic analogies to electricity and magnetism', *Nature 24* (1881), 360 ff.
64 'Ueber die Identität der Schwingungen des Lichts mit den electrischen Strömen', *Annalen der Physik und Chemie 131* (1867), 243 ff.
65 Ibid., 261-2.
66 Ibid., 262.
67 Ibid., 263.
68 'Resultate einer Untersuchungen über die Principien der Elektrodynamik', *Nachrichten von der Königlichen Gesellschaft der Wissenschaften und der Georg Augustus Universität zu Göttingen* (1868), 223-35. For Neumann's earlier attempt to derive the magnetic rotation of light from Weber's law of force see Knudsen (1976).
69 E.g., Helm, *Energetik*, 229-31.
70 P. M. Heimann, 'Helmholtz and Kant: the metaphysical foundations of *"Ueber die Erhaltung der Kraft"*', *Studies in History and Philosophy of Science 5* (1974), 205-38; Y. Elkana, 'Helmholtz's "Kraft": an illustration of concepts in flux', *Historical Studies in the Physical Sciences 2* (1970), 263-98.
71 'Erhaltung der Kraft', in Helmholtz, *Wissenschaftliche Abhandlungen*, 1:14; reprinted in Helmholtz, *Selected writings*, 5.
72 *Metaphysical foundations of natural science*, trans. J. Ellington (Indianapolis and New York, 1970), esp. chap. 2.

73 'Erhaltung der Kraft', in Helmholtz, *Wissenschaftliche Abhandlungen*, 1:17–19; reprinted in Helmholtz, *Selected writings*, 6–8.
74 Ibid., in Helmholtz, *Wissenschaftliche Abhandlungen*, 1:22; reprinted in Helmholtz, *Selected writings*, 11.
75 Ibid., in Helmholtz, *Wissenschaftliche Abhandlungen*, 1:14; reprinted in Helmholtz, *Selected writings*, 4.
76 'Ueber die Bewegungsgleichungen der Elektricität für ruhende leitende Körper', *Journal für die reine und angewandte Mathematik 72* (1870); reprinted in Helmholtz, *Wissenschaftliche Abhandlungen*, 1:567.
77 Ibid., 553.
78 Ibid., 565.
79 'Ueber die Theorie der Elektrodynamik', *Monatsberichte der Berliner Akademie* (1872); reprinted in Helmholtz, *Wissenschaftliche Abhandlungen*, 1:637.
80 'Bewegungsgleichungen der Elektricität', 1:625–8.
81 Ibid., 558.
82 'Theorie der Elektrodynamik', 1:639.
83 'Ueber die Gesetze der magnetischen und elektrischen Kräfte in magnetischen und dielektrischen Medien und ihre Beziehung zur Theorie des Lichtes', *Sitzungsberichte der Kaiserlichen Königlichen Akademie der Wissenschaften zu Wien 70* (1874), 596.
84 R. McCormmach, 'Heinrich Hertz', in *Dictionary of scientific biography*, vol. 6 (1972), 340–50.

10

'Subtler forms of matter' in the period following Maxwell

HOWARD STEIN
Department of Philosophy, University of Chicago, Chicago, Illinois 60637 USA

> This investigation is put forward as a confirmation of Professor Maxwell's electromagnetic theory of light, in which, though there are some points requiring further investigation, nevertheless the foundation has certainly been laid of a very great addition to our knowledge, and if it induced us to emancipate our minds from the thraldom of a material ether might possibly lead to most important results in the theoretic explanation of nature.
> *George Francis Fitzgerald (1878)*
>
> Luminiferous ether must be a substance of most extreme simplicity.
> *Lord Kelvin (1884; 1904)*
>
> Alas! Have all the barbers lived in vain
> That not one curl in nature has survived?
> *Wallace Stevens (1923)*

It will be useful to begin with two general remarks about the period to be considered in the present chapter. First, when Maxwell's phrase 'the subtler forms of matter' is brought to bear upon our period, a multiple pun can be seen: (1) 'Ether', conceived originally as 'subtle' in the sense *finely divided*, proves to require, for the understanding of its own attributes and its relationships to 'ponderable' matter, a more refined ('subtler') *system of concepts;* (2) the more finely divided forms (thus, in the original sense, 'subtler forms') of *ordinary* matter – molecules, ions, and eventually subatomic particles – come, during this period, to play an increasing role in physical theory (and in a manner that is closely linked with the theory of ether); (3) developments initiated in this period have led in the end to an unprecedentedly radical transformation and conceptual 'subtilisation' of the notion of 'ordinary matter' itself – so that in a very important sense *all* matter turns out to be 'subtler' than anyone could previously have imagined. Second, although the triumph of Maxwell's theory led ultimately to so basic a conceptual transformation that physicists now tend to regard the electromagnetic theory as having supplanted the 'mechanical' theory of light – as having in effect superseded attempts to construct a theory of the luminiferous (and now also electromagnetic) ether as a species of matter subject to the laws of elasticity and the laws of motion – this is a badly distorted view: There is in point of fact an enormous literature devoted to such attempts, to what may be called the 'classical ether-problem for the electromagnetic theory'. To survey this literature with any pretension to adequacy within the limits here available is out of the question. The primary aim of this chapter, therefore, is to sketch what I see as the most salient features of the developments referred to under the first head above – more particularly under (1), with some attention to (2) – with (it is hoped) historical responsibility, but with no claim at all to historical completeness.[1]

The MacCullagh–Fitzgerald dynamical ether

For an appreciation of the state of affairs in the early part of our period, it is important to remember what sort of theory it was that Maxwell himself left us. He outlined its general character with great clarity in article (3) of his decisive paper 'A dynamical theory of the electromagnetic field':

> The theory I propose may . . . be called a theory of the *Electromagnetic Field,* because it has to do with the space in the neighbourhood of the electric or magnetic bodies, and it may be called a *Dynamical* Theory, because it assumes that in that space there is matter in motion, by which the observed electromagnetic phenomena are produced.[2]

As to the detailed constitution, in mechanical terms – the actual motions and interconnections – of that matter 'by which the electromagnetic phenomena are produced', however, Maxwell's theory (in this paper and in the later *Treatise*) is deliberately noncommittal; except that in the *Treatise* he does distinctly embrace the view that the energy of the magnetic field is the kinetic energy of the medium, the energy of the electric field its potential energy.[3]

Twenty-three years after Maxwell's 'Dynamical theory', and fifteen years after his *Treatise,* Poincaré lectured at the Sorbonne on Maxwell's theory; these lectures were published as volume 1 of his *Electricité et optique.*[4] At the time of this course, Maxwell's theory was still widely regarded as a dark mystery (Hertz, for instance, declared: 'Many a man has thrown himself with zeal into the study of Maxwell's work, and . . . been compelled to abandon the hope of forming . . . an altogether consistent conception of Maxwell's ideas. I have fared no better myself'),[5] and Poincaré took it as his express aim to clarify the structure and content of the theory (for 'un lecteur français'). The course could not have been given at a more propitious time: at the end of his introduction, Poincaré remarks: 'Science has advanced with a rapidity that nothing could have allowed one to foresee at the moment I began this course. Since that time, the theory of Maxwell has received, in a brilliant manner, the experimental confirmation it had lacked'. Poincaré's next course on electrodynamics, given two years later and published as volume 2 of the same work, was devoted to the critical discussion of the epoch-making experiments of Hertz on electromagnetic waves.

The nub of Poincaré's diagnosis of the basic difficulty faced by Maxwell's readers, and of the clarification he offers, is expressed by him as follows:

> On opening Maxwell, a Frenchman expects to find a theoretical whole as logical and as precise as the physical Optics founded on the hypothesis of the ether; he thus prepares himself for a deception which I should wish to obviate for the reader . . .
>
> *Maxwell does not give a mechanical explanation of electricity and of magnetism; he restricts himself to demonstrating that that explanation is possible.*[6]

Poincaré goes on to make entirely plain what he means by a 'mechanical explanation', and what by a 'demonstration that mechanical explanation is possible'; the latter demonstration consists in the introduction of a kinetic and a potential energy function, in such a way that the resulting Lagrangian 'generalised dynamical equations' can be shown to be satisfied by the system in question; and Poincaré proves that when these conditions are met, it is always possible – and in infinitely many ways – to construct, in his sense, a 'mechanical explanation' of the processes of the system.[7] (Although this is no place for a full discussion, it should not go unremarked that this celebrated result is seriously defective: Poincaré's theorem is perfectly correct, but the sense he allows in it to the notion of 'mechanical explanation' is far too wide from the point of view of physical theory.) In short, the point made central by Poincaré is this: Maxwell demonstrated the possibility of a 'mechanical explanation' of electromagnetic phenomena, by subsuming the laws of these phenomena under the generalised mechanics of Lagrange.

Now, such a subsumption under the principles of generalised mechanics is certainly what Maxwell appears to aim at in part 4, chapters 5 and following, of the *Treatise;* it is therefore rather striking to find, on more careful examination, that neither Maxwell in the *Treatise* nor Poincaré in the *Electricité et optique* in actual fact applies Lagrangian dynamics to the electromagnetic field. Both, in effect, apply Lagrangian mechanics only to the case in which the sole electric currents of which account need to be taken are closed currents in well-defined curvilinear conducting circuits; in which in particular, then, the Maxwellian 'displacement currents' can be neglected. It is thus precisely to the case *not* characteristic of what we think of as 'Maxwell's theory' – the case, be it noted, that was not the subject of the 'brilliant experimental confirmation' achieved by Hertz – that both Maxwell and Poincaré demonstrate the applicability of generalised dynamics. In the *Treatise* this application occurs in chapters 6–8 of part 4; chapter 9 simply generalises the results obtained to the unrestricted case. (One is tempted to regard this step as a straightforward inference from the hypothesis that displacement currents obey all the same laws as ordinary currents; but this interpretation fails, because the preceding dynamical analysis will not go through for ordinary currents either, when the conductors cannot be regarded as one dimensional.) The situation is described by Lorentz (with his characteristic judiciousness and simplicity), in his pathbreaking paper of 1892 on the electrodynamics of moving bodies: 'The equations that determine the motion of electricity in bodies of three dimensions do not result, in the book of Maxwell, from a direct application of the laws of mechanics; they rest upon the results previously obtained for linear circuits'.[8]

Lorentz himself repairs the indicated defect; and we shall return to this point. But he was anticipated in this (although, as we shall later see, only partially); credit for the first demonstration that Maxwell's general field laws can indeed be brought within the scope of dynamical principles belongs to G. F. Fitzgerald, who accomplished this in 1878, in a work devoted to the electromagnetic theory of reflection and refraction.[9] This subject had not been pursued by Maxwell, either in 'A dynamical theory' or in the *Treatise;* he had contented himself with discussing the propagation of electromagnetic waves in homogeneous (although not necessarily isotropic) media. The first published intimation of an electromagnetic theory of the reflection and refraction of light (and of its advantages over the standard elastic-ether theories) was made in 1870 – thus three years before the publication of Maxwell's *Treatise* – by Helmholtz, in a few lines, in a footnote to the introductory section of the first of his remarkable series of comparative discussions of electrodynamic theories;[10] and Helmholtz's suggestion was worked out in detail by Lorentz in his doctoral thesis.[11] Fitzgerald's treatment was evidently elaborated in ignorance both of Lorentz's work and of Helmholtz's remarks.[12] His method, which is what chiefly concerns us here, is sketched at the beginning of the abstract of his investigation:

> In the first part of the paper the media are not assumed to be isotropic as regards electrostatic inductive capacity, so that the results are generally applicable to reflection and refraction at the surfaces of crystals. I use the expressions given by Professor J. Clerk Max-

well in his 'Electricity and Magnetism', vol. ii, part 4, chap 11, for
the electrostatic and electrokinetic energy of such media. By assuming three quantities, ξ, η, ζ, such that, t representing time, $d\xi/dt$, $d\eta/dt$, and $d\zeta/dt$ are the components of the magnetic force at any point, I have thrown these expressions for the electrostatic and electrokinetic energy of a medium into the same forms as M'Cullagh assumed to represent the potential and kinetic energy of the ether, in 'An Essay towards a Dynamical Theory of Crystalline Reflection and Refraction', published in vol. xxi of the 'Transactions of the Royal Irish Academy'. Following a slightly different line from his, I obtain . . . the same results as to wave propagation, reflection, and refraction, as those obtained by M'Cullagh . . . Of course, the resulting laws of wave propagation agree with those obtained by Professor Maxwell from the same equations by a somewhat different method. For isotropic media, the ordinary laws of reflection and refraction are obtained, and the well-known expressions for the amplitudes of the reflected and refracted rays.[13]

Fitzgerald, therefore, reduces the theory of electromagnetic propagation in dielectric media to the dynamical theory of ether that had been developed forty years earlier by James MacCullagh.[14] The vector with components ξ, η, ζ, referred to by Fitzgerald (let us henceforth call it **q**), is in MacCullagh's theory the displacement vector, at each given point and time, of ether – that is, the displacement from the given point of the ether particle whose 'normal' location is at that point; thus, Fitzgerald identifies the magnetic field intensity at any point with (roughly) the velocity $\dot{\mathbf{q}}$ of MacCullagh's ether at the same point (in conformity with the assumption that magnetic energy is kinetic). MacCullagh's basic dynamical assumption was that the potential energy of the optical medium is a quadratic form in the components of the vector field curl **q**. From this assumption there follows, as MacCullagh succeeded in showing, by purely dynamical reasoning (but under an approximating condition, often unremarked, that we shall have occasion to consider later), a system of laws of the propagation of ethereal disturbance in homogeneous media, and of boundary relations on opposite sides of a surface of discontinuity separating two such media, that agrees fully with the experimentally established facts of optics. Fitzgerald points out that if one puts the vector of magnetic field intensity proportional to $\dot{\mathbf{q}}$ and the vector of dielectric displacement proportional to curl **q,** then MacCullagh's expressions for the kinetic and potential energies coincide with Maxwell's, *and moreover* the fundamental relations of Maxwell's theory for regions free of charge and of conduction current all follow.[15] The success of MacCullagh's analysis of optical phenomena is thus fully transferred to the theory of Maxwell (in which, furthermore, the approximating condition just referred to is rigourously assumed); and, as it were incidentally, we have the subsumption of the general equations of the field (for charge- and current-free regions) under dynamical principles.

It should, perhaps, be noted that Whittaker (1951:144, n. 1) attributes this interpretation of MacCullagh's theory in electromagnetic terms to Heaviside (in a paper of 1891). It is strange that he gives the priority to Heaviside, for on the same page he refers to Fitzgerald as the first to have properly appreciated MacCullagh's work, and later (p. 286) he describes Fitzgerald's interpretation explicitly. (The careful reader should be warned that in Whittaker's footnote on p. 144 there is a substantive mistake, evidently a typographical error: In the notation there used, not **e**, as is written, but **è** 'corresponds to the magnetic force'.)

For a correct understanding of the significance of Fitzgerald's result, it is important to make clear some peculiar circumstances attaching to the work of MacCullagh. In the first place, that work had gained little attention and no acceptance. Larmor, for example, in a historical note added, in his collected papers, to the first of his series of articles 'A dynamical theory of the electric and luminiferous medium' (1893–7), states:

> When these papers were written the work of MacCullagh had fallen into almost complete discredit. In this country it had suffered under the destructive criticism of Stokes: while abroad it was assumed to be merely a belated version of the theory of crystalline optics of elastic-solid type, developed by F. E. Neumann. At present it has been restored in some degree to its proper position in the historical development of physical optics: see, for example, Rayleigh's obituary notice of Stokes, *Roy. Soc. Proc.* 1903, or F. Klein, *Entwicklung der Mathematik im 19 Jahrhundert* (1926).[16]

One receives a more poignant impression from the brief (unsigned) biography of MacCullagh in as late an edition of the *Encyclopaedia Britannica* as the eleventh (1911), where one reads: 'Overwork, mainly on subjects beyond the natural range of his powers, induced mental disease; and he died by his own hand in October 1847'; and: 'His methods . . . were altogether inadequate to the solution of the more profound physical problems to which his attention was mainly devoted, such as the theory of double refraction, &c. See G. G. Stokes's "Report on Double Refraction" (*B. A. Report*, 1862)'. (The same edition of the *Britannica* contains statements by Larmor and by Lorentz, placing a far higher value upon MacCullagh's work.)[17]

What Stokes had pointed out was that MacCullagh's potential-energy function is physically paradoxical: MacCullagh's medium would, on the one hand, offer no resistance whatever to irrotational distortion (which is strange, even if not strictly paradoxical); whereas, on the other hand, it would resist nondistorting rotation, with a quasi-elastic 'restoring torque' proportional (for 'small' displacements) to the total angle through which any infinitesimal element had been rotated from its 'normal' orientation in absolute space. (Note that it is not a question here of relative rotations of the parts of the medium with respect to one another, but of absolute rotations, as just specified.) It is quite clear that such a law not merely is different from any found to characterise 'ordinary' bodies, but is in conflict with the basic assumption of rotational invariance for the laws of nature. As is well known, violation of rotational invariance is associated with violation of the principle of the conservation of angular momentum. The objections to MacCullagh's theory were therefore by no means frivolous.

But none of this affects the standing of Fitzgerald's result. He was not in fact propounding a theory of the constitution of ether; in particular, he did not presume that the vector here called $\mathbf{q}$ represents ethereal displacement, but only that $\dot{\mathbf{q}}$ and curl $\mathbf{q}$ give the magnetic intensity and dielectric displacement, respectively. The existence of such a $\mathbf{q}$, under the indicated conditions, is implied by Maxwell's theory, as are the expressions for magnetic and electric energy. And MacCullagh's analysis had shown in complete rigour that the application of the principles of dynamics as if these were the kinetic and potential energies, respectively, leads to the ascertained optical laws. (No comparably complete success had been achieved by any of the more orthodox constitutive theories of the optical medium.)

In sum, MacCullagh had discovered a system of assumptions from which the laws of optics follow by dynamical reasoning; but these assumptions appeared (*malgré* the theorem of Poincaré) to be unrealisable by any material dynamical system. Fitzgerald, however, discovered that the electromagnetic field of Maxwell (more exactly, the 'free' field) is a system that satisfies exactly the assumptions of MacCullagh.

Fitzgerald's conditions for a material ether

The direct concern of Fitzgerald's investigation was not with questions about the dynamics of ether, such considerations having entered rather as ancillary to a method of attack upon his central problem; but his paper concludes with the striking passage quoted in the epigraph to the present chapter. The passage is undoubtedly pregnant, and in a measure prophetic; but some caution is required in its interpretation. When Fitzgerald here speaks of 'emancipating our minds from the thraldom of a material ether', what are we to take him to mean? The ensuing history (and Fitzgerald's part in it) can help to enlighten us.

A similar question can be raised about the statement of Lord Kelvin (or Sir William Thomson, as he then was), also quoted in my epigraph: What sort of thing did Kelvin mean by 'a substance of most extreme simplicity'? Here the clarification is immediate, for he tells us:

> We might imagine it to be a material whose ultimate property is to be incompressible; to have a definite rigidity for vibrations in times less than a certain limit, and yet to have the absolutely yielding character that we recognize in wax-like bodies when the force is continued for a sufficient time.[18]

And the passage immediately following makes it plain that the 'simplicity' that distinguishes, in Kelvin's opinion, ether from 'ponderable' matter is connected with its extreme fine-grainedness (its 'subtlety'):

> It seems probable that the molecular theory of matter may be so far advanced sometime or other that we can understand an excessively fine-grained structure and understand the luminiferous ether as differing from glass and water and metals in being very much more finely grained in its structure. We must not attempt, however, to jump too far in the inquiry, but take it as it is, and take the great facts of the wave theory of light as giving us strong foundations for our convictions as to the luminiferous ether.

In short, pending that eventual advance in the molecular theory, ether is to be treated as a continuous rather than a 'structured' medium, using the simplest applicable concepts of the standard theory of continuous elastic solid media, the whole account to be founded upon that 'natural history of the luminiferous ether' – 'an infinitely simpler subject than the natural history of any other body' – which is comprised in 'the great facts of the wave theory of light'.[19]

The general view of ether here advocated by Kelvin, and extensively developed in his *Baltimore lectures,* was first suggested by Stokes as a way to reconcile the demands of the wave theory of light with the fact of the ordinary unresisted motion of ordinary bodies (Whittaker, 1951:128). It is a view against which Fitzgerald argues, forcefully and repeatedly. Thus he writes in 1885:

> All theories of the ether that suppose it to be simply a jelly with matter spread through it, like grapes in a jelly, hardly seem to attribute sufficient importance to the difficulty of explaining upon any such simple hypothesis such phenomena as electricity and magnetism; and although the equations of motion of the jelly may fairly well represent the equations of motion of the ether, as regards its

propagation of light, yet the properties of a jelly prevent our supposing continuous rotation of its elements, which seems almost necessary in order that the same quantities which represent small motions in the light-propagation may represent known phenomena in electricity and magnetism.

Although Professor Stokes seems to think that there is no contradiction in supposing the ether to be a jelly, and at the same time sufficiently little rigid to permit the free motion of matter through it, nevertheless, there is no doubt that this is a serious stumbling-block in the way of a general acceptance of the hypothesis that the ether is, in all respects, like a thin jelly, and I hardly think the difficulty diminished when its strains, as a rigid body, are required to be capable of producing permanent electrical forces.[20]

In one of his last published pieces – a review of Larmor's book *Aether and matter* (1900) – Fitzgerald again puts forward this last consideration (with a pointed humour characteristic of him):

In discussing the result of Michelson and Morley's experiments, from which they concluded that the ether is carried along by the Earth in its motion, Mr. Larmor shows that such a hypothesis is quite inconsistent with the fact of aberration and with the tenability of Sir George Stokes's suggestion that ether is like a very soft jelly. How such a soft material could be the means by which tramcars are driven by shearing stresses seems an additional difficulty in the way of this suggestion.[21]

Returning to 1885, we have a brief communication to *Nature* on the subject of Kelvin's *Baltimore lectures* (the stenographic records of which had been made available).[22] In this note, Fitzgerald comments instructively on the differences between Kelvin's and Maxwell's views (with generosity towards both, but without concealing his own predilection for Maxwell). The note ends with the following paragraph:

I cannot conclude without protesting strongly against Sir William Thomson's speaking of the ether as *like* a jelly. It is in some respects *analogous* to one, but we certainly know a great deal too little to say that it is *like* one. May be Maxwell's conceptions as to its structure are not very definite, but neither are anybody's as to the actual structure of a jelly, and there is no real difficulty in supposing a medium whose condition is represented by symbols that obey the laws that Maxwell has shown should be the laws of symbols representing the condition of a medium that would explain electric and magnetic phenomena . . . It seems . . . likely that what

> he [Maxwell] called 'electric displacements' are changes in structure of the elements of the ether, and not actual displacements of the elements . . . so that I think the word 'displacement' was unfortunately chosen. I also think that Sir William Thomson, notwithstanding his guarded statements on the subject, is lending his overwhelming authority to a view of the ether which is not justified by our present knowledge, and which may lead to the same unfortunate results in delaying the progress of science as arose from Sir Isaac Newton's equally guarded advocacy of the corpuscular theory of optics.

In the light of these passages, it is clear that a part (at least) of the 'thraldom to a material ether' from which Fitzgerald wished our minds emancipated is just this doctrine of ether as a 'jelly'; it is moreover plain that he – in contrast to Kelvin – would have us include the relations of electromagnetism, alongside 'the great facts of the wave theory of light', within the 'natural history' of ether. But what sort of positive theory of ether might he have supposed would eventuate? Do his words of 1878 imply, in particular, that this ether might be in some sense *non*material? No conclusive evidence is at hand concerning Fitzgerald's more detailed intentions at the earlier date; but his subsequent remarks on the subject quite fail to support such a notion. In the paper of 1885 cited previously, after registering his objection to the view that 'the ether is, in all respects, like a thin jelly,' he proceeds as follows:

> There are, of course, many ways in which matter may move through the ether besides by displacing it; as, for instance, in the way in which a volume of liquid water might pass through ice, namely by dissolving in front, and by freezing as fast behind, and such hypotheses do not require any limit to be assigned to the rigidity of the ether. In all these cases it is, of course, evident, that when once it is shown that the energy of the medium depends on quantities which obey the laws of Maxwell's electric and magnetic induction and displacements, it follows that the forces on the places that represent the electrified and magnetized bodies must be the known electric and magnetic attractions and repulsions; and one great difficulty in framing hypotheses as to the connexion of the ether and matter is in explaining how the matter moves through the ether.[23]

Relating this suggestion – that ordinary matter might be fruitfully considered as, in effect, a state propagated through, rather than a substance interacting with and moving through, ether – to the 'vortex atom' hypothesis that had been advanced by Kelvin and explored by both Kelvin and J. J. Thomson,

Fitzgerald observes that 'there seems no doubt that the simplest theory as to the constitution of the ether is that it is a perfect liquid'; but, he continues,

> it seems almost impossible to explain electric and magnetic phenomena without some further hypothesis . . . Now, it seems certain that the only way in which a perfect liquid can become everywhere endowed with properties analogous to rigidity is by being everywhere in motion. The most general supposition of this kind would be, that it was what Sir William Thomson has called a vortex-sponge, *i.e.* everywhere endowed with vortex motion, but with this motion so mixed up as to have within any sensible volume an equal amount of vortex motion in all directions. There are many ways in which this supposition seems to be in accordance with what we know of the properties of the ether.[24]

Three points emerge rather clearly: (1) Fitzgerald has here no thought of ether as in any sense nonmaterial; (2) he *does* think it likely – or more than likely – that the constitution of the ether, and the physical properties derived by ordinary mechanical principles from that constitution, are very different from those of 'ordinary' matter; (3) a possibility to be reckoned with – although, for Fitzgerald, far more speculative than the preceding – is that the fundamental system of things, the *natura rerum,* may simply *be* ether, and what we perceive as the properties and interactions of 'ordinary bodies' may be the macroscopic aspect of what most basically are just processes within the ether (cf. the statement of Larmor: 'Matter may be and likely is a structure in the aether, but certainly aether is not a structure made of matter').[25]

The suggestion that ether may be a perfect liquid is less central to Fitzgerald's view; but his further suggestion that it may have the character of a 'vortex sponge' in the sense of Kelvin is based upon a more general consideration that is of great importance. Why does Fitzgerald reiterate (as he does) the need for physicists to search for 'such a mode of motion in space as will confer upon it the properties required'[26] to support electromagnetic phenomena? (Fitzgerald shows in general no especial predilection for the Cartesian principle that all material distinctions are to be grounded in motion.) Although he does not say so, the answer seems to be this: that by positing a medium that is a repository of fine-scale internal motion – and more particularly, of *angular momentum* distributed throughout its volume – Stokes's objection to MacCullagh's ether can be overcome. For if ether were full of microscopic rotational motion (tiny gyroscopes, or fluid vortices), the torque required to maintain what seems in the large like a constant deviation in the orientation of a portion of the medium might be seen as giving rise on the fine scale, through precessions of the rotating elements, to a continual transfer of angular momentum to that portion; and thus the conservation principle could be saved.

Kelvin's quasi-rigid ether

The vortex-sponge theory involved the notoriously difficult problem of the turbulent motion of a liquid, and was never developed beyond a rudimentary stage (for some details, and reference to further literature, Whittaker, 1951:295–301, may be cited); but the suggestion that quasi-elasticity based upon motion might serve the needs of the case was fully confirmed, some four years after Fitzgerald's proposal, by Kelvin, who devised a kinetic model (using cells in which gyroscopes were mounted) of what came to be called the 'quasi-rigid', or 'rotationally elastic', or 'gyrostatically loaded' ether.[27]

It is well known that Kelvin's references to Maxwell's theory are of a puzzling character. A brief sketch is given by Whittaker (1951:266–7) of the record of his fluctuating, ambivalent, at times downright hostile comments. The strongest favourable reference Whittaker cites is in Kelvin's preface to the English translation of Hertz's papers on electromagnetic wave propagation; there, Whittaker says, Kelvin 'appeared to accept "magnetic waves" '. (Yet if Kelvin, all things considered, seems odd on this subject, it must be said that Whittaker also seems a little strange; the phrase just quoted is a curious one to use of a passage in which Kelvin – referring to what he calls Hertz's 'experimental demonstration of magnetic waves' – says that 'for electricity and magnetism Faraday's anticipations and Clerk-Maxwell's splendidly developed theory have been established on the sure basis of experiment by Hertz's work'.)[28] It appears, in fact, that the favourable remarks in the Hertz preface signify a good deal more than (what one might suspect) an effort got up for the sort of ceremonial occasion on which 'a man is not upon oath': In the period from shortly before to a few years after 1890, stimulated presumably by a conjunction of the great results of Hertz, the influence of his compatriots Fitzgerald, Lodge, and Heaviside, and his own partial successes with the theory, Kelvin seems to have experienced a surge of enthusiasm for Maxwell's theory. Here, for instance, is what he told the Institution of Electrical Engineers, in his presidential address to that body in January 1889 (after referring to 'the velocity which is the conductance in electrostatic measure, and the resistance in electromagnetic measure of one and the same conductor', and stating that this velocity is 'not very different from that of light'):

> But its relationship to the velocity of light was brought out in a manner by Maxwell to make it really a part of theory which it never was before. Maxwell pointed out its application to the possible or

probable explanation of electric effects by the influence of a medium, and showed that that medium – the medium whose motions constitute light – must be ether. [*Note:* This is correctly transcribed from the source cited; but it seems to be a transposition – the intended sense, surely, is, 'that medium must be ether – the medium whose motions constitute light.'] Maxwell's 'electro-magnetic theory of light' marks a stage of enormous importance in electro-magnetic doctrine, and I cannot doubt but that in electro-magnetic practice we shall derive great benefit from a pursuing of the theoretical ideas suggested by such considerations.[29]

(A reference to 'Heaviside's way of looking at the submarine cable' follows as an example of such benefit.)

The connexion with Kelvin's own work on the theory of the quasi-rigid ether is manifest in his basic paper on that subject, although Witte (1906:47) is a little misleading when he describes this paper as containing an attempted mechanical explanation of the totality of electrical phenomena. Here is Kelvin's own summary of what he has and has not accomplished:

> We thus have simply *the undulatory theory of light,* as an inevitable consequence of believing that the displacement of an elastic solid by which, in my old paper [of 1847], I gave merely a 'representation' of the electric currents and the corresponding magnetic forces, is a reality. But to give anything like a satisfactory material realisation of Maxwell's electro-magnetic theory of light, it is necessary to show *electro-static force* in relation to the forcive (X,Y,Z) of my formulas; to explain the generation of heat according to Ohm's law in virtue of the action of this forcive when it causes an electric current to flow through a conductor; and to show how it is that the velocity of light *in ether* is equal to, or perhaps we should rather say, *is,* the number of electro-static units in the electro-magnetic unit of electric quantity. All this essentially involves the consideration of ponderable matter permeated by, or imbedded in ether, and a *tertium quid* which we may call electricity, a fluid go-between, serving to transmit force between ponderable matter and ether . . . I see no way of suggesting properties of matter, of electricity, or of ether, by which all this, or any more than a very slight approach to it, can be done, and I think we must feel at present that the triple alliance, ether, electricity, and ponderable matter is rather a result of our want of knowledge, and of capacity to imagine beyond the limited present horizon of physical science, than a reality of nature.[30]

This is a remarkable and trenchant statement. Much of our remaining discussion is concerned with a quite different line of research that succeeded in advancing knowledge of the matters pointed out by Kelvin as crucially problematic. But first let us consider some work in the immediate line of descent from Fitzgerald and Kelvin. Later I shall deal briefly with two other approaches to a theory of a materially constituted ether (one closely related to those previously discussed, the other not), and shall have especially to examine the nature of the difficulties that ultimately proved insuperable for all attempts to solve the 'classical ether-problem for the electromagnetic theory' (and proved to require, in Kelvin's words, a 'capacity to imagine beyond the limited present horizon of physical science' – farther beyond that horizon than Kelvin was ready to follow).

The problem of discrete electric charges: Larmor's electrons

Fitzgerald, we have seen, established the applicability of MacCullagh's dynamical analysis to Maxwell's electromagnetic field in uncharged dielectric media. The restriction is essential, for the identification of the dielectric displacement with curl **q** implies that this field is source-free. It is also, clearly, a serious restriction; from it, for example, it follows that the processes referred to by Kelvin as problematic – electrostatic forces, energy transformation in electric currents, and in fact most of the interactions by which energy is transferred between the electromagnetic field and ordinary matter – are outside the scope of this mechanical analysis.

Another consequence of the restriction is an inevitable ambiguity about how the translation between mechanical and electromagnetic terms ought really to be made. For in the charge-free and current-free case, there is complete symmetry in the field laws between electric fields and magnetic fields. Fitzgerald, following Maxwell, took magnetic energy to be kinetic, and accordingly interpreted the electric field as related to the twist of the medium; Kelvin, on the other hand, pursuing an analogy he had described as long before as 1847, preferred to associate the magnetic field with the twist: Therefore in his version it is the electric field that corresponds to the velocity of the medium.[31] The prospect thus arises that among the alternatives that stand on a par for free fields, one or another may offer an advantage for the extension to the interaction with charges and currents.

As to Kelvin's interpretation, a difficulty looms at once: Since electric charges are sources or sinks of the electric field, a theory identifying this field with the velocity of a material medium implies continual creation and destruction of the matter of this medium at the locations of charges (creation at the charges of one sign, destruction at those of the other). Perhaps this objection should not be regarded as necessarily fatal – perhaps, even (although this is very doubtful), Kelvin himself had this in mind when he spoke of a 'capacity to imagine beyond the limited present horizon of physical science' – but it would certainly be a serious departure from the notion of a 'classical' ether.

If we turn instead to the Fitzgerald version, the situation looks more encouraging (as an impressive array of investigators concluded: This mode of representation was pursued by Heaviside,[32] Sommerfeld,[33] Reiff,[34] and – to most notable effect – Larmor);[35] for the field of magnetic induction (assuming no 'true magnetic poles') is source-free, and can therefore be represented as the velocity field of an incompressible conserved substance. The crux of the problem becomes how to modify Fitzgerald's identification of the dielectric displacement **D** with curl **q,** and the associated identification of the electric field intensity **E** with the quasi-elastic restoring torque of the rotated element of the medium. Larmor proposed – first in an addendum, entitled 'Introduction of free electrons', to the initial paper of the series of three cited in n. 35 – to modify the theory of the quasi-rigid ether by postulating the presence of permanent (but mobile) *'centres of rotational strain'* in the medium, giving rise to uneliminable distributions of torque (upon which the additional torque proportional to additional twist is superposed).[36] These nuclei of 'intrinsic radial twist' would amount to elementary electric charges. Larmor names them *electrons* ('at the suggestion of G. F. Fitzgerald after G. J. Stoney's term'),[37] and goes on to sketch a 'purely electric theory of matter', based on the supposition that these electrons have no 'intrinsic inertia': that the laws of their motion are simply those of the dynamics of ether (in other words, that electrons have only what later came to be called 'electromagnetic mass'). He remarks that according to the theory such 'free electrons' could easily acquire velocities 'a considerable fraction of that of radiation', and suggests (this is in 1894) a possible connexion with the 'negative rays in vacuum tubes', remarking that he is informed by J. J. Thomson that these rays have been determined to have velocities of about 2×10^7 cm/sec.[38]

What we see emerge, as Larmor elaborates this theory, is a version of what has come to be, for us, the 'classical theory of electrons' – here developed out of speculation upon a constitution of ether that could provide a place in the very structure of this medium for what Kelvin calls the *tertium quid,* electricity. But the manner in which electricity serves for the transmission of force between 'ether' and 'ponderable matter', as it is envisaged in this theory, is not that of a go-between; instead there is projected an account of matter as in effect constituted out of electricity, which in its turn is nothing more than a particular (mobile) condition of a region of ether. This clearly entails exactly that view of the way matter can move through ether adumbrated by

Fitzgerald in his 1885 paper (quoted in at n. 23 of this chapter); and in his review of Larmor's book *Aether and matter* (1900), Fitzgerald repeats his own analogy: 'His theory practically assumes that matter can move through the ether much in the way that a drop of liquid water can move through a lump of ice, namely, by melting in front and freezing up again behind'.[39] Thus the framework appears to be present of a truly grand synthesis: a unified theory of matter, light, electricity, and magnetism, as manifestations of that single 'substance of most extreme simplicity', luminiferous ether.

Lorentz's particulate ether

Writing in 1901, in an obituary notice of Fitzgerald in the *Physical Review*, Larmor spoke of this work of his own in the following terms:

> In 1893 the formal development of electrodynamic theory from this point of view, inspired at the start by Fitzgerald's electric interpretation of MacCullagh's optical analysis, and by mechanical models based on Lord Kelvin's construction for a rotationally elastic aether, and resting directly on the single broad dynamical basis of the principle of Least Action . . . was initiated in Great Britain; it was practically complete early in 1897, running parallel in the main results, though not in method or mode of development, with Lorentz's final presentation in his tract of 1895.[40]

If, today, it is preeminently Lorentz, not Larmor, who is remembered as the great exponent of the theory of electrons, this has to be regarded as at least to some degree related to the difference in 'method or mode of development' that Larmor refers to – and to the fact that Lorentz's mode of thought proved in the end the more tenable.

The contrast may roughly be put thus: The centre of attention in the work we have been considering was upon the 'mechanical constitution' of the light-bearing medium, and the speculations of Fitzgerald, Kelvin, and Larmor were concerned especially with the fine-scale structure (or lack of structure) of ether; the centre of attention in the work of Lorentz was upon the electrical constitution of transparent media, and the speculations of Lorentz were concerned with the fine-scale structure of ordinary bodies.

We have already seen that Lorentz, like Fitzgerald, early addressed himself to the electromagnetic theory of reflection and refraction. In doing so, he was expressly following the lead that had been given by Helmholtz; and he employed Helmholtz's formulation of electrodynamic theory as the basis of his own investigation. Now, Helmholtz's procedure was the following: Starting from a theory – or, rather, a parametric family of theories – of charges and currents interacting at a distance, he investigated the laws that would result for the propagation of electromagnetic effects in a polarisable dielectric medium.[41] Helmholtz found that such a medium would, in general, carry both longitudinal and transverse electric oscillations, with different velocities (each characteristic of the medium). In the limiting case of what might be called 'complete polarisability', the velocity of the longitudinal wave becomes infinite, leaving in effect only transverse oscillations; and in this limit, Helmholtz's theory approaches that of Maxwell. (There are difficulties in the way of conceiving this limit as actually attained, so that it is only in a somewhat stretched sense that Helmholtz's account can be said to include Maxwell's 'as a special case'.) Electromagnetic waves, then, are represented in this picture as actual oscillations of 'electricity' in a polarisable medium; and Lorentz, pursuing within this general framework the question of the optical properties of bodies, came very naturally to deal with bodies through hypotheses about their electrical structure. In the early stages of this work, no special consideration of ether was required: It was simply a dielectric medium, comparable to air or glass.

Lorentz's thesis ends with a brief résumé of what has been treated and a review of problems that present interesting and promising subjects of further investigation.[42] Among these are 'the phenomenon of dispersion, the rotation of the plane of polarisation, and the way in which these phenomena are connected to molecular structure; then the mechanical forces that perhaps play a certain role in optical phenomena'; among them, too, is 'the influence exercised upon light by . . . the motion of the medium'.

The phenomenon of dispersion was treated by Lorentz in an early work: a paper published (in Dutch) in 1878 and in abridged form in German in 1880.[43] The most fundamental conclusion of this investigation is expressed by Lorentz in these words (my italics):

> If we accept the electromagnetic theory of light, there is nothing left, in my opinion, but to look for the cause of dispersion *in the molecules of the medium themselves*. And we can indeed obtain formulae from which a dispersion follows if we adopt the supposition that, in such a molecule, *as soon as an electric moment is excited, a certain mass is at the same time brought into motion*.[44]

This investigation of Lorentz's appears to have gone largely unnoticed for a considerable time. It is hard to understand why; but Gibbs, for example, in a series of papers dealing with related subjects (and with dispersion in particular), from 1882 to 1889, refers repeatedly to the German précis of Lorentz's thesis, but not once to Lorentz's theory of dispersion;[45] and a paper presented to the Berlin Academy in 1892 by Helmholtz seems to show that the latter remained, at that date, unaware of Lorentz's work on the subject. Helmholtz's paper opens with the remark that in his opinion a satisfying explanation of dispersion on the basis of the electromagnetic theory of light has not yet been given; he proceeds to explain the grounds of his dissatisfaction, and to develop the hypotheses on which a cogent account can be based [46] – which hypotheses turn out to be just those proposed fourteen years earlier by Lorentz.

The argument offered by Helmholtz is interesting. It was already understood from the elastic-ether theory of dispersion (to which Helmholtz had himself contributed) that a satisfactory account of this phenomenon seems to require the presence, in dispersive media, of particles capable of resonating to the optical oscillations at certain characteristic frequencies. In this paper, Helmholtz remarks that in Maxwell's theory of the forces exerted by the electromagnetic field upon neutral bodies, these forces are unchanged if the fields are reversed; these forces 'would therefore attain their greatest and smallest values each twice in each [optical] period of oscillation, so that they could not as a rule produce or sustain [mechanical] oscillations of the length of a simple period'. From this he concludes:

> Only if the ponderable particles contain charges of true electricity can the periodic alternations of the electric moments in the ether educe ponderomotive forces of the same period. The analogous assumption, that embedded atoms contain isolated northern or southern magnetism, I pass over as too improbable. On the other hand, the electrolytic phenomena, especially Faraday's law of electrolytic equivalents, have long since led to the assumption that electric charges of determinate magnitude attach to the valence positions of chemically bound ions – charges that can be now positive, now negative, but which must everywhere, for every valence position of every atom, have the same absolute magnitude.

This line of thought, therefore, leads to a break with the conception of electricity as a 'weightless fluid', and to its amalgamation instead with 'ponderable' matter – and, further, particulate matter. The working out of the theory demands a set of principles concerning the coupling of charged particles to the ether – that is, to the electromagnetic field – and this is not quite so straightforward a matter as it might be thought. So far as the electric field is concerned, to be sure, there can be no doubt: The charges of charged particles will be sources of the field of dielectric displacement, and they will subject the particles they belong to to the 'ponderomotive' force exerted, in an electric field, on charged bodies. But when Lorentz published his work of 1878, it was far less a matter of course to assume that a moving charged particle is to be regarded, in its relations to the magnetic field, as tantamount to a current. Lorentz explicitly introduces this as an assumption, with the remark:

'The experiments announced by Helmholtz as made by Rowland support this'.[47] More precisely, what Lorentz assumes is that a moving charged particle 'produces the same effects' as a current element: that is, that it is to be reckoned into the 'total current' term in Maxwell's equation connecting the current with the curl of the magnetic field. As to the 'ponderomotive' force on a moving charge, Lorentz has no occasion in this paper to consider anything but the electric contribution to that force; for this optical problem, the magnetic ponderomotive forces are negligible, and Lorentz does not even mention them.[48]

In his celebrated work of 1892 (cited in n. 8), Lorentz sets himself more far-reaching objectives. The problem of the electrodynamics of bodies in motion had just been treated by Hertz, on the tentative working assumption that ether within a moving body completely shares the motion of the body;[49] and Lorentz includes a chapter on Hertz's hypothesis. But Lorentz himself inclines to the extreme opposite view: that the motion of a body is to be regarded (at least in first approximation) as without any influence upon the state of motion of the ether through which it moves. This poses a very serious problem at the outset, for Maxwell's theory offers no clear guidance about the laws governing the connexion, or interaction, of the electromagnetic field with a body in motion through the ether; it was just this circumstance that led Hertz to adopt the working hypothesis he did: not because it was plausible (he points out that it is not), but because it was manageable.

Lorentz's solution of this difficulty is based upon two strategic moves. In the first place, he makes a general constitutive assumption about bodies: that they may be regarded as systems of charged particles, whose only interaction with ether is the interaction of charges (at rest or in motion) with the electromagnetic field. This means that the theory requires only, on the one hand, suitable assumptions about the binding of these charges into the system that is an ordinary body, and, on the other hand, the electrodynamic laws governing the relation to the field of a single charged particle – for it is presupposed that direct interaction with the field occurs independently for the individual charges of the system. Lorentz puts the matter thus:

> It has seemed to me useful to develop a theory of electromagnetic phenomena based on the idea of a ponderable matter perfectly permeable to the ether and able to move without communicating to the latter the least motion. Certain facts of optics may be invoked in support of this hypothesis and, although doubt is still permitted, it is certainly important to examine all the consequences of this view. Unfortunately, a quite serious difficulty presents itself from the beginning. How, in fact, is one to form a precise idea of a body

which, moving in the bosom of the ether and consequently traversed by this medium, is at the same time the seat of an electric current or a dielectric phenomenon? To surmount this difficulty, so far as I could, I have sought to reduce all the phenomena to a single one, the simplest of all, which is nothing else than the motion of an electrified body. One will see that, without any deeper study of the relation between ponderable matter and the ether, one can establish a system of equations suited to describe what passes in a system of such bodies. These equations accommodate very varied applications, which will be the object of the subsequent chapters.[50]

There remains the crucial problem of finding the laws of that 'simplest phenomenon of all,' the interaction of a single moving charge with ether; and to accomplish this, Lorentz makes his other strategic move. He alludes to this in the introduction to his paper,[51] where he contrasts his own procedure with that of Hertz. The latter had abstained, says Lorentz, from any 'mechanical' account of electromagnetic actions, and 'contented himself with a clear and succinct description, independent of any preconceived idea about what happens in the electromagnetic field' – a procedure (he says) that has its advantages (and, indeed, that Lorentz himself tends to follow in his later writings –after his own 'clear and succinct description' has been once attained). Nevertheless, he goes on, one is always tempted to return to mechanical explanations, and he will therefore make use of a generalisation of Maxwell's own method (in the *Treatise*) Then he adds:

> I have yet another motive to undertake these inquiries. In the memoir in which M. Hertz treats bodies in motion, he admits that the ether they contain moves with them. Now, optical phenomena have long since demonstrated that it is not always so. I therefore wished to know the laws that govern the electrical motions in bodies that traverse the ether without entraining it; and it seemed to me difficult to attain this end without having as guide a theoretical idea. The views of Maxwell [namely, on the application of mechanical principles to the field and charges as a 'connected mechanical system'] can serve as foundation of the desired theory.

The generalised dynamical assumptions Lorentz makes are (1) that the *configuration* – in the technical dynamical sense – of an electrodynamic system is specified jointly by the positions of the charged particles and the distribution of the electric field, and (2) that the potential and kinetic energies are as Maxwell had long since postulated. As what might be called 'constitutive conditions' or 'constraints' of the connected system, he assumes both the divergence equations of Maxwell (charges are the sources of the electric field;

the magnetic field is source-free) and Maxwell's equation connecting the curl of the magnetic field with the total current (where the latter is now understood to be the sum of the free-ethereal 'displacement current' of Maxwell and the 'convection current' constituted by motion of charges). From these assumptions Lorentz deduces, by standard variational methods of dynamics, first, Maxwell's equation for the curl of the electric field – which, of course, is old, but whose derivation constitutes the sought-for 'mechanical explanation' of the general laws of the field – and, second, the total 'ponderomotive' force exerted 'by the ether on a charged particle', a result of fundamental importance for Lorentz's theory and for all subsequent physics.[52]

Other ether models, c. 1900

From the point of view of one demanding a mechanically cogent theory, neither Lorentz's theory nor Larmor's was altogether immune to criticism, as we shall presently see. Two other hypotheses, still actively explored in the 1890s and 1900s, may be here briefly mentioned. One is an alternative conception of ethereal elasticity, put forward by Kelvin in 1888 (shortly before the 'quasi-rigid' ether), reviving a nearly fifty-year-old suggestion of Cauchy (cf. Whittaker, 1951:146 ff.), according to which ether has the character not of a 'jelly' but of a 'foam': 'a solid of such negative compressibility as should make the velocity of the condensational-rarefactional wave, zero or small'.[53] This theory (called that of the 'quasi-labile ether', 'contractile ether', or 'foam ether') has a remarkable relation to the theory of the quasi-rigid ether: In the foam ether, under given conditions of displacement of its particles (but assuming that the displacement vanishes outside a bounded region), the distribution of elastic forces will be the same as for the quasi-rigid ether; but there will be no volume distribution of (nonvanishing) torque. As Witte (1906:120) points out, the two theories ascribe to the strained ether stress tensors which differ, that of the foam ether being symmetric and that of the quasi-rigid ether not so, but which agree in the volume forces they determine. It follows that the actual motions in the foam ether will be the same (under the same initial conditions, and with the boundary condition mentioned previously) as in the quasi-rigid ether, so that the difficulties to be noted for the latter affect the former as well.

It is, perhaps, the eventual failure of these ether schemes that had looked so promising for electrodynamics that best explains the curious statement by Kelvin in his preface (1904) to the published version of his *Baltimore lectures,* cited by Whittaker (1951:267) as giving Kelvin's 'final judgment' on Maxwell's theory: 'The so-called "electro-magnetic theory of light" has not helped us hitherto'. (A further essential gloss upon this is another passage, near in date, in the same work – from appendix A, dated 1900: 'The so-called "electro-magnetic theory of light" does not cut away this foundation [namely, the elastic solid ether] from the old undulatory theory of light. It adds to that primary theory an enormous province of transcendent interest and importance; it demands of us not merely an explanation of all the phenomena of light and radiant heat by transverse vibrations of an elastic solid called ether, but also the inclusion of electric currents, of the permanent magnetism of steel and lodestone, of magnetic force, and of electrostatic force, in a comprehensive etherial dynamics'.)[54] At any rate, the theory that Kelvin in fact prefers, in the parts of the *Baltimore lectures* written last (and apparently until his death in 1907),[55] is a strange one indeed. In it (as for Lorentz) ether and 'ponderable matter' are perfectly permeable to one another; but the particles of matter and of ether act upon one another at a distance. Free ether he takes to be nearly incompressible (so that longitudinal waves in it have nearly infinite velocity); but the equilibrium of the ether in regions within 'ponderable matter', rearranged under the influence of the forces of attraction and repulsion there operating, he takes to be of the 'quasi-labile' kind, with nearly zero velocity for longitudinal waves. With this curious assortment of penetrable matters, actions at a distance, and both incompressible and 'contractile' elasticities,[56] Kelvin shows, in the parts of the *Baltimore lectures* written last, that he can account for much – but not quite all – of optics. He says in the preface: 'My object in undertaking the Baltimore Lectures was to find how much of the phenomena of light can be explained without going beyond the elastic-solid theory. We have now our answer: *every thing non-magnetic; nothing magnetic*'. (In fact this is slightly too optimistic: See his appendix B, on the 'two clouds' – neither of them essentially 'magnetic'.)

The second, and altogether different, view of ether referred to at the opening of this section has its roots partly in work of Kelvin's of around 1870, but especially in the investigations of C. A. Bjerknes (see Whittaker, 1951:284–6). It envisaged the ether as a perfect incompressible liquid, and was based upon the demonstrations by Kelvin and Bjerknes that in such a medium forces of attraction and repulsion would occur between immersed solid bodies in suitable vibratory states. Whittaker cites a work of Arthur Korn as containing a theory of gravitation based upon this general conception; but he fails to remark that the same work actually contains a development of the entire electromagnetic theory of Maxwell upon this foundation![57] But how is this possible? He who is curious will have to look into it; here it must suffice to say that Korn obtains a stock of parameters to manipulate by introducing special kinds of solid bodies, characterised by special intrinsic modes of vibration and by nonstandard boundary conditions for their surfaces of contact with the fluid ether. (As late as 1917–18, Korn was pursuing an analogous approach to an alternative to Einstein's general relativistic gravitational theory.)[58]

Another figure active in cultivating the point of view of Bjerknes was the latter's son, Vilhelm F. K. Bjerknes, himself a physicist of some distinction,[59] whose book *Fields of force,* devoted to the exposition of the analogies between hydrodynamics and electromagnetism, although without significant influence upon the latter theory, appears to have been of considerable importance for the former – and especially for meteorology.[60] It is a striking fact that this monograph appeared as publication number one of the Ernest Kempton Adams Fund for Physical Research, as the record of a course of lectures delivered in 1905 at Columbia University, and that publication number two of the same fund, containing lectures delivered at Columbia in 1906, was the first edition (1909) of Lorentz's famous book *The theory of electrons.*

The problem of ethereal motion

The difficulties alluded to at the beginning of the preceding section all have to do, although in somewhat different ways, with the idea of *ethereal motion*. We have already seen the problem that was presented for Kelvin's version of his quasi-rigid ether by the motion required in it in the simple case of electrostatic fields; and we have seen how the Fitzgerald–Larmor alternative seemed to promise a coherent solution. But it is necessary now to revert to a fine point, earlier slurred over. I said in the first section of the chapter that Fitzgerald identified the magnetic field intensity at any point 'with (roughly) the velocity $\dot{\mathbf{q}}$ of MacCullagh's ether at the same point'. The import of the qualification is this: The value at a given point (and time) of the vector field $\mathbf{q}$ is supposed to represent the displacement (then) of the ether particle 'belonging' there; but this particle is not in fact (then) at that point (except when $\mathbf{q}$ happens to vanish there); and the velocity $\dot{\mathbf{q}}$ is therefore, quite strictly, that of the ether, not at the position in question, but at the *displaced* position. This is an entirely familiar situation in the theory of elastic vibrations, where it is understood that the distinction just pointed to can be safely ignored, so long as the oscillations of the medium are 'small'. Analogously, one is able on similar grounds to disregard the difference between the rate of change (with time) of a quantity at a fixed location and the corresponding rate of change as one follows the motion of a single particle. Now, for the arguments establishing the agreement of Maxwell's electrodynamics with the properties of the quasi-rigid ether, it is necessary to neglect these differences; that is to say, for the theory as sketched to be tenable, it is necessary that the differences be negligible – that, in effect, the actual velocities and the actual displacements in the ether be 'small'. ('Smallness' of these quantities is actually required yet again, in this theory, for the cogency of the basic definition of 'rotational elasticity' itself; for it is only – in a sense that can be made precise – in the 'infinitesimal limit' that curl **q** can be said to represent the 'rotation' or 'twist' of the medium; for sizable values of **q**, a restoring torque proportional to curl **q** makes no physical sense.)

This, however, leads to almost immediate disaster. Once more it is the *non*optical case, the simple case of steady fields, that offers the obvious stumbling block; for if the intensity of the magnetic field represents the velocity **q̇** of ether, then in a steady magnetic field of sufficiently long duration, large displacements of the ether particles must ensue.

It would be misleading to suggest that this simple remark suffices to end the matter. Remedial measures might be sought, and were sought, for instance, by Larmor; and more elaborate discussion than is possible here would be required to do justice to the issue. But an accurate general statement can be made, and it is this: The 'rotational' theory of Larmor has as a consequence that in certain situations – those in which either high velocities or sizable displacements of ether particles occur – new observable effects, deviations from the orthodox theories of optics or electromagnetism, are to be expected. Such effects were sought for; but they were never found. One example at least is worth citing: Having modified his theory to take account of the difficulty mentioned, concerning curl **q** when **q** is of noticeable size, Larmor inferred from his revised dynamics of ether that in a steady magnetic field in a region initially neutral electrically, there will eventually be evolved a distribution of electric charge throughout the region.[61] (It is of some interest from a methodological point of view to note that these predictions do not constitute a point of 'falsifiability' for Larmor's theory, since the theory leaves unspecified the magnitude of the constants of proportionality relating electric and magnetic quantities to displacements and velocities in the ether: In any given case, a negative result can be explained away on the grounds that the effect was still too small to be observed. On the other hand, a positive outcome of the search for such an effect as the evolution of charge by a magnetic field would undoubtedly have meant a triumph for Larmor's theory comparable to the triumph supplied for Maxwell's by Hertz's production and detection of electromagnetic waves.)

In commenting, at the end of the first of his series of papers 'A propos de la théorie de M. Larmor' (see n. 36), upon an attempt by Lodge to detect one of the effects of ether motion suggested by Larmor's theory (namely, an alteration of the velocity of light by a strong magnetic field), Poincaré expresses himself as follows:

Thus this motion [of the ether] was so slow that the experiments of M. Lodge, although very precise, were yet not precise enough to detect it. To say all that I think, I believe that if these experiments had been a hundred or a thousand times more precise, the result would still have been negative.

In support of this opinion I have nothing to offer but a subjective conviction [*des raisons de sentiment*]; if the result had been positive, one would have been able to measure the density of the ether, and – if the reader will forgive me the vulgarity of this
expression – it is repugnant to me to think that the ether is *si arrivé que cela*.[62]

(I have left the last phrase untranslated, for its full colloquial savour; a rough equivalent would be 'that big a success'.)

In the second and third papers of this series, Poincaré turns his attention to the problem of the electrodynamics of moving bodies (which Larmor himself, in the preliminary article that had occasioned Poincaré's reflections, had not yet discussed); and, reviewing the salient evidence and the major theories thus far proposed, he concludes that of all these theories Lorentz's is 'the least defective'.[63] Yet Lorentz's theory does have, in his view, a major defect, which makes it impossible for one to accept it as definitive: In this theory, the principle of the equality of action and reaction is violated.

The nub of the difficulty can be seen in the formula for, or indeed in the very concept of, the 'Lorentz force'; for this, the only 'ponderomotive' force in Lorentz's electrodynamics, is exerted only upon charges, and only by ether. A comparison with the more orthodox Maxwellian tradition is instructive: There the 'displacement current' of Maxwell is regarded as subject to the exercise of magnetic force, just as is the ordinary current in a conductor; and this means that 'ponderomotive' forces will be exercised upon ether itself, as a seat of displacement currents.[64] (Thus we note, incidentally, an important break, in the theory of Lorentz, with Maxwell's principle of the 'displacement current': The latter is *not* in all electrodynamic respects 'equivalent to a current'; it is so in its relation to the distribution of fields, but not as regards the determination of moving force; and just this break leads to the failure of the principle of reaction. It is also clear, of course, that Lorentz has departed quite fundamentally from his earlier adherence to Helmholtz's theory of the 'electrically polarisable ether': Of all the investigators, Lorentz is the one most radically 'emancipated from the thraldom of a material ether'.)

In his estimate of the theoretical situation, Poincaré goes further than to register his unhappiness at the violation of the principle of reaction in Lorentz's theory; he also deprecates the notion that forces upon ether should be required in order to maintain that principle, saying: 'It seems to me very hard to admit that the principle of reaction is violated, even in appearance, and that it is no longer true if one envisages solely the actions suffered by ponderable matter and leaves aside the reaction of this matter upon the ether'.[65] He then cites the fact that a series of experiments have strongly suggested the impossibility of detecting the relative motion of 'ponderable matter' with respect to ether (Lodge's experiments are one example; 'a recent experiment of M. Michelson' is another), connecting this with his remark about action and reaction – and, by implication, with the 'subjective conviction' stated in his comment on the experiments of Lodge – and expressing the hope that the same repair of theory that succeeds in explaining why motion with respect to ether can in principle not be detected will also repair the breach of the principle of reaction, restoring this principle for ponderable matter considered alone and apart from ether. One might fairly sum up Poincaré's conviction as that of the impotence of ether.

One consideration would seem to make this view puzzling. Maxwell had shown that on his electromagnetic theory of light a reflecting or absorbing surface is subject to *pressure* from the light that falls on it;[66] quite elementary arguments then show that if the principle of reaction – and therefore that of conservation of momentum – is to be maintained, a beam of light must contain, and transport, momentum; and this *must* be reckoned into the balance if conservation is not to be 'violated in appearance'. How, then, could Poincaré have held the hope and expectation that he did?

The answer seems to be simply this: that for Poincaré in 1895, Maxwell's light pressure was a theoretical notion of not significantly higher standing than ether winds or the influence of magnetic fields on the velocity of light. He does not say this; but it is extremely hard to make sense of his position otherwise. And indeed the pressure of radiation *was* at the time a dubious matter; it was first detected experimentally by Lebedev in 1899, despite attempts dating back to 1873 (Whittaker, 1951:267). It is true that Fitzgerald suggested in 1882 that the pressure of sunlight might be responsible for comets' tails;[67] but this was very speculative. It is also true that pressure of radiation had played a crucial part in the theoretical investigations by Boltzmann (1884) and Wien (1893) of black-body radiation (Whittaker, 1951:374, 379–80); but these, too, were speculative and, at the time, somewhat out-of-the-way pieces of work (although within another few years they had proved to have played a role in transforming the foundations of physics). The clearest possible indication of how precarious a status belonged to light pressure is provided by this passage from a paper of Larmor in 1892:

> A formula has been given by Maxwell for the intensity of the pressural force produced by electric undulations in the aether striking against a plate of conducting matter, a force which has apparently not been detected for the case of light-waves. If the notions here suggested have any basis, this force may likely be non-existent.[68]

To this there is a note, added in the edition of 1929: 'The pressure of radiation is now abundantly verified, and has become a keystone of theory'. (Note in passing that anyone concerned to form a just estimate of the stance of Kelvin, and impressed by Lodge's 1898 statement that 'Kelvin doesn't even believe in Maxwell's light pressure' [Whittaker, 1951:267] must take into consideration this passage from Larmor.)

Now, once one accepts the pressure of radiation as genuine, any theory of radiation as consisting in the transmission of motions through a system of interacting masses has to attribute to this underlying material carrier of the radiation not merely fine-scale oscillations, but net streaming motions. And this means that the difficulties that were obvious for the Kelvin and Larmor theories in the case of steady fields are really present in the 'purely optical' case as well.

What, then, in this situation, was the position of Lorentz? First, Lorentz's theory definitely implies that light exerts a pressure: yet another manifestation of the breakdown, in this theory, of momentum conservation 'if one envisages solely . . . ponderable matter'. Second, Lorentz accepts the evidence suggesting the undetectability of 'streamings' of ether, and systematically abstains from any attempt to characterise ether as a system of masses capable of motions and interacting by forces. As we have seen, this by no means prevents him from applying, as Fitzgerald had done, to ether – and further, as Fitzgerald had not done, to ether and charged particles (in arbitrary motion) as a single system – the principles of generalised dynamics.

An account of the developments that clearly established Lorentz's point of view as the most fruitful – and, at the same time, did indeed solve the problems of momentum conservation and of undetectability of motion with respect to ether, although not in the way anticipated by Poincaré – would require a treatment of the history of the special theory of relativity, a subject that would demand another chapter. Limitations of space compel the present discussion to end with no more than the briefest suggestion of the shape of the results. These may be described as a striking confirmation of half of what Poincaré expected, and an equally striking disconfirmation of the other half. So far as the question of the velocity of ether is concerned, Poincaré's 'principle of impotency' is true: In the definitive form of the theory, the concept 'state of motion of the ether at a given point' does not occur at all. So far as the momentum of ether is concerned, Poincaré's 'principle of impotency' is false: None of the fundamental conserved quantities of classical physics – momentum, angular momentum, energy – is in fact conserved, unless the part associated with 'ether', that is, with the electromagnetic field, is reckoned into the balance. That such dynamical quantities as momentum and angular momentum are to be ascribed to the field, without definite masses moving with definite velocities as their bearers, is a very basic result, comparable in significance to the earlier extension of the concept of energy beyond the kinematically grounded *vis viva* of the older mechanics.

When this conceptual transformation has been effected, it is very striking how some older notions and principles remain, shining through the new forms, whereas others simply vanish. A striking example has already been given: The ether has momentum; but it has no velocity. Another notable instance is the tensor of stress in the field that had been introduced by Maxwell, but seemed to have fallen out of the Lorentz theory (since Lorentz's force was not derived from the Maxwell stress; had it been so derived, it would have had to satisfy the principle of reaction). And lo! the clarifying explanation: The Maxwell stresses indeed have a role in the theory, intimately associated with the conservation principle, for they define the *flow of momentum;* but since momentum is no longer entirely 'kinetic', only a part of this momentum flow is through 'ponderomotive forces' that transfer momentum to moving bodies. On the other hand, the angular momenta posited by the various 'gyroscopic' or 'vortex-sponge' theories do *not* survive this transformation, for contrary to these theories, there is not a store of angular momentum in a pure magnetic field (or in a pure electric field); again, contrary to all the 'classical' ether theories, the momentum vanishes both for pure electric and for pure magnetic fields.

It might be supposed that this attribution of dynamical quantities like momentum and angular momentum to fields is merely a kind of 'tailoring' of a theory to make it shapely, or (*vide* Stevens) a kind of 'barbering' of natural appearances to make them comely. Perhaps the most convincing demonstration that it is not so would be a discussion of the role these notions have played in the development of the quantum theory (a discussion, however, that would demand, not another chapter, but another volume). Let one example suffice. Arnold Sommerfeld, in his very important book *Atomic theory and spectral lines,* discusses the Zeeman effect, an effect of magnetism upon the emission of light, for the discovery and – partial – explanation of which he and Lorentz shared a Nobel Prize in 1902 (cf. Kelvin's phrase of 1904, 'nothing magnetic', and his associated comment on the unhelpfulness of 'the so-called "electro-magnetic theory of light" '). In the course of his discussion,

Sommerfeld remarks that a previous attempt to develop a quantum-theoretic account of this effect had had to end with the statement that 'Bohr's energy equation . . . can never account for the polarisations'; but we now see, he tells us, that it is only necessary to add the equation of angular momentum balance between atom and radiation to obtain a satisfactory treatment.[69]

The example may also remind us that the same Sommerfeld worked on ether models in the 1890s and on quantum-theoretic models of atoms and molecules in the 1910s and 1920s. The two inquiries were comparably well motivated, and were executed with comparable skill; the former has left no trace upon the physics of our own time, the latter has been profoundly important. Some 'curls' survive; which do is determined by something other than barbering.

Notes

1 No historical account known to me approaches adequacy. The later chapters of Whittaker (1951) provide a useful point of departure; but they are deficient both in comprehensiveness and (as we shall have occasion to see) in reliability. Another secondary source of considerable value is Witte (1906); this monograph, which originated as a doctoral dissertation under the sponsorship of Planck, is conceived as a critical discussion of the existing state of theory rather than as a historical account, but it contains a great deal of historical information and much valuable elucidation of the relationships among theories.
2 J. C. Maxwell, *The scientific papers of James Clerk Maxwell*, 2 vols. (Cambridge, 1890), 1:527.
3 J. C. Maxwell, *A treatise on electricity and magnetism*, 3rd ed., 2 vols. (Oxford, 1892), 2:276 (art. 638).
4 H. Poincaré, *Electricité et optique*, vol. 1, *Les théories de Maxwell et la théorie électromagnétique de la lumière* (Paris, 1890). (The lectures were delivered in the spring of 1888, not of 1889 as the title page indicates: See vol. 2 of the same work, *Les théories de Helmholtz et les expériences de Hertz* [Paris, 1891)], vii–viii.)
5 H. Hertz, *Electric waves*, trans. D. E. Jones (London, 1893), 20.
6 *Électricité et optique*, 1:vii. See also H. Poincaré, *Science and hypothesis* (reprinted, New York, 1952), 214–15.
7 *Électricité et optique*, 1:ix–xiv.
8 H. A. Lorentz, 'La théorie électromagnétique de Maxwell et son application aux corps mouvants', *Archives Néerlandaises* 25 (1892), §3; reprinted in H. A. Lorentz, *Collected papers*, 9 vols. (The Hague, 1934–9), 2:168.
9 H. A. Lorentz, two papers with the same title, 'On the electromagnetic theory of the reflection and refraction of light', the first an abstract of the second, in *Proceedings of the Royal Society* (1879), and *Philosophical Transactions of the Royal Society* (1880), respectively; G. F. Fitzgerald, *The scientific writings of the late George Francis Fitzgerald*, ed. Joseph Larmor (Dublin and London, 1902), 41–4, 45–73.
10 H. Helmholtz, 'Ueber die Theorie der Elektrodynamik: erste Abhandlung: ueber die Bewegungsgleichungen der Elektricität für ruhende leitende Körper', *Journal für die Reine und Angewandte Mathematik* 72 (1870); reprinted in H. Helmholtz, *Wissenschaftliche Abhandlungen von Hermann Helmholtz*, 2 vols. (Leipzig, 1882–3), 1:558–9, n. 1.

11 H. A. Lorentz, 'Over de Theorie der Terugkaatsing en Breking van het Licht', Leiden, 1875; reprinted in Lorentz, *Collected papers*, 1:1-192 (French translation, ibid., 193-383). A précis was published in German, in three parts: 'Ueber die Theorie der Reflexion und Refraction des Lichtes', *Zeitschrift für Mathematik und Physik* 22 (1877), 1-30, 205-19, and 23 (1878), 197-210.
12 Cf. Larmor's statement to this effect in *Scientific writings of Fitzgerald*, xlii.
13 *Scientific writings of Fitzgerald*, 41-2.
14 *Transactions of the Royal Irish Academy 21* (1846); MacCullagh's paper is dated 9 Dec. 1839. (Several writers refer to this volume as issued in 1848; but the copy I have consulted in the library of Columbia University bears the date MDCCCXLVI.)
15 *Scientific writings of Fitzgerald*, 45-9.
16 J. Larmor, *Mathematical and physical papers*, 2 vols. (Cambridge, 1929), 1:415.
17 'MacCullagh, James (1809-1847)', *Encyclopaedia Britannica*, 11th ed. (1911), 17:207 'Aether', ibid., 1:296 (article signed by Sir James [sic] Larmor; through some inadvertence, MacCullagh's first initial is here given as 'T'.); 'Light', ibid., 16:622 (subsection 'Nature of Light' signed by Hendrik Antoon Lorentz).
18 W. Thomson, *Baltimore lectures on molecular dynamics and the wave theory of light* ('founded on Mr. A. S. Hathaway's stenographic report of twenty lectures delivered in Johns Hopkins University, Baltimore, in October 1884') (London, 1904), 12.
19 Ibid., for all quoted phrases in this paragraph.
20 G. F. Fitzgerald, 'On a model illustrating some properties of the ether', *Scientific Proceedings of the Royal Dublin Society* (read 19 Jan. 1885); reprinted in Fitzgerald, *Scientific writings of Fitzgerald*, 153.
21 G. F. Fitzgerald, 'The relations between ether and matter', *Nature* (19 July 1900); reprinted in Fitzgerald, *Scientific writings of Fitzgerald*, 508.
22 G. F. Fitzgerald, 'Sir W. Thomson and Maxwell's electromagnetic theory of light', *Nature* (7 May 1885); reprinted in Fitzgerald, *Scientific writings of Fitzgerald*, 170-3.
23 *Scientific writings of Fitzgerald*, 155-6.
24 Ibid., 154.
25 J. Larmor, *Aether and matter* (Cambridge, 1900), vi n.
26 *Scientific writings of Fitzgerald*, 162.
27 W. Thomson, 'Motion of a viscous liquid; equilibrium or motion of an elastic solid; equilibrium or motion of an ideal substance called for brevity *ether;* mechanical representation of magnetic force', in W. Thomson, *Mathematical and physical papers*, 6 vols. (Cambridge, 1882-1911), 3:436 ff.
28 Hertz, *Electric waves,* Kelvin's preface to the English ed., xiii.
29 Thomson, *Papers,* 3:490.
30 Ibid., 465.
31 For a very interesting discussion of alternative electromagnetic interpretations of a 'rotationally elastic ether' see O. Heaviside, *Electromagnetic theory*, 3 vols. (reissued, London, 1922), 1:243-56.
32 Ibid., 1:127-31, 243-56. See also O. Heaviside, *Electrical papers*, 2 vols. (London, 1892), 1:467.
33 A. Sommerfeld, 'Mechanische Darstellung der elektromagnetischen Erscheinungen in ruhenden Körpern', *Annalen der Physik und Chemie, n.s. 46* (1892), 139-51.
34 R. Reiff, *Elasticität und elektricität* (Freiburg, 1893).
35 J. Larmor, in the three-part series of papers 'A dynamical theory of the electric and luminiferous medium', *Philosophical Transactions of the Royal Society 185* (1894), 719-822, *186* (1895), 695-743, and *190* (1897), 205-300; reprinted, with amplifying commentary, in Larmor, *Papers*, 1:414-535, 543-97, and 2:11-132. Extensive

abstracts of these papers, from the *Proceedings of the Royal Society,* are also given in Larmor's *Papers:* 1:389–413, 536–42, 625–39. See also Larmor, *Aether and matter*.

36 Larmor's style is not of the most transparent; in one of his reviews of *Aether and matter*, Fitzgerald comments that it is 'in many places so condensed and general in its language as to be very difficult to follow'. *Scientific writings*, 515. Helpful accounts are given by Witte (1906), 150 ff., and by Poincaré, 'A propos de la théorie de M. Larmor', *L'Eclairage Electrique 3* (1895), 5–13, 289–95, and *5* (1895), 5–14, 385–92; reprinted in H. Poincaré *Oeuvres de Henri Poincaré*, 11 vols. (Paris, 1936–56), 9:369–426; also reissued, slightly revised and with an addendum – most useful for our present concern – not included in the *Oeuvres*, in the 2nd ed. (Paris, 1901) of *Électricité et optique* (for the addendum, 'Forme définitive de la théorie de Larmor', see 627–32).

37 *Papers*, 1:536, historical fn. (For Larmor's first introduction of the term, see ibid., 514, 516.)

38 Ibid., 521–4. The assumption of 'no intrinsic inertia' appears on 522, the phrase 'a purely electric theory of matter' in a marginal note on 523, and the comparison with cathode rays on 524, n. 1.

39 *Scientific writings of Fitzgerald*, 513.

40 Reprinted in ibid., liii–liv.

41 The paper of Helmholtz cited in n. 10 and the sequels to that paper merit more attention than they seem to have received. Helpful accounts are given in Poincaré, *Électricité et optique*, vol. 2; and in Emil Wiechert, *Grundlagen der Elektrodynamik*, issued as pt. 2 of the *Festschrift zur Feier der Enthüllung des Gauss–Weber–Denkmals in Göttingen* (Leipzig, 1899), 64–70.

42 *Collected papers*, 1:382–3.

43 H. A. Lorentz, 'Concerning the relation between the velocity of propagation of light and the density and composition of media', in Lorentz, *Collected papers*, 2:1–119 (this is an English translation of the full original work: 'Ueber die Beziehung zwischen der Fortpflanzung des Lichtes und der Körperdichte', *Annalen der Physik und Chemie, n.s. 9* [1880], 641–65).

44 *Collected papers*, 2:79–80.

45 For the papers in question, see J. W. Gibbs, *The collected works of J. Willard Gibbs* 2 vols. (New York, 1928), 2:pt. 2, 182–252; references to Lorentz occur on 220, 221, 238, 252.

46 H. Helmholtz, 'Elektromagnetische Theorie der Farbenzerstreuung', *Sitzungsberichte der Berliner Akademie* (Dec. 1892), 1093–1109; reprinted in *Annalen der Physik und Chemie, n.s. 48* (1893), 389–405; and in Helmholtz, *Wissenschaftliche Abhandlungen*, vol. 3 (Leipzig, 1895), 505–25. Whittaker (1951), although he refers in passing to Lorentz's memoir of 1878 (392, n. 6), gives the erroneous impression that it was only in 1892 that a theory of dispersion was developed by Lorentz.

47 *Collected papers*, 2:5.

48 See ibid., 81, equations (19).

49 See Hertz, *Electric waves*, 241 ff. (note in particular Hertz's expression of scepticism towards his own working hypothesis, 242–3).

50 'La théorie électromagnétique de Maxwell et son application aux corps mouvants', beginning of chap. 4; reprinted in Lorentz, *Collected papers*, 2:228.

51 Ibid., §3; reprinted in Lorentz, *Collected papers*, 2:168.

52 Lorentz's fundamental hypotheses and derivations are to be found in ibid., §§75–80; reprinted in Lorentz, *Collected papers*, 2:230–8. The formula for the force on a moving charge was obtained by similar reasoning in a paper of Heaviside in 1889: see his *Electrical papers*, 2:505–6. To see in Lorentz's derivation a strong influence

of the distant-action theory of Clausius, as Whittaker (1951), 393 ff., esp. 395, does, appears rather farfetched.
53 Thomson, *Baltimore lectures,* 351 (where the paper of 1888 is quoted – this lecture having itself been 'written afresh 1902' – see 324).
54 Ibid., 468 n.
55 See W. Thomson, 'On the motions of ether produced by collisions of atoms or molecules, containing or not containing electrions' (1907), in Thomson, *Mathematical and physical papers,* 6:235–43.
56 The basic assumptions of this theory are stated in the rewritten lecture 19 (1903) of the *Baltimore lectures,* 411–14.
57 A. Korn, *Eine Theorie der Gravitation und der elektrischen Erscheinungen auf Grundlage der Hydrodynamik,* 2nd ed. (Berlin, 1898).
58 A. Korn, 'Mechanische Theorie des elektromagnetischen Feldes', a series of papers in the *Physikalische Zeitschrift 18, 19,* and *20* (1917–19).
59 Cf. Hertz, *Electric waves,* 17.
60 See *Encyclopaedia Britannica,* 11th ed., 18:286.
61 'A dynamical theory of the electric and luminiferous medium', pt. 3, §17; reprinted in Larmor, *Papers,* 2:42–4.
62 *Oeuvres,* 9:381–2.
63 Ibid., 409.
64 Cf., e.g., Heaviside, *Electromagnetic theory,* 1:108.
65 *Oeuvres,* 9:412.
66 *Treatise,* 2:440–1(arts. 792–3).
67 *Scientific writings of Fitzgerald,* 108–9.
68 *Papers,* 1:286.
69 A. Sommerfeld, *Atomic structure and spectral lines,* trans. H. L. Brose (New York, 1923), 304.

SELECT BIBLIOGRAPHY OF SECONDARY SOURCES

Aiton, E. J. 1969. 'Newton's aether-stream hypothesis'. *Annals of Science 25:* 255–60.
– 1972. *The vortex theory of planetary motions*. London.
Aronson, S. 1964. 'The gravitational theory of George-Louis Le Sage'. *Natural Philosopher 3:* 53–73.
Bechler, A. 1974. 'Newton's law of forces which are inversely as the mass'. *Centaurus 18:* 184–222.
Berkson, W. 1974. *Fields of force: the development of a world view from Faraday to Einstein*. London.
Bromberg, J. 1967. 'Maxwell's concept of electric displacement'. Unpublished doctoral dissertation. University of Wisconsin.
– 1968*a*. 'Maxwell's displacement current and his theory of light'. *Archive for History of Exact Sciences 4:* 218–34.
– 1968*b*. 'Maxwell's electrostatics'. *American Journal of Physics 36:* 142–51.
Brush, S. G. 1970. 'The wave theory of heat: a forgotten stage in the transition from caloric theory to thermodynamics'. *British Journal for the History of Science 5:* 145–67.
– 1976. *The kind of motion we call heat*. 2 vols. Amsterdam.
Buchwald, J. Z. 1977. 'William Thomson and the mathematization of Faraday's electrostatics'. *Historical Studies in the Physical Sciences 8:* 101–36.
Caneva, K. L. Forthcoming. 'Ampère, the etherians, and the Oersted connection'.
Cantor, G. N. 1970. 'The changing role of Young's ether'. *British Journal for the History of Science 5:* 44–62.
– 1971. 'Henry Brougham and the Scottish methodological tradition'. *Studies in History and Philosophy of Science 2:* 69–89.
– 1975. 'The reception of the wave theory of light in Britain: a case study illustrating the role of methodology in scientific debate'. *Historical Studies in the Physical Sciences 6:* 109–32.
Cohen, I. B. 1956. *Franklin and Newton: an inquiry into speculative Newtonian experimental science and Franklin's work in electricity as an example thereof*. Philadelphia.
Corson, D. W. 1974. 'The Newtonian aether: an historical and critical study of its conceptual development'. Unpublished doctoral dissertation. Cornell University.
D'Agostino, S. 1968. 'La pensée scientifique de Maxwell et la developpement de la théorie du champ electromagnétique dans le mémoire "On Faraday's lines of force"'. *Scientia 103* (supp.): 155–64.

Dobbs, B. J. T. 1975. *The foundations of Newton's alchemy or "the hunting of the greene lyon"*. Cambridge.

Doran, B. G. 1975. 'Origins and consolidation of field theory in nineteenth century Britain: from the mechanical to the electromagnetic view of nature'. *Historical Studies in the Physical Sciences* 6:133–260.

Duhem, P. 1902. *Les théories electriques de J. Clerk Maxwell*. Paris.

Fox, R. 1971. *The caloric theory of gases from Lavoisier to Regnault*. Oxford.

– 1979. 'The science of fire: J. H. Lambert and the study of heat', in *Université de Haute Alsace: colloque international et interdisciplinaire Jean-Henri Lambert. Mulhouse, 26–30 September 1977*, pp. 325–42. Paris.

French, R. K. 1969. *Robert Whytt, the soul, and medicine*. London.

Goldberg, S. 1970. 'In defense of ether: the British response to Einstein's special theory of relativity, 1905–11'. *Historical Studies in the Physical Sciences* 2:89–125.

Goldfarb, S. 1977. 'Rumford's theory of heat: a reassessment'. *British Journal for the History of Science* 10:25–36.

Guerlac, H. 1963. 'Francis Hauksbee: éxperimentateur au profit de Newton'. *Archives Internationales d'Histoire des Sciences* 16: 1113–28. Reprinted 1977. In H. Guerlac, *Essays and papers in the history of modern science*. Baltimore and London.

– 1967. 'Newton's optical aether: his draft of a proposed addition to his *Opticks*'. *Notes and Records of the Royal Society of London* 22:45–57. Reprinted 1977. In H. Guerlac, *Essays and papers in the history of modern science*. Baltimore and London.

Hall, A. R., and Hall, M. B. 1967. 'Newton and the theory of matter'. *Texas Quarterly* 10: 54–68. Reprinted 1970. In *The 'Annus Mirabilis' of Sir Isaac Newton*, ed. R. Palter. Cambridge, Mass., and London.

Hawes, J. L. 1968a. 'Newton and the electrical attraction unexcited'. *Annals of Science* 24:121–30.

– 1968b. 'Newton's revival of the aether hypothesis and the explanation of gravitational attraction'. *Notes and Records of the Royal Society of London* 23:200–12.

– 1970. 'Concepts of force in Britain from 1700–50'. Unpublished doctoral dissertation. University of London.

Heilbron, J. L. 1979. *A history of electricity in the 17th and 18th centuries*. Berkeley, Calif.

Heimann, P. M. 1970. 'Maxwell and the modes of consistent representation'. *Archive for History of Exact Sciences* 6:171–213.

– 1971. 'Faraday's theories of matter and electricity'. *British Journal for the History of Science* 5:235–57.

– 1972. 'The *unseen universe:* physics and the philosophy of nature in Victorian Britain'. *British Journal for the History of Science* 6:73–9.

– 1973. ' "Nature is a perpetual worker": Newton's aether and eighteenth-century natural philosophy'. *Ambix* 20: 1–25.

Heimann, P. M., and McGuire, J. E. 1971. 'Newtonian forces and Lockean powers: concepts of matter in eighteenth-century thought'. *Historical Studies in the Physical Sciences* 3:233–306.

Hesse, M. 1961. *Forces and fields: the concept of action at a distance in the history of physics*. London.

– 1967a. 'Action at a distance and field theory'. In *The encyclopedia of philosophy*, ed. P. Edwards, 1:9–15. 8 vols. New York.

– 1967b. 'Ether'. In *The encyclopedia of philosophy*, ed. P. Edwards, 3:66–9. 8 vols. New York.

– 1973. 'Logic of discovery in Maxwell's electromagnetic theory'. In *Foundations of scientific*

Select bibliography

method: the nineteenth century, eds. R. N. Giere and R. S. Westfall, pp. 86–114. Bloomington, Ind.

Hirosige, T. 1966. 'Electrodynamics before the theory of relativity, 1890–1905'. *Japanese Studies in the History of Science* 5:1–49.

– 1968. 'Theory of relativity and the ether'. *Japanese Studies in the History of Science* 7:37–53.

– 1969. 'Origins of Lorentz' theory of electrons and the concept of the electromagnetic field'. *Historical Studies in the Physical Sciences* 1:151–209.

– 1976. 'The ether problem, the mechanistic worldview, and the origins of the theory of relativity'. *Historical Studies in the Physical Sciences* 7:3–82.

Holton, G. 1969. 'Einstein, Michelson, and the "crucial experiment"'. *Isis* 60:133–97.

Home, R. W. 1972. 'Franklin's electrical atmospheres'. *British Journal for the History of Science* 6:131–51.

– 1974. 'Some manuscripts on electrical and other subjects attributed to Thomas Bayes F.R.S.'. *Notes and Records of the Royal Society of London* 29:81–90.

– 1977a. ' "Newtonianism" and the theory of the magnet'. *History of Science* 15:252–66.

– 1977b. 'Boerhaave's "fire" and the foundations of eighteenth-century electrical theory'. In *Abstracts of scientific section papers, XVth Congress of the History of Science*, p. 116. Edinburgh.

Kargon, R. 1969. 'Model and analogy in Victorian science, Maxwell's critique of the French physicists'. *Journal of the History of Ideas* 30:423–36.

Klein, M. J. 1973. 'Mechanical explanation at the end of the nineteenth century'. *Centaurus* 17:58–82.

Knudsen, O. 1972. 'From Lord Kelvin's notebook: ether speculations'. *Centaurus* 16:41–53.

– 1976. 'The Faraday effect and physical theory'. *Archive for History of Exact Sciences* 15:235–81.

Kottler, M. J. 1974. 'Alfred Russel Wallace, the origin of man, and spiritualism'. *Isis* 65:145–92.

La Rosa, M. 1912. *Der Aether: Geschichte einer Hypothese*. Leipzig.

Laudan, L. 1970. 'Thomas Reid and the Newtonian turn of British methodological thought'. In *The methodological heritage of Newton*, eds. R. Butts and J. David, pp. 103–31. Toronto.

Lawrence, C. Forthcoming. 'Joseph Black: the natural philosophical background'. In *Joseph Black, 1728–1799: a commemorative volume*, ed. A. D. C. Simpson.

Lémeray, E. M. 1922. *L'éther actuel et ses précurseurs*. Paris.

Love, R. 1972. 'Some sources of Hermann Boerhaave's concept of fire'. *Ambix* 19:157–74.

– 1974. 'Hermann Boerhaave and the element-instrument concept of fire'. *Annals of Science* 31:547–69.

McCormmach, R. 1970. 'H. A. Lorentz and the electromagnetic view of nature'. *Isis* 61:459–79.

McGuire, J. E. 1967. 'Transmutation and immutability: Newton's doctrine of physical qualities'. *Ambix* 14:69–95.

– 1968. 'Force, active principles and Newton's invisible realm'. *Ambix* 15:154–208.

– 1970. 'Atoms and the "Analogy of nature": Newton's third rule of philosophizing'. *Studies in History and Philosophy of Science* 1:3–58.

– 1974. 'Forces, powers, aethers and fields'. In *Methodological and historical essays in the natural and social sciences*, eds. R. S. Cohen and M. Wartofsky, pp. 119–59. Dordrecht and Boston.

– 1977. 'Neoplatonism, active principles and the *Corpus Hermeticum*'. In R. S. Westman and G. McGuire, *Hermeticism and the scientific revolution*. Los Angeles.

McMullin, E. 1978. *Newton on matter and activity*. Notre Dame, Ind.

Mendelsohn, E. 1964. *Heat and life: the development of the theory of animal heat*. Cambridge, Mass.

Metzger, H. 1938. *Attraction universelle et religion naturelle chez quelques commentateurs Anglais de Newton*. Paris.

Morse, E. W. 1972. 'Natural philosophy, hypothesis and impiety: Sir David Bewster confronts the undulatory theory of light'. Unpublished doctoral dissertation. University of California, Berkeley.

Moyer, D. F. 1977. 'Energy, dynamics, hidden machinery: Rankine, Thomson and Tait, Maxwell'. *Studies in History and Philosophy of Science 8:* 251–68.

– 1978. 'Continuum mechanics and field theory: Thomson and Maxwell'. *Studies in History and Philosophy of Science 9:* 35–50.

Olson, R. 1975. *Scottish philosophy and British physics, 1750–1880*. Princeton, N.J.

Quinn, A. J. 1970. 'Evaporation and repulsion: a study of English corpuscular philosophy from Newton to Franklin'. Unpublished doctoral dissertation. Princeton University.

Ritterbush, P. C. 1964. *Overtures to biology: the speculations of eighteenth century naturalists*. New Haven and London.

Rosenfeld, L. 1956. 'The velocity of light and the evolution of electrodynamics'. *Nuovo Cimento*, supp. 4, pp. 1630–69. Reprinted 1979. In *Selected papers of Léon Rosenfeld*, eds. R. S. Cohen and J. J. Stachel. Dordrecht, Boston, and London.

– 1965. 'Newton and the law of gravitation'. *Archive for History of Exact Sciences 2:* 365–85. Reprinted 1979. In *Selected papers of Léon Rosenfeld*, eds. R. S. Cohen and J. J. Stachel. Dordrecht, Boston, and London.

– 1969. 'Newton's views of aether and gravitation'. *Archive for History of Exact Sciences 6:* 29–37. Reprinted 1979. In *Selected papers of Léon Rosenfeld*, eds. R. S. Cohen and J. J. Stachel. Dordrecht, Boston, and London.

Schaffer, S. 1977. 'Halley's atheism and the end of the world'. *Notes and Records of the Royal Society of London 32:* 17–40.

– 1978. 'The phoenix of nature: fire and evolutionary cosmology in Wright and Kant'. *Journal for the History of Astronomy 9:* 180–200.

Schaffner, K. F. 1969. 'The Lorentz electron theory of relativity'. *American Journal of Physics 37:* 498–513.

– 1972. *Nineteenth-century aether theories*. Oxford.

Schofield, R. E. 1970. *Mechanism and materialism: British natural philosophy in an age of reason*. Princeton, N.J.

Shapin, S. 1980. 'Social uses of science, 1660–1800'. In *The ferment of knowledge: changing perspectives of eighteenth-century science*, eds. R. S. Porter and G. S. Rousseau, pp. 93–139. Cambridge.

Siegel, D. M. 1975. 'Completeness as a goal in Maxwell's electromagnetic theory'. *Isis 66:* 361–8.

Silliman, R. H. 1963. 'William Thomson: smoke rings and nineteenth-century atomism'. *Isis 54:* 461–74.

Smith, C. 1978. 'A new chart for British natural philosophy: the development of energy physics in the nineteenth century'. *History of Science 16:* 231–79.

Spencer, J. B. 1970. 'On the varieties of nineteenth-century optical discoveries'. *Isis 61:* 34–51.

Stein, H. 1970. 'On the notion of field in Newton, Maxwell and beyond'. In *Historical and philosophical perspectives of science*, ed. R. H. Stuewer, pp. 264–86, 299–310. Minneapolis.

Sudduth, W. M. 1978. 'Eighteenth-century identifications of phlogiston with electricity'. *Ambix* 25:131–47.
Swenson, L. S. 1972. *The ethereal aether: a history of the Michelson–Morley–Miller aether-drift experiments, 1880–1930*. Austin, Tex., and London.
Talbot, G. R. 1967. 'Origins and solutions of some problems in heat in the eighteenth century'. Unpublished doctoral dissertation. University of Manchester.
Thackray, A. 1970. *Atoms and powers: an essay on Newtonian matter-theory and the development of chemistry*. Cambridge, Mass., and London.
Tonnelat, M. A. 1959. 'De l'idée de milieu á la notion de champ'. *Archives Internationales d'Histoire des Sciences* 12:337–56.
Turner, F. M. 1974. *Between science and religion: the reaction to scientific naturalism in late Victorian England*. London.
Turner, J. 1955. 'Maxwell's method of physical analogy'. *British Journal for the Philosophy of Science* 6:226–38.
Walker, D. P. 1972. *The ancient theology: studies in Christian Platonism from the fifteenth to the eighteenth century*. London.
Werkmeister, W. H. 1975. 'Kant and modern physics'. In *Reflections on the philosophy of Immanuel Kant*, ed. W. H. Werkmeister, pp. 69–92. Tallahassee, Fla.
Westfall, R. S. 1971. *Force in Newton's physics: the science of dynamics in the seventeenth century*. London.
Whiteside, D. T. 1964. 'Newton's early thoughts on planetary motion: a fresh look'. *British Journal for the History of Science* 2:117–37.
– 1970. 'Before the *Principia*: the maturing of Newton's thoughts of dynamical astronomy'. *Journal for the History of Astronomy* 1:5–19.
Whittaker, E. T. 1910. *A history of the theories of aether and electricity*. Vol. 1. London.
– 1926. *A history of the theories of aether and electricity*. Vol. 2. London.
– 1943. 'The aether: past and present'. *Endeavour* 2:117–20.
– 1951. *A history of the theories of aether and electricity*. New ed. Vol. 1. London and New York.
Whyte, L. L. 1960. 'A forerunner of twentieth century physics: a re-view of Larmor's *Aether and matter*'. *Nature* 186:1010–4.
Wilde, C. 1980. 'Hutchinsonianism, natural philosophy and religious controversy in eighteenth-century Britain'. *History of Science* 18:1–24.
Wilson, D. B. 1971. 'The thought of late Victorian physicists: Oliver Lodge's ethereal body'. *Victorian Studies* 15:29–48.
– 1972. 'George Gabriel Stokes on stellar aberration and the luminiferous ether'. *British Journal for the History of Science* 6:57–72.
Wise, M. N. 1977. 'The flow analogy to electricity and magnetism: Kelvin and Maxwell'. Unpublished doctoral dissertation. Princeton University.
– 1980. 'The Maxwell literature and British dynamical theory: an essay review'. *Historical Studies in the Physical Sciences* 11:49–83.
Witte, H. 1906. *Ueber den gegenwärtigen Stand der Frage nach einer mechanischen Erklärung der elektrischen Erscheinungen*. Berlin.
Woodruff, A. E. 1962. 'Action at a distance in nineteenth-century electrodynamics'. *Isis* 53:439–59.
– 1968. 'The contributions of Hermann von Helmholtz to electrodynamics'. *Isis* 59:300–11.

Wynne, B. 1977. 'C. G. Barkla and the J phenomenon: a case study in the sociology of physics'. Unpublished master's thesis. University of Edinburgh.
- 1979. 'Physics and psychics: science, symbolic action and social control in late Victorian England'. In *Natural order: historical studies of scientific culture*, eds. S. B. Barnes and S. A. Shapin, pp. 169–86. London.

Zahar, E. 1973. 'Why did Einstein's programme supersede Lorentz's?' *British Journal for the Philosophy of Science 24:* 95–123, 223–62.

Ziemacki, R. L. 1974. 'Humphry Davy and the conflict of traditions in early nineteenth century British chemistry'. Unpublished doctoral dissertation. University of Cambridge.

INDEX

aberration of light, 52
Achinstein, P., 261
action at a distance, 19–24, 39–44, 189, 192–6, 270, 300–2
active principles, 6–7, 16–17, 20, 62–9
aer, 3
aether, 2–3, 7–9
Airy, George Biddell, 34, 47
aither, 2–9, 16
Aiton, E. J., 61
Ampère, André-Marie, 197, 200–6, 243, 249–51, 277–8
Anaxagoras, 4
Anaximenes, 3
anima mundi, 7, 10, 13, 16, 26
animal heat, 75–6, 103
animism, 113–14, 125–32
Arago, Dominique François Jean, 52, 204–5
Argyle, Duke of, 91
Aristotle, 2, 4–6, 36, 111–12
Aronson, S., 30
Avogadro, Amadeo, 194–6

Bajollet, J., 126
Barton, Richard, 148–9
Battie, William, 123–4
Baxter, Andrew, 121
Beattie, James, 96
Beccaria, Giambatista, 194–5
Becquerel, Antoine-César, 204
Beer, August, 280
Bellini, Lorenzo, 127
Berkeley, George, 33, 147–8
Bernoulli, Daniel, 61
biblical cosmology, 141–2
Biot, Jean-Baptiste, 47, 192, 196, 200, 217
Bjerknes, Carl Anton, 292, 330
Bjerknes, Vilhelm F. K., 331

Black, Joseph, 75, 86, 101–3
Boerhaave, Hermann, 2, 10, 11, 14, 24–6, 69, 73–4, 113
  Elementa chemiae, 10, 24, 69, 73–4
  fire, 24–6, 69–70, 73–4
  Institutiones et experimenta chemiae, 24
Bordeu, Théophile de, 131–2
Borelli, Giovanni Alfonso, 115
Boscovich, Roger Joseph, 32, 44, 188–9
Boyle, Robert, 15–17, 24–5, 36–7
Bradley, James, 52
Brewster, David, 34, 136, 138–9, 228
Briot, Charles, 232
Bromberg, J., 252
Brown, Thomas, 170
Brush, S. G., 50
Buchwald, J. Z., 34, 240, 285
Buffon, Georges-Louis Leclerc, Comte de, 118

caloric, 27–8
Cambray, N., 124
Canton, John, 192–4
Cantor, G. N., 28, 34, 35, 63, 79, 218
Carnot, Nicolas Léonard Sadi, 45
Cauchy, Augustin-Louis, 35, 47, 222–7, 232–3, 329
Cavallo, Tiberius, 193–4
Cavendish, Henry, 32, 63, 76
Challis, James, 47
Chambers, Ephraim, 24
Charleton, Walter, 3
chemistry, 22, 28, 46, 73–6, 85–106
Cheyne, George, 113
Children, John George, 197
Christie, J. R. R., 28, 34, 153
Clarke, Samuel, 66, 150–1
Cleghorn, William, 102, 106
Cohen, I. B., 61, 68, 71

cohesion of bodies, 97–8
Colden, Cadwallader, 72
Comte, Auguste, 33
Cook, John, 136
Corson, D. W., 57
Cowper, William, 123
Crawford, Adair, 101–4
Crombie, Alexander, 143
Cudworth, Ralph, 18–19
Cullen, William, 74–6, 86–103, 117–18
cultural role of science, 85–96

Dallowe, Timothy, 24, 69
Davy, Humphry, 37, 78–80, 196–7, 201–2
De la Rive, Auguste, 204
Democritus, 4, 7
Demonferrand, J. F., 200
Desaguliers, Jean Theophilus, 23
Descartes, René, 9, 11–17, 43, 113
  La dioptrique, 12
  Le monde, 11–12
  Les météores, 12
  magnetism, 17–18
  optics, 15
  planetary motions, 14–15
  Principia philosophiae, 12, 17–18
  subtle matter, 12, 16, 61–2
  theory of light, 12, 31
Dijksterhuis, E. J., 11
Dirac, P. A. M., 53–4
Dobbs, B. J. T., 140
Donovan, A. L., 102
Doran, B. G., 51–2, 240
Duhem, Pierre, 11–12

Earnshaw, Samuel, 230
Edinburgh, Philosophical Society of, 90, 91
Edinburgh, Select Society of, 90, 91
Edinburgh University, 86
Einstein, Albert, 52–4, 289, 335–6
elective affinities, 96–104
electric charge, 190–4
electrical polarisation, 193–5, 299
electricity, theories of, 23, 27, 38, 45, 70–2, 116, 190–207, 239–68, 309–37
electromagnetism, theories of, 38, 50–2, 199–207, 239–68, 269–303, 309–37
electrostatic action, 38, 190–9, 202, 278
Empedocles, 4
energeticists, 275, 303
energy, theory of, 39–42, 246–54, 274–5
Engels, Friedrich, 32
Epicurus, 7, 142
epistemology, 157–85
Erman, Paul, 203
ether, problem of definition, 1–2, 187
Euler, Leonhard, 61, 159, 165

Faraday, Michael, 31, 37, 80, 141, 188–90, 195, 201, 204–7, 239–44, 271, 278
Fechner, Gustav Theodor, 278, 283–7, 296
field, concept of, 35–44, 187–90, 271
fire, 3, 8, 15, 25–6, 61, 69–70, 73–4, 117
Fitzgerald, George Francis, 51, 239, 258–9, 261, 309, 310–20, 335
Fontana, Felix, 116
force, concepts of, 19–25, 35–44, 47–50
Franklin, Benjamin, 27, 30, 36–7, 190
French, R. K., 28, 31, 34, 150
Fresnel, Augustin Jean, 31, 34–5, 47, 49, 52, 173, 215, 219–22

Galen, 27, 111–12
Galvani, Luigi, 116
Galvanism, 196–9
gases, 28, 102–6
Gassendi, Pierre, 3
Gauss, Carl Friedrich, 241, 287
Gilbert, William, 17–18
Glasgow Literary Society, 99
Glasgow University, 86, 91
Glisson, Francis, 113
Goldberg, S., 53
Good, John Mason, 150
Gooding, D., 31
Gottsched, Johannes, 118–19
Green, George, 241
Greene, Robert, 67
Gregory, John, 96
Grosseteste, Robert, 9
Grotthuss, Theodor von, 197
Guerlac, H., 23

Hales, Stephen, 68, 113
Hall, T. S., 130–1
Haller, Albrecht von, 116, 121
Halley, Edmund, 143–4
Hamilton, William Rowan, 227–8, 260–1
Hansteen, Christopher, 204
Hare, Robert, 31
Hartley, David, 2, 32, 87, 96, 123, 145–6, 159–64
Harvey, William, 6, 113
heat, theories of, 27–8, 45, 49, 75–80, 97–106, 197
Heaviside, Oliver, 320–1
Heilbron, J. L., 27, 31, 41–3, 271
Heimann, P. M., 20, 26, 34, 37–8, 42–4, 147
Helmholtz, Hermann von, 51, 255, 271, 275, 288, 292, 295–303, 324–7, 333
Helmont, Johann Baptista van, 10
Heraclitus, 3
Herbart, Johann Friedrich, 273, 284, 288
Hermes Trismegistus, 9
Herschel, John Fredrick William, 35, 173–6

Index

Herschel, William, 32
Hertz, Heinrich, 302–3, 312, 320, 328
Hesse, M. B., 40, 42, 252
Higgins, Bryan, 76
Hippocratic medicine, 113–14
Hirosige, T., 52–3, 297, 303
Hobbes, Thomas, 142
Hoffmann, Friedrich, 28, 114, 120
Holbach, Paul Henri Thiery, Baron d', 150
Holton, G., 52
Home, Henry, Lord Kames, 91
Home, R. W., 61, 71
Hume, David, 87–93
Hunter, John, 118
Hutcheson, Francis, 93, 127
Hutchinson, John, 141, 149
Hutton, Charles, 78
Hutton, James, 76–8, 79–80
Huygens, Christian, 11, 14, 37, 215–16

imponderable fluids, 61–80
Irvine, William, 101–4

Joule, James Prescott, 80, 244

Kant, Immanuel, 188–9, 273, 274, 296–7
Keill, John, 67
Kelland, Philip, 228–30
Kelvin, Lord, *see* Thomson, William
Kepler, Johannes, 11, 17
Kirchoff, Gustav Robert, 292
Kirwan, Richard, 76
Knight, Gowin, 71, 79
Knudsen, O., 239
Kottler, M., 147
Koyré, Alexandre, 11
Kuhn, T. S., 44–5

Lagrange, Joseph Louis, 260–1, 311
Lamarck, Jean Baptiste, 118
La Mettrie, Julien Offray de, 150
Langrish, Browne, 121–2
Laplace, Pierre-Simon, 47, 142, 217
Larmor, Joseph, 51, 239, 262–3, 314, 317, 322–4, 332, 334–5
latent heat, 102
Laudan, L. L., 30, 31, 34, 152
Lavoisier, Antoine Laurent, 28, 86, 104–6
Leibniz, Gottfried Wilhelm von, 13, 14, 65–6
LeSage, George-Louis, 30, 158, 164–9, 256
Leslie, Peter Dugud, 75–6
Leucippus, 4, 7–8
Leyden jar, 196
light
  absorption of, 228–30
  diffraction of, 218–19
  dispersion of, 224–8

polarisation of, 220–1, 230, 232, 233–4
theories of, 2, 9, 12–15, 21–3, 45–7, 49–52, 71–2, 75–80, 138–9, 173–81, 203–4, 215–37, 269, 277–8, 301–2
Lloyd, Humphrey, 35, 225–6
Locke, John, 91
Lodge, Oliver, 146–7, 239, 258–9, 332–3
Lorentz, Hendrik Antoon, 51–4, 302–3, 312, 323–9
Lorenz, Ludwig Valentin, 293–4
Love, R., 24
Lovett, Richard, 137
Lucretius, 3, 7–8, 142, 150, 256

McCormmach, R., 52, 303
MacCullagh, James, 51, 232, 310–15
McEvoy, J. G., 37
McGuire, J. E., 22, 37–8, 42–4, 61, 65
Maclaurin, Colin, 72–3, 87, 93
McMullin, E., 21, 58
magnetism, theories of, 17–18, 21–3, 61–2, 278
Malus, Etienne-Louis, 47, 217–18
Martin, Benjamin, 68
Martine, George, 86
Martinez, Martin, 119–20
Marum, Martinus van, 191
materialism, 142–3, 150, 283
mathematics, 39–42, 47, 219–25, 241–2, 248–54, 281–3, 287–91, 298–9
Maxwell, James Clerk, 35, 38–41, 50–1, 140, 165, 188, 239–68, 309–16, 332–7
Mayer, Julius Robert, 272–6
Mead, Richard, 123
mechanical explanation, 44–50, 276–9, 311
'mechanical philosophy', 15–19, 23–4, 35–7, 81
medicine, 96
methodology of science, 91–5, 102, 157–85
Metzger, H., 141
Michell, John, 32
Michelson, Albert Abraham, 52, 259, 334
Mill, John Stuart, 31, 173–80
Milner, Thomas, 193
Monro, Alexander, 93
moral theory, 89
More, Henry, 3, 18–19
Morley, Edward Williams, 52
Morse, E. W., 139
Moyer, D. F., 291

natural theology, 136–9
naturphilosophie, 198, 271–6, 288, 294, 296
nebular hypothesis, 144
neo-Platonism, 7, 9, 13, 16, 17–18, 67, 147–8
Neumann, Carl Gottfried, 294–5
Neumann, Franz Ernst, 298, 314

Newton, Isaac, 1–2, 11, 13, 14, 17–24, 25–6, 35–7, 47–8, 58, 61–8, 73–5, 86, 88
  'active principles', 20, 62–9
  'De aere et aethere', 20
  ether letter to Boyle, 20, 26, 64, 70
  gravitational theory, 1, 22, 35–6, 48, 57, 64–6, 143
  'Hypothesis explaining the properties of light', 19–24, 64, 70, 73, 139
  Opticks, 20, 21, 22, 23, 26, 30, 31, 36–7, 61–9, 121–4, 144, 145
  optics, 20, 216
  planetary motion, 22
  Principia, 20, 26, 36–7, 57, 62–8
Noxon, J., 87

O'Brien, Matthew, 230
Olson, R., 63
Ørsted, Hans Christian, 190, 198–200, 203–6
Ostwald, Wilhelm, 33
Otto, Everardus, 127

Paley, William, 136
Paracelsus, 10, 17
Parmenides, 4
Paul, Saint, doctrine of 'spiritual body', 8, 19, 146
Pemberton, Henry, 67–8
phlogiston, 75
physiology, 28–9, 111–34
Pike, Samuel, 141–2
Pitcairne, Archibald, 118
Plato, 3, 17
Playfair, John, 63, 78, 143
Plotinus, 7
Plummer, Andrew, 93
pneuma, 2–10, 16, 26, 112
Pohl, G. F., 204
Poincaré, Jules Henri, 34, 310–11, 333–7
Poisson, Siméon-Denis, 271
potentials, 39–42
Powell, Baden, 35, 47, 225–8, 231
Priestley, Joseph, 31, 36–7, 76, 118, 135, 159
Proclus, 7, 17
psychology, 87–8, 114, 285–6

quantum mechanics, 54
quinta essentia, 5
Quesnai, François, 124–5

Ramsay, Michael, 147–8
Rankine, William John Macquorn, 49, 245
Regnault, Victor, 241
Reid, Thomas, 32, 96, 133–4, 135, 146, 170–2
relativity theory, 52–4, 335
Riemann, Bernhard, 276, 288–92

Ritterbush, P. C., 116, 147
Robinson, Bryan, 26, 68–70, 96–8, 123, 147, 158, 182
Robison, John, 32, 63, 102
Roget, Peter Mark, 202
Rumford, Count, *see* Thompson, Benjamin

Salisbury, Lord, 33
Santorini, Giovanni Domenico, 115–16
Sarrau, Emile, 232
Sauvages, de la Croix, François Boissier de, 127–9
Savart, Félix, 200
Savary, Félix, 200
Schaffer, S., 143
Schaffner, K. F., 52, 303
Schelling, Frederick Wilhelm Joseph, 273
Schofield, R. E., 35–7, 61, 63, 71, 80
secularisation, 93, 95
Seebeck, T. J., 202
Shapin, S., 153
Shaw, Peter, 24, 26, 69, 73–4
Siegel, D. M., 33, 34, 41–3, 51, 140, 271, 292, 299
Silliman, R. H., 51, 140
Smellie, William, 118
Smith, Adam, 87–93
Sommerfeld, Arnold, 336
soul, 2–4, 112–18, 125–32
spirit, 10, 28, 97, 111–34, 147–51
spiritualism, 146–7
Stahl, Georg Ernst, 28, 120
states of matter, 27–8, 96–106
Stefan, Josef, 302
Stein, H., 51–2, 57
Stewart, Balfour, 140
Stokes, George Gabriel, 314, 319
Stoics, 2, 6–7

Tabor, John, 117
Tait, Peter Guthrie, 140, 255, 258
Talbot, G. R., 75
Thackray, A., 68, 73
theology, 28–9, 45, 66–7, 93–4, 135–53, 285
Thompson, Benjamin, 36, 78–80
Thomson, George, 123
Thomson, Joseph John, 256, 323
Thomson, William, 38–41, 51, 140, 239–68, 271, 309, 316–24
Tovey, John, 228–30
Tucker, Abraham, 147
Turner, F. M., 147

Vince, Samuel, 142–3
vision, theories of, 145
Volta, Alessandro Giuseppe Antonio Anastasio, 191

Index

vortex model of the atom, 50–1, 245–7, 254–60, 292

Walker, Adam, 78
Walker, D. P., 148
Wallace, Robert, 93
Watts, Isaac, 43–4
Weber, Ernst Heinrich, 283, 285–6
Weber, Wilhelm Eduard, 276–301
Westfall, R. S., 65
Whewell, William, 31, 137–9, 144, 173–180
Whittaker, E. T., 51–3, 252, 314, 320, 329, 334, 335
Whytt, Robert, 113

Wilde, C., 141
Willis, Thomas, 114
Wilson, Benjamin, 70, 158
Wilson, D. B., 147
Wise, M. N., 33, 34, 41–3, 51
Witte, H., 337
Wollaston, William Hyde, 196, 200–1
Woodruff, A. E., 276, 297
Wynne, B., 147

Young, Thomas, 46–7, 52, 79, 140–1, 218

Zeno of Citium, 6
Ziemacki, R. L., 82–3

LIBRARY OF DAVIDSON COLLEGE

Books on regular loan may be checked out for **two weeks**. Books must be presented at the Circulation Desk in order to be renewed.

A fine is charged after date due.

Special books are subject to special regulations at the discretion of the library staff.

APR 30. 0			